Peter A. Reinhardt
K. Leigh Leonard
Peter C. Ashbrook

Pollution Prevention and Waste Minimization in Laboratories

LEWIS PUBLISHERS

Boca Raton   New York   London   Tokyo

Library of Congress Cataloging-in-Publication Data

Reinhardt, Peter A.
  Pollution prevention and waste minimization in laboratories /
Peter A. Reinhardt, K. Leigh Leonard, Peter C. Ashbrook.
    p.   cm.
  Includes bibliographical references (p.   ) and index.
  ISBN 0-87371-975-1 (alk. paper)
  1. Laboratories--Waste disposal. 2. Pollution. 3. Waste
minimization. I. Leonard, K. Leigh. II. Ashbrook, Peter C.
III. Title.
TD899.L32R45   1995
628.4′2--dc20   95-340
                                                                           CIP

This book contains information obtained from authentic and highly regarded sources. Reprinted material is quoted with permission, and sources are indicated. A wide variety of references are listed. Reasonable efforts have been made to publish reliable data and information, but the author and the publisher cannot assume responsibility for the validity of all materials or for the consequences of their use.

Neither this book nor any part may be reproduced or transmitted in any form or by any means, electronic or mechanical, including photocopying, microfilming, and recording, or by any information storage or retrieval system, without prior permission in writing from the publisher.

All rights reserved. Authorization to photocopy items for internal or personal use, or the personal or internal use of specific clients, may be granted by CRC Press, Inc., provided that $.50 per page photocopied is paid directly to Copyright Clearance Center, 27 Congress Street, Salem, MA 01970 USA. The fee code for users of the Transactional Reporting Service is ISBN 0-87371-975-1/96/ $0.00+$.50. The fee is subject to change without notice. For organizations that have been granted a photocopy license by the CCC, a separate system of payment has been arranged.

CRC Press, Inc.'s consent does not extend to copying for general distribution, for promotion, for creating new works, or for resale. Specific permission must be obtained in writing from CRC Press for such copying.

Direct all inquiries to CRC Press, Inc., 2000 Corporate Blvd., N.W., Boca Raton, Florida 33431.

© 1996 by CRC Press, Inc.
Lewis Publishers is an imprint of CRC Press

No claim to original U.S. Government works
International Standard Book Number 0-87371-975-1
Library of Congress Card Number 95-340
Printed in the United States of America 1 2 3 4 5 6 7 8 9 0
Printed on acid-free paper

The Editors

The editors: K. Leigh Leonard, Peter A. Reinhardt, and Peter C. Ashbrook

Peter A. Reinhardt is the Assistant Director for Chemical & Environmental Safety at the University of Wisconsin-Madison's Safety Department. Since 1979 he has held positions in radiation safety and as the Hazardous Waste Officer and the Chemical Safety Supervisor there. He directs a staff of ten who perform hazardous waste management, industrial hygiene, environmental engineering, chemical analysis, site remediation, and environmental audits to manage the University's chemical and environmental risks. His program oversees the University's compliance with environmental and safety laws pertaining to hazardous waste, Superfund, chemical emergency planning and response, toxic substances, water, air, and other chemical uses. He inspired and edited the University's *Chemical Safety and Disposal Guide* (1984,1994) and assists other Wisconsin state agencies as a member of the Hazardous Materials Management Council.

He is an active member of the American Chemical Society and its division of Chemical Heath and Safety. He was appointed to the ACS's task force on Laboratory Waste Management in 1989 and was selected as Chair in 1994.

Since 1992 he has been elected to chair the Dane County Local Emergency Planning Committee. He had been elected Vice Chair since the Committee's inception in 1987.

He is an appointed member of the National Committee on Clinical Laboratory Standards' subcommittee on waste management, which prepared the 1993 NCCLS guideline on Clinical Laboratory Waste Management.

He coauthored *Hazardous Waste Management at Educational Institutions* (1987), *Infectious and Medical Waste Management* (1991), and *Laboratory Waste Management* (1994), and has written several other technical publications, chapters, and articles. He is also a frequent speaker, instructor, and presenter on the topics of hazardous waste and chemical safety.

In 1993 he was appointed by the National Research Council to subcommittees on Mixed Waste and Pollution Prevention to help draft parts of *Prudent Practices for Handling, Storage, and Disposal of Chemicals in the laboratory (1995).*

Mr. Reinhardt has a Bachelor's degree in Biochemistry with a concentration in Toxicology from the University of Wisconsin-Madison (1977) and a Master's degree in Public Policy and Administration (Environmental Risk Policy focus) from UW-Madison's Robert M. La Follette Institute of Public Affairs (1991).

K. Leigh Leonard, CHMM (Certified Hazardous Material Manager), is the Associate Environmental Health and Safety Manager for the University of Wisconsin System, which includes 26 campuses. She provides policy development and technical oversight for UW's hazardous waste management program. She also works with a multi-agency committee to administer the State of Wisconsin's mandatory hazardous waste services contract.

In this capacity, she monitors the hazardous waste market, conducts hazardous waste facility inspections, and recently worked on a rebid of the contract that resulted in a 50 percent cost savings to the State. Ms. Leonard's role in the State's contract earned her a State of Wisconsin award for Excellence in Purchasing in 1995. Other professional specialties include laboratory waste minimization, management of high-hazard wastes, chemical safety, safety in the arts, and biosafety.

Leigh has worked in environmental health and safety at the University for over eight years. Prior to that she worked for the Oswego County Environmental Management Council in New York, where she co-managed the County's first household toxics collection program in 1985.

Ms. Leonard co-founded in Madison the first chapter of the Federation of Environmental Technologists (FET), a Wisconsin-based organization dedicated to the education of environmental compliance professionals. Through FET she has organized numerous seminars on several topics including an environmental primer, SARA compliance, management of the "ten most unwanted wastes," and large-quantity hazardous waste generator training. In 1995, Ms. Leonard received an award from FET in appreciation for outstanding and dedicated service during her term as senior chapter officer.

Ms. Leonard frequently speaks on environmental health and safety-related topics for Wisconsin professional associations, Wisconsin secondary school teachers, the College and University Hazardous Waste Conference, and the Certified Hazardous Materials Manager certification course.

Ms. Leonard holds a Masters degree from the University or Wisconsin-Madison in Urban and Regional Planning with a concentration in environmental planning (1989) and a Bachelor's degree from Oberlin College in Biology (1982) where she graduated Phi Beta Kappa.

Peter C. Ashbrook is the Head of Hazardous Waste Management at the University of Illinois at Urbana-Champaign. In this capacity, he has directed the development of a comprehensive hazardous waste management program including the preparation of the University's Part B permit and the construction of a new building designed for chemical waste management. In 1990, he obtained a two-year grant to evaluate the design and implementation of laboratory waste minimization activities.

Mr. Ashbrook graduated in 1973 from Carleton College, Northfield, Minnesota in 1973. and obtained his Master's degree in Environmental Health from the University of Minnesota in 1976.

Mr. Ashbrook is an active member of the American Chemical Society (ACS), having served as an officer for the Division of Chemical Health and Safety, as well as being appointed a member of both the Laboratory Waste Management Task Force and the Committee on Chemical Safety. He is the co-author of a regular column on laboratory waste minimization in *Chemical Health and Safety,* which is published jointly by the ACS and its Division of Chemical Health and Safety. Mr. Ashbrook served on two subcommittees of the National Research Council to assist in the preparation of *Prudent Practices for Handling, Storage, and Disposal of Chemicals in the Laboratory.* He was an invited participant in two workshops addressing the impact of hazardous waste regulations on laboratory research sponsored by the National Research Council's Government, University, and Industry Research Roundtable. He has also served on the National Association of College and University Business Officers' (NACUBO) Hazardous Waste Management Task Force.

He was co-editor of *Safe Laboratories: Principles and Practices for Design and Remodeling* (1990) and is a frequent speaker, instructor, and presenter on topics related to chemical waste management in the laboratory.

Acknowledgments

Words cannot express our tremendous gratitude to the book's contributors for their chapters. Their perspectives and expertise give this book exceptional breadth and depth. All chapter authors were timely and most cooperative in their contributions, yet maintained the book's high standard of quality and integration. Readers will enjoy their insights and revelations.

The idea for this book was born following a symposium on this subject at the American Chemical Society 1992 Spring Meeting. We wish to credit the Society, its Laboratory Waste Management Task Force, its Chair — Russell Phifer, and the ACS Department of Government Relations and Science Policy for allowing this idea to germinate.

The national Survey of Laboratory Chemical Disposal and Waste Minimization would not have been possible without the consultation and assistance of Linda J. Penaloza, Associate Director, and Jane Campbell, Senior Research Specialist, of the University of Wisconsin — Extension, Continuing Education Extension, Wisconsin Survey Research Laboratory. We also thank W. Emmett Barkley and Cheryl A. Warfield of the Howard Hughes Medical Institute Office of Laboratory Safety for their support and assistance in conducting the survey. We thank the American Chemical Society Division of Chemical Health and Safety for permission to use their member directory. We very much appreciate the support of David Schleicher of the ACS Office of Government Relations and Science Policy for including the survey in their newsletter.

We greatly appreciate David Drummond and Patricia Kandziora, who permitted the flexible work schedules we needed for this idea to blossom into a book.

We would especially like to thank Helen and Steve for their patience and support while completing the survey, writing our chapters, and editing the final text. They endured the company of sleepy midnight writers and weekend editors lost in their manuscripts at the kitchen table. We love them and promise to spend more time with them.

The editors: Peter A. Reinhardt
K. Leigh Leonard
Peter C. Ashbrook

A Note of Caution

Chemicals, biological agents, and radioactive materials used in laboratories are inherently hazardous. In many circumstances, laboratory procedures, waste minimization practices, and methods of pollution prevention are also inherently hazardous. As a result, users of this book need to take additional precautions to protect health, property, and the environment. Because there exists much uncertainty and paucity of information in the field of safety, users of this book must assume the inherent risks of laboratory materials and practices, even when additional precautions are taken.

To use this book, you must have a thorough understanding of the hazards of chemicals, biological agents, and radioactive materials you use. We have included safety information in this book or referenced other sources. Unfortunately, much is not known about the hazards of laboratory materials and practices; we are continuously learning of new circumstances that lead to hazardous situations and of previously unknown hazardous characteristics of laboratory materials. To promote safety, you need to review the safety information in this book (and referenced information) before implementing any waste minimization or pollution practices. We highly recommend that you also read the following references:

- Material Safety Data Sheets for the materials you use.
- *Prudent Practices for Handling, Storage and Disposal of Hazardous Chemicals in Laboratories* (National Academy of Sciences, Washington, D.C., 1983; new edition pending).
- *Safety in Academic Chemistry Laboratories, 6th edition* (American Chemical Society: Washington, D.C., 1995).
- *Handbook of Laboratory Safety,* 3rd ed. (CRC Press: Boca Raton, Florida, 1990).
- *General Laboratory Safety,* Tentative Guideline (National Committee for Clinical Laboratory Standards: Villanova, PA, 1993) for medical laboratories.
- *Handbook of Reactive Chemical Hazards,* 5th ed., L. Bretherick (Royal Society of Chemistry: London, 1995).

This book is meant for use by people who work in laboratories. Procedures apply only to laboratory-scale quantities of materials and waste. Users of this book should not modify or alter any equipment, procedures, or practices described in this book in a way not intended by the authors.

Waste minimization and pollution prevention activities must comply with all federal, state, and local laws. In particular, the reader is directed to

- Title 29 of the Code of Federal Regulations, including those parts pertaining to the Laboratory Standard of the U.S. Occupational Health and Safety Administration. Waste minimization or pollution prevention practices need to be implemented within the scope of your laboratory's chemical hygiene plan.

- Title 40 of the Code of Federal Regulations, including those sections for the management of medical and hazardous waste.
- Title 10 of the Code of Federal Regulations, including those parts describing the management of radioactive materials and wastes.

Necessary measures have been taken to ensure the accuracy and reliability of the information contained in this book. However, the editors and authors make no representation, warranty, or guarantee, and thereby disclaim any liability or responsibility for loss or damages resulting from the use of any information in this book, or for the violation of any law with which any of the information may conflict.

Contributors

Milly Archer
Environmental Compliance
University of Vermont
Burlington, Vermont

Margaret-Ann Armour
Department of Chemistry
University of Alberta
Edmonton, Alberta, Canada

Peter C. Ashbrook
Division of Environmental Health
and Safety
University of Illinois at
Urbana-Champaign
Urbana, Illinois

Bruce D. Backus
Department of Environmental Health
and Safety
University of Minnesota
Minneapolis, Minnesota

John Brandon
Division of Environmental Health
and Safety
University of Illinois at
Urbana-Champaign
Urbana, Illinois

Robert B. Charbonneau
Environmental Protection Services
University of California
Oakland, California

Jeff Christensen
Department of Risk Management
University of Arizona
Tucson, Arizona

Gerald A. Denys
Microbiology, Serology, and
Virology
Methodist Hospital
Indianapolis, Indiana

Judith Gordon
Gordon Resources Consultants, Inc.
Reston, Virginia

Jennifer L. Hernandez
Beveridge and Diamond
San Francisco, California

Cynthia Klein-Banai
Division of Environmental Health
and Safety
University of Illinois at
Urbana-Champaign
Urbana, Illinois

K. Leigh Leonard
University of Wisconsin System
Administration
Madison, Wisconsin

Hector Mandel
University of Illinois at
Urbana-Champaign
Urbana, Illinois

John A. Mangravite
Department of Chemistry
West Chester University
West Chester, Pennsylvania

John F. O'Brien
Mayo Clinic
Rochester, Minnesota

Ronald Pike
Department of Chemistry
Merrimack College
North Andover, Massachusetts

Peter A. Reinhardt
Safety Department
University of Wisconsin
Madison, Wisconsin

Roger R. Roark, Jr.
B/R Instrument Corporation
Easton, Maryland

Roger R. Roark, Sr.
B/R Instrument Corporation
Easton, Maryland

Richard A. Senn
Agracetus, Inc.
Middleton, Wisconsin

Mono M. Singh
Department of Chemistry
Merrimack College
North Andover, Massachusetts

Gregory D. Smith
Mayo Clinic
Radiation Safety
Rochester, Minnesota

Ralph Stuart
Department of Risk Management
University of Vermont
Burlington, Vermont

Zvi Szafran
Merrimack College
North Andover, Massachusetts

Linda J. Tanner
3M Environmental Engineering
and Pollution Control
St. Paul, Minnesota

Fay M. Thompson
Department of Environmental Health
and Safety
University of Minnesota
Minneapolis, Minnesota

Paul VanTriest
B/R Instrument Corporation
Easton, Maryland

Richard J. Vetter
Mayo Clinic
Rochester, Minnesota

Lloyd M. Wundrock
Department of Risk Management
University of Arizona
Tucson, Arizona

Contents

PART 1: INTRODUCTION AND OVERVIEW

Chapter 1
Why Pollution Prevention? 3
K. Leigh Leonard, University of Wisconsin System Administration
   This chapter discusses how pollution prevention and waste minimization are defined in the laboratory setting and details the reasons that pollution prevention and waste minimization make sense.

Chapter 2
A Survey of Laboratory Waste Management and Minimization Practices 15
K. Leigh Leonard, University of Wisconsin System Administration; and Peter A. Reinhardt, University of Wisconsin-Madison
   This chapter describes the findings of a survey of U.S. laboratories on their use of the sanitary sewer, normal trash, chemical treatment methods, and waste minimization.

Chapter 3
Investigating Waste Minimization Possibilities 31
Cynthia Klein-Banai, University of Illinois at Urbana-Champaign
   This chapter describes the basic concepts in solving waste minimization problems and discusses what can be done. Also described are the informational and compliance programs that must be in place to explore waste minimization.

Chapter 4
The Law on Waste Minimization in Laboratories 55
Jennifer L. Hernandez, Beveridge & Diamond
   This chapter describes EPA's pollution prevention and waste minimization policies and regulations that govern recycling and other waste minimization activities. What do they mean for a laboratory? This chapter explains how to stay in compliance while preventing pollution.

PART 2: EFFECTING POLLUTION PREVENTION AND WASTE MINIMIZATION

Pollution prevention doesn't just happen. Optimal implementation requires planning, policymaking, resources, and finding ways to overcome impediments. Implementation is facilitated when waste minimization is integrated into the scientific curriculum.

Chapter 5
Planning and Development of a Model Waste Minimization Program 69
Robert Charbonneau, University of California System
   This chapter describes waste audits, tracking waste sources and pathways, and institutional strategies and plans to achieve waste minimization.

Chapter 6
Institutional Policy, Commitment, and Support 89
Fay M. Thompson, University of Minnesota
> This chapter describes the management of a pollution prevention program at the administrative level of larger organizations, including policymaking, budgeting and provision of laboratory support services for training, chemical management, and waste disposal. Also discussed are the importance of executive-level commitment and the roles of the institutional safety committee and the chemical hygiene officer.

Chapter 7
Overcoming Impediments to Waste Minimization 97
Peter A. Reinhardt, University of Wisconsin-Madison
> This chapter describes how to cope with liability concerns, economic disincentives for waste minimization, and institutional bureaucracy. The content and potential of the Laboratory Waste Coalition's Laboratory Equity and Waste Minimization Act is also reviewed.

PART 3: APPROACHES BY MEDIA, SOURCE, AND WASTE TYPE

Preventing pollution and minimizing waste can be approached from several perspectives. This part describes how experts look at the problem by media, source, and waste type.

Chapter 8
Management of Laboratory Air Emissions 123
Ralph Stuart and Milly Archer, University of Vermont
> Releases to the atmosphere may be the most significant environmental impact from laboratories. Can fume hood and fugitive emissions from laboratory operations be reduced? This chapter describes how to contain and abate routine emissions of air pollutants.

Chapter 9
Management of Laboratory Effluents to the Sanitary Sewer 141
Lloyd Wundrock and Jeff Christensen, University of Arizona
> Every laboratory has a sink, yet allowable releases to the sanitary sewer vary considerably by locale. This requires each laboratory to understand the capabilities, limitations, and rules of its treatment works. This chapter describes how laboratories can manage the problem of minimizing releases of hazardous chemicals to the sewer.

Chapter 10
Pollution Prevention in Clinical Laboratories 153
Richard J. Vetter, John F. O'Brien, and Gregory D. Smith, Mayo Clinic
> Clinical laboratories manage many hazardous materials, including infectious agents, blood, chemicals, and radioactive materials. As a result, they are challenged to reduce a great variety of pollution and waste. Procedures and methods in clinical labs tend to be standardized, which can facilitate feasibility analysis

for pollution prevention. Microscale samples and methods have long been a trend. Training and professional staff aid implementation.

Chapter 11
Minimization of Waste Generation in Medical Laboratories 163
**Judith G. Gordon, Gordon Resource Consultants, Inc.; and
Gerald A. Denys, Methodist Hospital**
Biomedical and clinical laboratories are faced with the problem of managing multiple infectious and medical wastestreams that are, in many locales, subject to new regulation. Waste reduction is one solution, but its potential for success depends upon waste type, available technology, and laboratory practices. Pressure to improve productivity of the medical laboratory raises the importance of employee training and easy-to-use systems.

Chapter 12
Minimization of Low-Level Radioactive Wastes from Laboratories 195
**Peter C. Ashbrook, John Brandon, and Hector Mandel,
University of Illinois at Urbana-Champaign**
Much progress has been made in reducing low-level radioactive waste from laboratories. Because few offsite disposal sites have been available, generators have refined their techniques for onsite management. Storage for decay and source separation by waste type and radionuclide are two important waste minimization methods. One of the most successful examples of material substitution is here.

PART 4: WHAT INDIVIDUAL LABORATORIES CAN DO

Perhaps it is most effective when individuals take responsibility for pollution prevention and initiate new laboratory practices. The methods in this part can be implemented by people who work in laboratories with minimal organizational support. In most cases, all it takes is personal initiative. These methods can also be implemented at the organizational level to optimize pollution prevention for all of their laboratories. For example, some institutions take advantage of the economies of scale by distilling solvents at a central facility. The chapters are in approximate order of the waste minimization hierarchy.

Chapter 13
The Microscale Chemistry Laboratory 207
Ronald M. Pike, Zvi Szafran, and Mono M. Singh, Merrimack College
This chapter describes the success of microscale equipment and methods in reducing waste in chemistry laboratories.

Chapter 14
At the Lab Bench: Finding the Right Balance in Source Separation 221
Peter A. Reinhardt, University of Wisconsin-Madison
Keeping normal trash out of the hazardous wastestream is the first step in waste minimization. Also important is maximizing disposal and treatment opportunities by keeping dissimilar wastes separate.

Chapter 15
Solvent Recycling by Spinning Band Distillation: Theory, Equipment,
and Limitations 239
**John A. Mangravite, West Chester University; and Roger R. Roark, Jr.,
B/R Instrument Corporation**
> *This chapter describes the available equipment, the problem of azeotropes, theoretical limitations, and safety considerations for distillation and reuse of organic solvents.*

Chapter 16
Chemical Treatment Methods to Minimize Waste 275
Margaret-Ann Armour, University of Alberta
> *This chapter describes elementary neutralization and other chemical treatment methods that reduce and/or destroy hazardous laboratory waste at the point of generation.*

PART 5: WHAT ORGANIZATIONS CAN DO

Organizations can do much to facilitate and support pollution prevention and waste minimization in their laboratories. The success of a surplus chemical exchange accelerates with the number of participating laboratories. Bulking and commingling is especially effective when many laboratories cooperate, but is also a way for all laboratories to save substantial disposal costs. Although purchasing is usually done centrally, any laboratory can benefit from working with vendors and commercial services to effect pollution prevention.

Chapter 17
Recruiting Vendors to Achieve Waste Minimization and
Pollution Prevention 295
K. Leigh Leonard, University of Wisconsin System Administration
> *Purchasing practices, commercial services, and other offsite options can facilitate pollution prevention. Purchasing procedures can control acquisition of degradable and hazardous chemicals, and carry out material substitution policies. Buying fewer and smaller containers avoids waste and surpluses. Commercial recycling firms are available for organic solvents, mercury, batteries, etc., as are cooperative and commercial chemical exchange brokers and gas cylinder reconditioning firms. Low- or no-cost options include voluntary exchanges between organizations and return of surplus compressed gases to their manufacturers. Precautions for contractor selection and contracting are discussed.*

Chapter 18
Surplus Chemical Exchange: Successes and Potential 311
Jeff Christensen, University of Arizona
> *Surplus and unwanted laboratory chemicals can often be redistributed to other chemical users within an institution. This chapter describes the elements of a*

successful redistribution program, including information exchange, quality control, and transportation that is safe and follows DOT rules.

Chapter 19
Cost Savings and Volume Reduction By Commingling Wastes 319
Peter C. Ashbrook, University of Illinois at Urbana-Champaign
This chapter describes safe methods of bulking and commingling organic solvents, mineral acids, and other wastes in order to reduce volume and achieve significant cost savings.

PART 6: CASE STUDIES AND APPLICATIONS

This part explains the details of applying pollution prevention methods and the specifics of implementation in certain organizations.

Chapter 20
Applications for Waste Solvent Recovery Using Spinning
Band Distillation 329
John A. Mangravite, West Chester University; Roger R. Roark, Jr., Roger R. Roark, Sr., and Paul VanTriest, B/R Instrument Corporation
Which laboratory waste solvents can be successfully distilled and reused? This chapter details the applications of solvent distillation.

Chapter 21
The Implementation of Waste Minimization Strategies in a
Biotechnology Research and Development Laboratory 377
Richard A. Senn, Agracetus, Inc.
The waste minimization program at a biotechnology firm is described.

Chapter 22
Waste Minimization and Management at 3M R&D/Laboratories 389
Linda J. Tanner, 3M
The waste minimization efforts at a major industrial research and development laboratory is described.

Chapter 23
Implementation of Waste Minimization Strategies 405
Bruce D. Backus, University of Minnesota
Waste minimization activities at a large academic institution are explained.

PART 7: CONCLUSION

Chapter 24
Thoughts on the Future of Pollution Prevention in the Laboratory 425
Peter C. Ashbrook, University of Illinois at Urbana-Champaign
This chapter explores the direction of environmental safety in laboratories.

APPENDICES

Appendix A
Guidance to Hazardous Waste Generators on the Elements of a Waste
Minimization Program 435
The EPA's Notice and Interim Final Guidance.

Appendix B
National Survey of Laboratory Chemical Disposal and Waste Minimization:
Survey Design, Methods, and Analysis 443
For the survey results detailed in Chapter 2, this Appendix describes how the survey was conducted and other background information.

Appendix C
State and Local Clearinghouses for Pollution Prevention and
Waste Minimization Resources and Information 457

Glossary 459

Index 465

Pollution Prevention and Waste Minimization in Laboratories

Part 1:
Introduction and Overview

CHAPTER 1

Why Pollution Prevention?

K. Leigh Leonard

CONTENTS

The Scope and Purpose of this Book 3
Setting the Stage: The Evolution of U.S. Environmental Policy 4
    A Whirlwind Tour of U.S. Environmental Policy 5
        The Clean Water Act 5
        The Clean Air Act 6
        The Resource Conservation and Recovery Act (RCRA) 6
        The Comprehensive Environmental Response,
        Compensation, and Liability Act (CERCLA) 7
        Other Statutes Controlling Toxic and Hazardous Substances 7
        Summary of Two Decades 7
        The Pollution Prevention Act 8
    So, Why Pollution Prevention? 8
    The "Greening" of Corporate America 9
EPA Defines the Terms 9
    Pollution Prevention 9
    Waste Minimization 10
    The "EPA Hierarchy" 10
The Philosophical Approach Behind this Book 11
How to Make the Best Use of this Book 12
Endnotes 13

THE SCOPE AND PURPOSE OF THIS BOOK

This book is about laboratories and the various types of wastes they generate. Its purpose is to provide information on efficient waste management, with the major emphasis on reducing or minimizing laboratory wastestreams. Chapters 8 and 9 illustrate other approaches to laboratory pollution prevention such as reducing wastewater discharges and fume hood emissions.

The scope of this book is comprehensive. A wide range of waste types is addressed, including hazardous, infectious, medical, and radioactive. Solid waste minimization is not addressed per se, but the subject is addressed peripherally in several chapters. The information in this book pertains to all laboratory operations where chemicals or other hazardous materials are used, including academic instructional and research labs, medical and clinical laboratories, analytical labs, and labs devoted to research and product development. It will also pertain to large institutions with multiple laboratories.

Who may benefit from reading this book? This book is, to a great extent, a "nuts and bolts" book. Specific waste minimization and pollution prevention techniques are detailed that will work in particular types of labs for specific wastestreams. The laboratory manager, principal investigator, or technician will want to apply ideas presented in some chapters directly to their laboratory operation. Environmental health and safety staff whose realm of responsibility includes laboratories will find this book to be valuable for advising laboratory staff, and also for developing a broader, institutional framework to plan and set priorities for pollution prevention in labs. Parts of this book, especially Chapters 3, 5, 6, and 7, may also be of interest to the corporate executive or institutional administrator for understanding the vital role of management support in achieving pollution prevention and waste minimization goals.

So, why is a book like this needed? Laboratory operations are notoriously difficult to prescribe waste minimization solutions for. Their waste types are numerous, and usually small in volume. Some wastes are extremely hazardous to handle due to the need for highly reactive agents in some types of research. These challenges are further compounded in pure research labs where there is a high frequency of change in experimental procedures and reagents used. Just when you have found a way to minimize a wastestream from a particular protocol, the study or project comes to an end.

So, in laboratories, much more so than with industrial operations, it is critical to have a "toolbox" of waste minimization and pollution prevention techniques and concepts readily at hand, so that effort can be concentrated on getting them in use, rather than identifying them in the first place. That, precisely, is this book's *raison d'être*.

SETTING THE STAGE: THE EVOLUTION OF U.S. ENVIRONMENTAL POLICY

There is little doubt that environmental quality is important to the public in the U.S. As environmental issues go, few are as emotionally gripping as waste management — especially where hazardous, radioactive, PCB, and infectious wastes are concerned. Environmental nightmares of the past are etched on the public psyche. Two extreme examples are the catastrophic methyl isocyanate gas release at Bhopal, India, which injured or killed thousands and the Love Canal site in New York where a residential neighborhood was built on a heavily contaminated chemical disposal site with sobering consequences. Public reaction to situations such as these has fueled a cascade of federal legislation and supporting rules pertaining to environmental protection, so much so that today it is virtually impossible for an environmental compliance professional, let alone a lay person, to understand the entire complex of U.S. environmental regulation in depth.

It is not the point of this book to describe all of the laws and regulations pertaining to laboratory waste management, though appropriate references to codes are included with the discussion of specific practices. However, the general history of U.S. environmental law is a useful framework for understanding why pollution prevention and waste minimization are regarded by both the regulators and the regulated community as the culmination of U.S. environmental policy and the logical summit of an evolution in thinking about contaminant control and environmental protection.

A WHIRLWIND TOUR OF U.S. ENVIRONMENTAL POLICY

The discipline of environmental law was virtually unrecognized prior to 1970. Not that environmental problems weren't already at the forefront of current events. Tort[1] litigation alleging damages or harm from pollution formed a common law foundation for many of the statutes that followed. Historically, general environmental protection concerns were not viewed to be the constitutional purview of the federal government. However, as interstate disputes over pollution control heightened along with public concern, the federal government saw fit to initiate legislation under the interstate commerce clause of the Constitution.

There are many historical sources of law that impact environmental quality (e.g., land use law, preservation law). The sources that most directly affect the management of residuals (wastes, air contaminants, water pollutants) are those that deal with resource remediation following damage or exploitation and those stemming from public health field.[2] The key environmental statutes that affect the management of hazardous chemicals and wastes in laboratories are shown in Table 1 and highlighted here.

The Clean Water Act

The federal government's constitutional right to regulate navigable waters makes the area of water pollution control the elder of environmental law. Antecedents to the Clean Water Act included the Rivers and Harbors Act (1899) and the Federal Water Pollution Control Act (1948) and its amendments. However, it wasn't until the 1972 amendments that national water quality standards were set forth. The 1977 Clean Water Act emphasized toxic pollutants as well as the more conventional water quality concerns.

Unless they dispose of no chemical substances down the drain, laboratories are generally affected by the Clean Water Act in one of two ways. Most labs discharge to a sanitary sewer system that flows to a local POTW,[3] which may restrict the types and amounts of contaminants in the laboratory's effluent. Alter-

Table 1 Significant U.S. Environmental Laws

Year enacted	Title of statute
1972, 1977	Clean Water Act and Amendments
1970, 1977, 1990	Clean Water Act and Amendments
1976, 1984	Resource Conservation and Recovery Act (RCRA) and Hazardous and Solid Waste Amendments (HSWA)
1980, 1986	Comprehensive Environmental Response, Coompensation and Liability Act (CERCLA or "Superfund") and Superfund Amendments and Reauthorizaiton Act (SARA)
1990	Pollution Prevention Act

natively, some labs may be part of a large facility that has its own on-site wastewater treatment system which may be permitted to discharge directly to a water body, or to the sanitary system. In this case, the lab's discharges would be subject to the capabilities and permit requirements of the treatment system. What this amounts to, in either case, is that allowable levels of contaminants are determined locally by the requirements of the off-site POTW or the on-site treatment system. So, there is much variability from lab to lab as to what contaminants each is allowed to put down the drain.

The Clean Air Act

The Clean Air Act of 1963 empowered federal district courts to take action to abate emissions, but left standard setting to the states. In 1970, the Clean Air Act amendments established national standards and set forth the policy of "nondegradation" of ambient air quality. Since then, the Clean Air Act has seen substantial revisions in 1977 and 1990, making this one of the hotbeds of regulatory development.

The Clean Air Act has great potential to affect labs, as provisions of the 1990 statute are codified in regulations. The more intensive regulation of nearly 200 Hazardous Air Pollutants (HAPs), including many commonly used laboratory chemicals (e.g., acetaldehyde), is certain to affect the use of specific chemicals in lab fume hoods and is likely to spawn further technology developments in the capture and treatment of lab fume hood emissions.

The Resource Conservation and Recovery Act (RCRA)

This act, together with its amendments, probably has the greatest direct impact on laboratories because it dictates strict management standards for hazardous wastes. While antecedents date back to 1965, the law that first defined hazardous wastes and set forth a "cradle-to-grave" management system for them was the Resource Conservation and Recovery Act of 1976. This law and its supporting regulations are what causes laboratories to ship hazardous waste out to licensed TSD facilities at a premium cost. The 1984 *Hazardous and Solid Waste Amendments* (HSWA) to RCRA banned the land disposal of untreated hazardous waste, and generally tightened the requirements for hazardous waste management. However, the most significant development relative to this book is that HSWA included the first statutory reference for hazardous waste minimization. Congress stated that the reduction or elimination of hazardous waste generation at the source should take priority over the management of wastes after they are generated.[4] This policy became reflected in a regulatory requirement that generators certify that they have a waste minimization program in place. (For detailed discussion of this requirement, see Chapter 4.)

The Comprehensive Environmental Response, Compensation, and Liability Act (CERCLA)

CERCLA was passed in 1980 to deal with the damage from past disposal practices. The Act, also known as the Superfund Law, allows the federal government to sue "potentially responsible parties" (PRPs) for cleanup costs. PRPs can include generators whose waste was disposed at the site, transporters who brought waste to the site, and past and present site owners. Its "joint and several" liability provisions allow the government to sue any PRP for the entire cost of the cleanup — a prospect that has deeply affected how U.S. institutions and firms, including laboratories, manage hazardous waste.

Superfund was amended in 1986 through the Superfund Amendments and Reauthorization Act (SARA). These amendments are significant because they are the first federal requirements to comprehensively address the use of chemicals at a facility, and to require public disclosure of amounts of chemicals maintained on site and (legally) released to the environment. At this time, these requirements do not impact many laboratory operations because the quantity of chemicals managed falls below reporting thresholds, and because there are exceptions for the supervised use of small quantities of chemicals. However, SARA's chemical inventory tracking, emergency planning, and toxic release reporting requirements presage future requirements that may have a greater impact on laboratory operations.

Other Statutes Controlling Toxic and Hazardous Substances

There are other statutes that may affect particular types of laboratory operations. The Toxic Substances Control Act (TSCA) regulates the use and disposal of PCBs, dioxin, asbestos, and Chlorofluorocarbons (CFCs). It sets forth strict requirements for registering and testing newly synthesized chemicals, or for using existing chemicals in a new manufacturing use. The Hazardous Materials Transportation Act (HMTA) regulates the transport of hazardous materials delivered to the lab, and hazardous wastes shipped from the lab. In addition, there are other federal requirements for the management of asbestos, pesticides, radioactive materials, and wastes.

Summary of Two Decades

To summarize, roughly 20 years of regulatory development for environmental protection has been based mainly on pollution treatment and control "at the pipe," that is, after the pollution is created. Under the 1984 amendments to RCRA, there was the first glimpse of a federal mandate for waste minimization. Under the 1986 Superfund amendments, the federal government compelled institutions to track the inventories of certain hazardous substances, estimate annual releases of specific toxic substances, and disclose this information to the public. This is a subtle shift in regulatory emphasis from management of pollution after it is created to managing hazardous materials in a manner that reduces the impact of spills and minimizes releases. This was coupled with an emerging ethic that the public has a right to know how companies and institutions are carrying out their business with respect to national environmental protection goals. So, what will the third decade bring?

The Pollution Prevention Act

This act, passed in 1990, confirms that the U.S. is entering a new phase of environmental regulatory development. In this act, Congress stated that it is national policy to prevent or reduce pollution at the source, wherever feasible, and that disposal or other release into the environment is to be used as a last resort. It specifically requires companies that file a Toxic Release Inventory to meet SARA requirements to document their efforts to reduce those releases. (The Pollution Prevention Act and its applicability to labs is discussed further in Chapter 4.)

Though most labs are not yet affected by this act, the writing is on the wall. Environmental professionals presume that, when RCRA is reauthorized, it will include measures that support or mandate waste minimization and pollution prevention. In the meantime, EPA Secretary Carol Browner has placed waste minimization and pollution prevention at the top of her agency's priority list. In 1993, EPA released in the May 28th *Federal Register* its interim final guidance on the elements of a waste minimization plan (see Appendix A). As of 1994, according to Browner's policy, the EPA plans to publicly disclose the location of all large quantity hazardous waste generators. The intent of this is to bring public pressure to bear on generators to reduce their hazardous waste, in the absence, as yet, of a clear regulatory mandate.

SO, WHY POLLUTION PREVENTION?

We have established that pollution prevention is not just on the horizon; it's knocking at the doors of institutions that use hazardous substances in the process of doing business. So, is this a good thing? EPA thinks so. Publications describing EPA's pollution prevention philosophy cite the following benefits:[5]

- Reduce costs for raw materials and waste management
- Increase process efficiency and productivity
- Maintain or increase competitiveness
- Decrease exposure to long-term liability
- Reduce present and future regulatory burdens, and lessen compliance costs

- Improve workplace safety by reducing worker exposures
- Improve environmental quality and ensure community safety
- Maintain or improve corporate image

There are also significant obstacles to implementing pollution prevention: initial capital requirements, regulatory obstacles, immediate production concerns, lack of staff time and technical expertise, and institutional inertia.

THE "GREENING" OF CORPORATE AMERICA

If pollution prevention is such a good idea, why isn't it already being done, despite the obstacles? Actually, it is being done. Pollution prevention has been referred to as "business planning with environmental benefits."[6] Companies, businesses, and institutions are beginning to talk about concepts such as "sustainable development" and "full cost pricing," and some are beginning to incorporate these ideas into their approach to business. The underlying premise is "meeting the needs of the future without compromising the ability of future generations to meet their own needs."[7] Some companies are showing a tremendous ability to change their corporate cultures.

One of the most mature efforts of nearly 20 years is 3M's *Pollution Prevention Pays* program, emphasizing source reduction. By the early 1990s, 3M had cut pollution per unit of production in half, prevented more than 600,000 tons of emissions, and saved $573 million.[8] Another front runner is S.C. Johnson, who eliminated CFCs from its aerosols in 1975, and has transferred the CFC-free technology to every manufacturer they have acquired throughout the world.[9]

These developments are now beginning to extend beyond singular entities. The Chemical Manufacturers Association (CMA) recently developed the Responsible Care® initiative to guide companies in taking a life-cycle view of their products when accounting for environmental impacts. The Business Council for Sustainable Development, a worldwide coalition of business leaders, was founded in 1991 to foster the integration of sustainability into business practices and seek government and international policy changes that would catalyze sustainable growth. The growing realization is that, not only is pollution prevention a positive activity, it is beginning to be perceived as the key to the survival of economic vitality.[10]

EPA DEFINES THE TERMS

So, what exactly are pollution prevention and waste minimization? This section will detail EPA's formal definitions of these terms. The following section will critique EPA's definitions and elaborate on the philosophy behind this book.

POLLUTION PREVENTION

EPA defines pollution prevention to be "any in-plant practice that reduces or eliminates the amount and/or toxicity of pollutants which would have entered any wastestream or would otherwise have been *released into the environment prior to management techniques such as recycling, treatment, or disposal*. It includes the design of products or processes that will lead to less waste being produced by the manufacturer or end user" (emphasis added).[11]

Notice that this definition does not include recycling, reuse, or reclamation as a step following waste generation, nor does it include treatment or destruction. Also, pollution prevention applies, not only to management of all types of wastes, but also to the management of releases to all media (air, water, soil, waste, and so on).

According to the Pollution Prevention Act, pollution prevention does include *in-process recycling*, which means the direct use, reuse, or reclamation of a waste material within a process. The primary emphasis, though, is on *source reduction*, which includes any practice that reduces the amount of contaminant entering a wastestream or the environment prior to recycling, treatment, or disposal, while reducing the hazards to public health and the environment. These practices may include equipment or process modifications, reformulation or redesign of products, substitution of less-toxic raw materials, and improvements in operating procedures, maintenance protocols, training, or inventory control practices.

WASTE MINIMIZATION

EPA defines waste minimization to include any pollution prevention measures that reduce RCRA hazardous waste. In addition, the term includes *environmentally sound recycling*. This means the material must be used, reused, or reclaimed according to the definitions of RCRA.[12]

Therefore, pollution prevention and waste minimization may be conceptualized as two overlapping circles. Many activities would be classified as both pollution prevention and waste minimization, but a few activities may be classified as only one or the other. What activities would qualify neither as pollution prevention nor as waste minimization?

- Separating wastes at the source of generation (e.g., preventing hazardous waste from being mixed with solid waste, thereby making the entire quantity hazardous)
- Transferring pollutants from one media to another (e.g., collection of organic emissions from a fume hood on a carbon filter that will ultimately have to be treated or disposed)
- Concentrating or commingling wastes for the sole purpose of volume reduction
- Other waste treatment (e.g., elementary neutralization)
- Appropriate and legal waste disposal

None of these techniques qualify as waste minimization or pollution prevention, even though they may be generally considered to be beneficial.

THE "EPA HIERARCHY"

The federal shift in emphasis from regulating pollution control to cultivating management systems that prevent the production of waste or residuals has an implicit hierarchy. Source reduction that prevents pollution is the highest form of environmental protection. If source reduction is not feasible, then pollution should be recycled in an environmentally safe manner. If recycling isn't feasible, then pollution should be treated in an environmentally safe manner. Disposal or other release into the environment should only be undertaken as a last resort, and should still be undertaken in an environmentally safe manner.

THE PHILOSOPHICAL APPROACH BEHIND THIS BOOK

The ultimate end of the EPA approach as presently expressed in national policy is a world with no waste and no other externalities. This is an agreeable long-term goal, but institutions have present constraints that make this goal impracticable to attain. Internal costs, that is, what it will cost a firm or institution in terms of staff effort, production adjustments, and retraining, are not very well identified in the policy. Energy costs are also not explicitly addressed. New equipment or processes that create less pollution may well be more energy intensive when decentralized.

For example, if a process creates nickel and chromium wastes as a by-product (e.g., the steel industry) and if these wastes cannot be further reduced following pollution prevention policy, a company would incorporate an on-site smelter into their production process to reclaim these metals. However, for smaller-scale operations, the efficiency of doing this is nil. It will probably always be more efficient to send these wastes off site to a facility that has specifically developed the technology to undertake metals reclamation most efficiently on a large scale. Along the same lines, there are few companies that could possibly undertake the retorting and triple distillation of mercury, though the use of mercury in fluorescent lamps is not likely to be surpassed by new technology in the near future. EPA does make allowances for what is "feasible" and what is "economically practicable," but to the author's knowledge has not defined these criteria to any extent.

The point of this is not to quibble with basic soundness of the EPA hierarchy. The philosophical approach of this book is to support the rationale of the EPA hierarchy while recognizing the present realities of practicing it. Responsible, efficient waste management involves elements not considered to be pollution prevention or even waste minimization by EPA. Yet, some of these elements merit attention because they are part of a total (long-term) approach to attaining the goals of less waste, reduced risk, and a more cost- and energy-efficient laboratory process. This is because these elements:

- Provide data or information vital to targeting pollution prevention or waste minimization initiatives
- Make waste management more efficient, thus freeing up staff time for waste minimization and pollution prevention
- Reduce risks to employees, waste handlers, and the public
- Save money while not compromising environmental protection — money that may help defray start-up costs for waste minimization or pollution prevention initiatives

For these reasons, this book addresses topics such as benchtop treatment, safe and appropriate use of the sanitary sewer, bulking and commingling of wastes, and source separation as well as "purer" topics such as microscale procedures, product substitution, and solvent distillation. It is the authors' intention to impart a comprehensive understanding of managing laboratory waste efficiently, economically, and in a manner that maximizes opportunities for achieving genuine source reduction and waste minimization.

HOW TO MAKE THE BEST USE OF THIS BOOK

This book is organized into seven major parts. Your role with respect to laboratories should help to define which parts would be of most interest to you. Where applicable, the chapters within a part are sequenced in order of EPA's hierarchy, with source reduction techniques discussed first.

Part 1 provides an overview to the subjects of hazardous waste management in laboratories, concepts of waste minimization, and government policies and requirements for waste minimization and pollution prevention. It will be of general interest to most readers.

Part 2 broadly discusses organizational approaches and impediments to achieving waste minimization and pollution prevention. This part may be of most interest to laboratory managers, environmental health and safety staff, executives, administrators, and policy makers.

Part 3 gets specific: its chapters discuss methods to be used to minimize pollution by media (e.g., air, water), by type of lab (e.g., clinical), and waste type (e.g., infectious, medical, radioactive). The content of each chapter will rather narrowly define its readership.

Part 4 discusses specific techniques that can be implemented at the laboratory level, regardless of whether overarching administrative support for waste minimization exists. This part may be of particular interest to those working in labs on a day-to-day basis.

Part 5 describes what can be done at the organizational level, across a number of laboratories that are part of the same institution or firm. This part will be of most interest to administrators or environmental health and safety staff who have a wide span of influence in a large organization.

Part 6 presents several case studies and applications that have been tried and tested at particular laboratory operations. These chapters should be of general interest.

Finally, Part 7 concludes the book and offers thoughts on the future direction of environmental compliance and waste management for laboratories.

Chapter 1 having set the stage, the reader should be prepared now to delve more deeply into the subject of pollution and prevention and waste minimization for laboratories. Though much of what follows is technical and detailed, we hope the reader finds that this book also suggests a sound philosophical approach to achieving waste minimization and pollution prevention, as well as specific tools to engage in the process.

ENDNOTES

1. Tort: A civil wrong that is compensable in a court of law (e.g., "negligence").
2. Robinson, N., The origins and framework of environmental law in the United States, *Earth Law J.*, 1975, 323–325.
3. POTW: publicly owned treatment works, the legal federal term for a municipal wastewater treatment plant.
4. 58 *Federal Register* 31114.
5. For example: University of Wisconsin — Extension, Solid and Hazardous Waste Education Center, July 1993, *Pollution Prevention: A Guide to Program Implementation.*
6. From University of Wisconsin — Extension, July 1993, p. 5.
7. Jane Hutterly, S.C. Johnson Company, March 8, 1994, *Environment '94* conference, Federation of Environmental Technologists, p. 8.
8. Jane Hutterly, S.C. Johnson Company, March 8, 1994, *Environment '94* conference, Federation of Environmental Technologists, p. 6.
9. Jane Hutterly, S.C. Johnson Company, March 8, 1994, *Environment '94* conference, Federation of Environmental Technologists, p. 15.
10. For detailed treatment of this subject, readers may refer to Hawken, P., *The Ecology of Commerce*, HarperCollins, New York, 1993.
11. From University of Wisconsin — Extension, July 1993, p. 3.
12. 40 CFR 261.1.
13. RCRA is codified at 42 USC § 6901 *et seq.;* herein referenced as "RCRA § 1001 *et seq.*"

CHAPTER 2

A Survey of Laboratory Waste Management and Minimization Practices

K. Leigh Leonard and Peter A. Reinhardt

CONTENTS

Introduction 16
Research Methods 16
    Survey Instrument 16
    Sampling Approach 17
        Howard Hughes Medical Institutes (HHMI) 17
        Network News Subscribers (NET) 17
        ACS Chemical Health and Safety Division (CHAS) 17
    Some Caveats 17
Results 18
    Profiles of Groups Surveyed 18
    Laboratory Chemical Waste Management Practices 19
        Off-Site Management Practices Used 20
        On-Site Management Practices Used 20
        Management Practices Not Used 21
        Use of the Sewer and Normal Trash 23
        Annual Cost of Laboratory Waste Management 24
        Laboratory Waste Management Problems 24
    Laboratory Hazardous Waste Minimization Practices 25
        Is Waste Minimization Happening? 25
        Popularity of Waste Minimization Practices 25
        Most Beneficial Waste Minimization Techniques and Why 27
        Cost Savings from Waste Minimization 27
Conclusions 28

INTRODUCTION

Laboratory staff face many challenges when mounting a serious effort to minimize their wastes. Laboratory wastes are numerous and diverse, but each type is normally generated in small volumes that are difficult to manage efficiently or economically. Some laboratory wastes, such as organic peroxides and compressed gases, can be highly dangerous to handle, and extraordinarily costly to properly dispose. Research laboratory processes are often of limited duration, making it a challenge to discover and apply waste minimization techniques before the procedure ends. Protocols may also be constrained by government regulations or national standards that proscribe waste-intensive research procedures. For example, some EPA environmental research protocols, which must be adhered to by laboratories receiving grant money from EPA, call for glassware washing with multiple solvents prior to the start of the research procedure.

In light of these and other constraints, how often do laboratory staff actually pursue waste minimization? Where laboratory waste minimization is being achieved, which techniques are proving most successful? What tangible benefits accrue from laboratory waste minimization? EPA's pollution prevention policy emphasizes techniques that reduce the amount of waste being generated. It also favors on-site recycling over off-site. How closely do laboratory practices reflect the EPA hierarchy?

Because there is scant literature addressing these questions, the authors embarked on a national survey to illuminate current laboratory waste management practices. The results of this survey help to anchor in reality the contents of this book. Laboratory waste minimization is taking place, and institutions and firms are garnering tangible benefits, including significant cost savings. These findings are detailed and explained in this chapter.

RESEARCH METHODS

A written survey was mailed to several hundred laboratories, the majority located in the U.S., within a period of about 1 year. The survey was mailed to laboratories from each of three target populations. We received a total of 180 responses. A detailed explanation of research methods and error rate calculations, as well as a copy of the survey instrument, is given in Appendix B.

SURVEY INSTRUMENT

We used a six-page written survey instrument divided into seven sections: (1) respondent profile, (2) laboratory waste management practices, (3) attitude toward sewering of laboratory chemicals, (4) attitude toward waste neutralization and treatment, (5) waste minimization practices and benefits, (6) management of selected laboratory chemicals, and (7) problems/costs of laboratory chemical waste. This chapter draws on data from sections 1, 2, 5, and 7 of the survey.

SAMPLING APPROACH

Three available subpopulations were selected for sampling: host institutions of the Howard Hughes Medical Institute, members of the Chemical Health & Safety Division of the American Chemical Society, and subscribers to *Network News*, a newsletter published by the Government Relations and Science Policy Office of the American Chemical Society. A detailed description of the sampling approach used for each group is described below.

Howard Hughes Medical Institute (HHMI)

A survey was sent to the environmental health and safety office of each host institute. There are 53 total host institutes. If the survey was not returned within a period of several weeks, we sent a second survey. After two survey rounds we received 46 responses, for a response rate of 87%.

Network News Subscribers (NET)

One uncoded survey form was sent to each subscriber as an attachment to the ACS newsletter *Network News*. There was no mechanism to follow up on nonrespondents, although the newsletter editor did include a reminder in a later issue requesting readers to fill out and return the survey. We received 60 responses out of 327 subscribers for a response rate of 18%. Because of the low response rate, we have used the results from this source primarily for qualitative purposes.

ACS Chemical Health and Safety Division (CHAS)

Using the CHAS mailing list, we selected members who we thought had a high likelihood of being associated with private, nonacademic entities that include laboratories as part of their operations. From this set of 352, we randomly selected a sample of 200. The survey forms were coded so that we could follow-up with reminders and additional survey forms to those who had not responded. After two rounds of reminder post cards, and one round in which we resent the survey form, we obtained 74 respondents. We eliminated from the sample 25 respondents that were not laboratories, leaving 175 remaining in the total sample. The response rate was 42%.

SOME CAVEATS

When reviewing data summaries, it is helpful to keep in mind that each result is dependent on the individuals filling out the survey. Based on the profile information supplied, we are confident that the majority of responses are supplied by a person knowledgeable of the operations within his or her realm (an environmental health and safety staff person, a laboratory director or principal investigator, or a laboratory staff person). Also, the survey instructions specified that we were seeking answers that pertain on an institution-wide level, but exclude consideration of operations other than laboratories (e.g., physical plant). We recognize that there may be some knowledge limitations on the part of individuals filling out the survey, but addressed this to the extent possible within the confines of a mail survey.

As already mentioned, these are three very specific subsets of laboratories. Each subset may well have an inclination toward environmental management principles. The Howard Hughes Medical Institute operates under central leadership that values environmentally sound laboratory practices. As for the other two subsets, their affiliation with those two particular ACS functions (*Network News* and Division CHAS) reflects an inherent interest in environmental management. Consequently, one must be cautious in extrapolating the results of our survey to all U.S. laboratories.

However, due to their response rates, the CHAS and HHMI survey results do reflect the behaviors of laboratory staff within each group, and we can therefore make accurate generalizations about them. These findings are important because they shed light on laboratory waste management practices, and document significant waste minimization activities within these two groups. Although the *Network News* data are more limited for the purpose of analysis, they also provide a few insights into laboratory waste management and minimization practices. In summary, while this data cannot be generalized to all U.S. laboratories, this research still comprises perhaps the most comprehensive and rigorous study to date on laboratory waste management and minimization practices.

RESULTS

This section describes the findings of the survey that are relevant to the subjects addressed in this book. The profiles of each group are presented and contrasted first. Each group reflects a different mix of laboratory types and annual waste generation rates. The next section presents an overview of waste management practices and costs within each group as a backdrop against which waste minimization practices may be considered. The last section details the waste minimization practices used by laboratories, why they are used, and the benefits laboratories are accruing from participating in waste minimization.

PROFILES OF GROUPS SURVEYED

Profile information includes the business sector of the institution, the state in which is it located, the number of laboratory employees (and students) it has, the responsibilities of the person filling out the survey, and the generation rate of regulated hazardous waste from laboratories. We did not ask for information on the generation of acute hazardous wastes. For ease of discussion we assign generator categories, but these do not necessarily correlate with regulatory categories. An institution whose laboratories generate less than 100 kg of regulated chemical hazardous waste per month is referred to as a *very small quantity generator*. One that generates between 100 and 1000 kg/month is referred to as a *small quantity generator*; one that generates more than 1000 kg/month is referred to as a *large quantity generator*. These categories are intended to convey an idea of the overall volume of regulated waste that must be handled by each laboratory rather than to describe what regulatory requirements may apply.

The survey responses indicate that the HHMI set of respondents are predominantly (95%) large academic and medical institutions. They are located in 21 states and all 10 EPA regions. Eighty percent of them are employers of greater than 1000 employees and students. (Thirty-seven percent have greater than 5000.) The group has the highest proportion (57%) of large quantity generators followed by (37%) small quantity generators. The majority (91%) of responses were supplied by the institution's environmental health & safety staff.

The survey responses indicate that the CHAS set of respondents are predominantly (93%) nonacademic, nonmedical private sector laboratories. Companies that responded are located in 29 states, and there was one response from Puerto Rico. All ten EPA regions were represented. Eighty-nine percent of them are employers of 1 to 500 employees. The group has the highest proportion of very small quantity generators (59%) followed by small quantity generators (24%). Responses were filled out by environmental health & safety staff (47%), laboratory directors/managers (30%), and laboratory staff (15%).

To summarize, the HHMI set represents mainly large academic and medical institutions that may generally be assumed to hold not-for-profit status. The CHAS set represents mostly small private sector (for profit) laboratory operations. The HHMI respondents are generally large quantity generators, while more than half the CHAS respondents are very small quantity generators.

LABORATORY CHEMICAL WASTE MANAGEMENT PRACTICES

The second section of the survey instrument provides a comprehensive list of laboratory waste management practices and directs respondents to identify whether each practice is conducted on-site, off-site, or not used at all. Section 7 requests information on the annual cost of laboratory waste management. It also includes an open-ended question that asks respondents to identify their most difficult problem in managing laboratory waste. The results of these two sections of the survey are summarized below. Only the highlights of this analysis are provided because the emphasis of this book is on laboratory waste minimization.

For both HHMI and CHAS respondents, some management practices were much more likely to be used off-site than on-site, such as hazardous waste incineration, fuel blending, use of hazardous waste landfills, and medical/pathological incineration. Conversely, elementary neutralization, redistribution, long-term storage, and evaporation are predominantly on-site methods. For both off-site and on-site management, HHMI laboratories use a greater range of management techniques for laboratory chemical waste than CHAS laboratories.

Table 1 Off-Site Waste Management Practices Used

Management practice	HHMI (%)	CHAS (%)
Hazardous waste incineration	98	68
Fuel blending	80	51
Hazardous waste landfill	74	55
Med/path incineration	65	23
Recovery (not solvents)	46	30
Chemical treatment	35	9
Elementary neutralization	33	4
Open burn/detonate	28	7
Storage	9	7
Redistribution	4	11
Wastewater (not POTW)	4	5
Solvent distillation	4	5
Evaporation	0	0
Other atmospheric release	0	0

Off-Site Management Practices Used

These survey results are summarized by Table 1. The most commonly used off-site waste management techniques for each data set are destruction in a permitted hazardous waste incinerator, disposal in a permitted hazardous waste landfill, and fuel blending for energy recovery in cement kilns. For these practices, a lower percentage of the CHAS respondents use each practice with the greatest difference being for the practice of fuel blending: of the HHMI respondents, 80% utilize fuel blending while 51% of the CHAS respondents do.

Recovery, not including solvent distillation, constitutes a significant off-site activity for both the HHMI respondents (46%) and the CHAS respondents (30%). These activities might include off-site reclamation of mercury, other precious metals, and perhaps compressed gas cylinders. Off-site solvent distillation is not used extensively by either the HHMI or CHAS respondents (5% or less). This finding may reflect, among other factors, the difficulty of attracting commercial interest in the small volumes of solvent mixtures usually generated by laboratories. It also correlates with the high prevalence of fuel blending, a less expensive alternative for drum-quantity solvent volumes.

Redistribution of unwanted chemicals is not used extensively by either the CHAS respondents or the HHMI group (11 and 4%, respectively).

On-Site Management Practices Used

These survey results are summarized by Table 2. The data show that, in general, members of the HHMI sample are utilizing a greater range of on-site management techniques for laboratory chemical wastes than are members of the CHAS sample. More than 50% of the HHMI respondents report using elementary neutralization, redistribution, recovery, and storage. In comparison, only one on-site management technique, elementary neutralization, was used by more than 50% of the CHAS respondents.

On-site redistribution was used prevalently (76%) by the HHMI set, while this practice was used by only 31% of the CHAS respondents.

Table 2 On-Site Waste Management Practices Used

Management practice	HHMI (%)	CHAS (%)
Elementary neutralization	72	68
Redistribution	76	31
Recovery (not solvents)	54	36
Storage	61	26
Solvent distillation	44	26
Chemical treatment	43	22
Med/path incineration	35	5
Evaporation	15	19
Other atmospheric release	9	8
Fuel blending	9	8
Wastewater (not POTW)	9	5
Incineration	4	3
Landfill	0	7
Open burn/detonate	0	3

Elementary (simple acid-base) neutralization was used by a high percentage of both the HHMI (72%) and the CHAS (68%) respondents. This is to be expected, since corrosives are a common waste, and elementary neutralization is easy to practice and is the one type of treatment clearly allowed under federal hazardous waste regulations. Chemical treatment (other than elementary neutralization) is used by 43% of the HHMI respondents, and by 22% of the CHAS respondents.

On-site solvent distillation is practiced by 44% of the HHMI respondents and 26% of the CHAS group. Recovery of materials from laboratory chemical wastes, not including solvent distillation, ranks high in both groups. Examples of these activities might include recovery of metallic silver from photographic solutions and reuse of barely contaminated research solvents in applications where purity is not as important. Of the HHMI respondents, 54% report using recovery, as compared with 36% of the CHAS respondents.

As surprising result is that a small, but significant, percentage of each set of respondents (HHMI: 15%, CHAS: 19%) report the use evaporation as an on-site management technique for laboratory chemical wastes. We speculate that this includes the evaporation of water from inorganic salt solutions to reduce the volume (and hence the cost) of hazardous waste that must be shipped off-site.

Management Practices Not Used

Table 3 summarizes data indicating the extent to which management practices are not used at all (either on-site or off-site). These data do not include blanks, where the respondent did not fill in any boxes to indicate whether a management practice is used on-site, off-site, or not used at all. To be included in this summary, respondents actively indicated that their institution's laboratories do not use a particular management technique.

There is a marked contrast between behaviors of the HHMI and CHAS groups. Except for three management practices, a much greater percentage of the CHAS respondents reports that they do not use a certain management technique than do the HHMI respondents. Specific findings from Table 3 are selectively discussed below.

Table 3 Waste Management Practices not Used

Management practice	HHMI (%)	CHAS (%)
Other atmospheric release	89	87
Evaporation	85	77
Wastewater (not POTW)	83	78
Open burn/detonate	70	89
Distillation	54	65
Storage	33	66
Chemical treatment	33	64
Med/path incineration	13	69
Redistribution	22	55
Landfill	26	38
Recovery (not solvents)	17	35
Fuel blending	11	39
Elementary neutralization	15	27
Incineration	2	28

Both groups are significantly averse to disposal of laboratory chemical wastes in a licensed hazardous waste landfill. Of the HHMI respondents, 26% do not use such landfill disposal, while 38% of the CHAS respondents do not. This may reflect policy choices by these institutions to limit long-term liability, as well as the broad scope of the federal land ban restrictions, which effectively prohibit the landfill of the majority of laboratory wastes without pretreatment.

Only 2% of the HHMI group report that they do not use incineration. Because incineration is a widely accepted means of disposing hazardous wastes with minimal long-term liability, this result is to be expected. Of the CHAS respondents, 28% report that they do not use incineration, a surprisingly high figure. Whether due to an aversion to the cost of incineration, a lack of wastestreams that require incineration for effective management, or some other factor is not clear.

The CHAS respondents appear to have a greater aversion to fuel blending than do the HHMI respondents. Only 11% of the HHMI group do not use fuel blending, as compared with 39% of the CHAS group. There may be many reasons for this. These results may reflect reluctance among some private laboratories concerning the use of fuel blending, a somewhat controversial management technique, or, considered with the data on incineration, this may also indicate that some laboratories in the CHAS group may not generate solvent wastestreams, and hence would have less need for incineration or fuel blending.

The use of distillation is similar for both groups. Of the HHMI respondents, 54% report that they do not use distillation, as compared with 65% of the CHAS respondents.

A significant percentage of both groups (HHMI: 15%, CHAS: 27%) do not use elementary neutralization as a management technique. It is difficult to believe that these percentages of laboratory institutions do not generate any acid/base laboratory wastes. It may be that a number of laboratories are electing to avoid the use of elementary neutralization. This is surprising, given the legality of the practice in most states, and the ease and cost effectiveness with which it can be carried out.

LABORATORY WASTE MANAGEMENT AND MINIMIZATION PRACTICES

Table 4   Use of the Sewer and Normal Trash

Management practice	HHMI (%)	CHAS (%)
Uses sewer[a]	65	54
Does not use sewer	35	43
Uses normal trash[a]	59	49
Does not use normal trash	39	49

[a] The percentage for use of sewer or normal trash includes all respondents who report either on-site or off-site use, or both. Numbers may not add up to 100% because some respondents did not answer the question.

Chemical treatment (other than elementary neutralization) is also avoided by a sizable proportion of each group, especially by CHAS respondents. Of the HHMI group, 33% report that they do not use chemical treatment, as compared with 64% of the CHAS respondents.

Both the CHAS respondents and the HHMI group generally avoid the use of evaporation as a waste management technique. Of the CHAS respondents, 77% report that they do not use evaporation as compared with 85% of the HHMI respondents.

Use of the Sewer and Normal Trash

The use of the sanitary sewer and normal trash for disposal of certain laboratory chemical wastes is highlighted very briefly here since these practices are not a major emphasis of this book.

Nonhazardous chemical wastes disposed through the normal trash generally present no hazard to solid waste handlers. Wastes normally selected for drain disposal are quickly degraded through the normal wastewater treatment process. Responsible use of the normal trash and sanitary sewer for disposal of laboratory chemicals is much less costly, and more efficient, than other management options.

However, even where strict controls are used, the perception that "harmful" chemicals are being disposed inappropriately may prevail. On occasion, these perceptions may be expressed through sweeping local regulations prohibiting disposal of any chemical waste via the sewer or normal trash. Even in the absence of local restrictions, the threat of public criticism or the risk of a waste separation mistake may completely inhibit laboratories from appropriately using these avenues.

Table 4 summarizes the use of the sewer and normal trash by both groups. It is interesting to note the extent to which these practices are not used at all. Of the HHMI respondents, 35% report that they do not use the sanitary sewer for the disposal of laboratory chemical wastes, as compared with 43% of the CHAS respondents. For use of the normal trash, 39% of the HHMI respondents report that they do not use this management option as compared with 49% of the CHAS respondents. These are significant proportions of both groups. Apparently, many laboratories are either prohibited or strongly inhibited from utilizing these options for the management of laboratory chemical wastes.

Annual Cost of Laboratory Waste Management

Tables 5 and 6 summarize data on laboratory hazardous waste disposal costs. The survey requested respondents to provide the approximate annual cost for off-site commercial disposal of laboratory chemical wastes. Table 5 shows the minimum, the maximum, and the average annual cost for waste management. The average cost is much higher for the HHMI group ($223,511) than for the CHAS group ($52,163). Two respondents in the CHAS sample reported that they spend zero dollars on off-site disposal, implying that they are able to manage their wastes on-site.

Table 6 presents average cost for each category of waste generation. As expected, both groups show a similar trend of increasing total cost with increasing rate of generation.

Laboratory Waste Management Problems

There was a wide variety of answers to this open-ended question in section 7 of the survey. It is beyond the purpose of this chapter to present detailed analysis of this data. However, to provide the reader with a feel for the type of waste management problems reported by laboratories, a few examples are listed as follows.

Table 5 Annual Cost for Off-Site Waste Disposal ($)

Minimum	Maximum	Average
Howard Hughes Medical Institutes (98% reporting)		
12,500	2,100,000	223,511
Chemical Health and Safety Division (77% reporting)		
0	500,000	52,163

Table 6 Average Off-Site Waste Disposal Cost by Generation Rate

Generation rate (kg/month)	Average cost ($)
Howard Hughes Medical Institutes (98% reporting cost)	
>1000	294,423
100–1000	99,719
<100	28,750
Chemical Health and Safety Division (77% reporting cost)	
>1000	185,625
100–1,000	38,833
<100	25,153

From all respondents (including the *Network News* sample), compressed gases, mercury and other heavy metal wastes, mixed wastes, and unknowns were commonly identified problem wastestreams. There were numerous administrative problems identified as well. Examples include difficulty training laboratory staff, getting compliance with waste collection procedures, and lack of funding or personnel resources for waste program.

LABORATORY HAZARDOUS WASTE MINIMIZATION PRACTICES

To this point, this chapter has examined survey results relating to management practices, costs, and problems associated with laboratory chemical waste disposal. This section summarizes laboratory waste minimization practices reported through the survey.

Is Waste Minimization Happening?

Table 7 provides an overview of the extent to which waste minimization is practiced and documented. Survey results show that laboratory waste minimization is widely practiced. Of the HHMI respondents, 98% report that their institutions make an effort to minimize laboratory chemical wastes, as do 91% of the CHAS respondents. Of the *Network News* subscribers, 100% report that they engage in waste minimization. These findings are very encouraging, and demonstrate that laboratories are minimizing wastes despite the difficulty of doing so.

A much lower proportion of laboratories has written documentation for waste minimization though. Of those HHMI respondents who report waste minimization activities, 56% report that their institution has a waste minimization plan or policy, and the proportion for CHAS respondents is even lower at 39%. Even fewer of the respondents have written plans that set specific goals within their waste minimization plans: of the respondents who have plans/policies, 36% (HHMI) and 39% (CHAS) report that they have goals.

What are some examples of specific laboratory waste minimization goals? A few of the goals reported by individual respondents from both groups include achieving a 10% reduction in chemical waste by January 1994, distilling alcohol wastes by 1994, achieving a 20% reduction in waste generation by 1995, and evaluating each area for reduction within 2 years. One of the *Network News* respondents reported as a goal to reduce lab pack wastes to 50% of total laboratory wastes.

Popularity of Waste Minimization Practices

The proportion of respondents who reported using particular waste minimization techniques is summarized by Table 8. Percentages shown in Table 8 represent the ratio of the number who report the use of a particular technique to the total number reporting that their institution engages in waste minimization. First discussed are methods most commonly used and least commonly used by both

Table 7 Extent of Waste Minimization Programs

Measure of program	HHMI (%)	CHAS (%)
Practice waste minimization	98	91
Have waste minimization plan[a]	56	39
Have specific goals for plan[b]	36	39

[a] The percent of those that report waste minimization practices who also report that they have a written plan.

[b] The percent of those who report that they have a written plan that also have specific written goals for waste minimization.

Table 8 Popularity of Waste Minimization Practices

Minimization practice	HHMI (%)	CHAS (%)
Substitute nonhazardous material	91	94
Bulk/commingle	96	72
Purchase less	76	85
Change equipment or procedure	67	82
Reduce scale (includes microscale)	76	69
Recovery or recycling (not distill)	87	52
Redistribute surplus chemicals	78	48
Chemical inventory management	49	70
Chemical treatment	71	42
Other process change to use less	49	69
Purchase controls	27	69
Distillation	44	28
Volume reduction (not bulking)	33	39
Other beneficial use	36	33
Computer simulation	11	18

groups, followed by discussion of methods where the groups differ substantially in their behaviors.

Substitution with less hazardous or nonhazardous materials, a true source reduction measure, receives the highest average ranking (HHMI: 91%, CHAS: 94%). Purchasing less to minimize surplus chemicals is also highly ranked (HHMI: 76%, CHAS: 85%). Also favored by both groups are changes in equipment or procedure (CHAS: 82%; HHMI: 67%) and reducing scale, including microscale techniques (HHMI: 76%, CHAS 69%).

Least commonly used is computer simulation or modeling to replace wet laboratories though significant percentages of both groups report that they do use this method: 11% for HHMI and 18% for CHAS. Other beneficial use and volume reduction (not bulking) are other methods not commonly used by either group.

Though second highest ranked on average, commingling of similar wastes to reduce volume is practiced more by the HHMI respondents (96%) than the CHAS respondents (72%).

The HHMI respondents use recovery/recycling (not distillation) (87%) much more extensively than the CHAS respondents (52%).

Distillation is used to a greater extent by the HHMI respondents (44%) than the CHAS respondents (28%), a finding that correlates with on-site waste management data. Chemical treatment (which may include elementary neutralization) is also used to a greater extent by the HHMI respondents (71%) than by the CHAS respondents (42%).

Some very interesting differences between the two groups occur for three practices: redistribution of surplus chemicals, chemical inventory management, and purchasing controls. The HHMI respondents are much more likely to engage in redistribution than the CHAS respondents, (78% for HHMI, 48% for CHAS). This may be because larger institutions have greater opportunities for chemical redistribution. In contrast, the CHAS respondents are much more likely to use purchasing controls (69%) and chemical inventory management (70%) than are the HHMI respondents (27 and 49%, respectively). These findings may be related and are discussed in detail in the conclusions.

Most Beneficial Waste Minimization Techniques and Why

Survey respondents were asked to identify their institution's most beneficial waste minimization method for laboratory chemical waste and to state why it is the most beneficial.

For the HHMI respondents reporting, commingling of wastes to reduce volume is most often (22%, or ten respondents) cited as the most beneficial waste minimization technique, because of associated cost savings, and its ease and effectiveness. In contrast, commingling is cited by only four CHAS respondents.

Reducing the scale of procedures is also frequently cited by HHMI respondents (20%, or nine respondents) as the most beneficial waste minimization technique, because it reduces waste volumes, employee exposures, and chemical storage needs, and because it is "advanced" and can be "applied universally." Fourteen CHAS respondents (21%) also favor reducing scale. In addition to reasons already mentioned, these respondents cite conformance with EPA protocols and cost savings.

Redistribution of excess chemical products is reported by 16% of the HHMI respondents as the most beneficial waste minimization technique because of associated cost savings, because it reduces wastes, and is "advanced" and effective. Redistribution was mentioned by only four of the CHAS respondents.

Purchasing-related techniques were also reported by CHAS respondents (21%) as the most beneficial waste minimization approach. These include purchasing less, as well as controlling purchases and chemical inventory management. Only 13% of HHMI respondents identify purchasing-related techniques as most beneficial.

Except for waste commingling, techniques regarded as most beneficial in both groups are management strategies that achieve true waste reductions. The practice of commingling obviously has substantial monetary and efficiency benefits that cannot be disregarded, even though this practice does not achieve true reductions in the amount of waste generated.

Cost Savings from Waste Minimization

Those respondents who report that their institution minimizes waste were asked to report the estimated annual net savings from their most beneficial waste minimization technique. These data are summarized in Table 9.

Table 9 Annual Savings From Most Beneficial Waste Minimization Method

Method	Savings reported ($)
Howard Hughes Medical Institutes	
Silver recovery	4,000
Microscale	10,000
Redistribution	15,000
Neutralization	5,000–20,000
Solvent distillation	4,000–30,000
Commingling	10,000–200,000
Chemical Health and Safety Division	
Process change	1,000
Distillation	25,000
Substitution	50,000
Purchasing-related methods	500–100,000
Reducing scale	2,300–80,350
Redistribution	500–400,000

Of the HHMI respondents who engage in waste minimization, only 44% could identify estimated annual net savings associated with their stated most beneficial waste minimization method. Figures were most readily provided for commingling, with savings up to $200,000 reported. Significant savings were also reported for solvent distillation, neutralization, and redistribution.

Even fewer (21%) of the CHAS respondents who engage in waste minimization were able to identify savings associated with their stated most beneficial technique. Redistribution garnered the highest savings (up to $400,000). Purchasing-related methods (inventory control, purchasing controls, purchasing less) saved up to $100,000. Significant savings were also reported for reducing scale, substitution, and distillation.

Respondents were also asked to report the annual cost of, and savings from, laboratory waste minimization if known. Only a handful of respondents from each group have a handle on total savings from waste minimization. Four HHMI respondents report annual costs for waste minimization ranging from zero to $20,000, and net savings ranging from $10,000 to $135,000. Five CHAS respondents report annual costs for waste minimization ranging from zero to $12,000, and net savings ranging from minus $9,500 (costs exceeded savings in this year) to $105,439.

CONCLUSIONS

Laboratories use a wide range of laboratory waste management techniques. For these two groups, the CHAS data set reflected generally lower percentages of waste management methods used (and higher percentages of methods not used) than the HHMI data set. The explanation for this is not obvious, but we offer a bit of speculation. The private sector laboratories (CHAS) appear to use a narrow range of waste management techniques, possibly reflecting fewer wastestreams as may be generated by some types of commercial laboratories (e.g., QA/QC laboratories, narrowly targeted analytical laboratories). The large academic institutions (HHMI) appear to use a wide variety of management techniques, perhaps due to the complexity and number of their laboratory operations, and the diversity of their wastestreams. In future research it would be helpful to request information on types of wastes generated to confirm this hypothesis.

To highlight the numerous findings outlined in this chapter, key findings from both the waste management and waste minimization data are discussed and tied together:

Distillation — Off-site solvent distillation is not widely used. As already stated, this may reflect unfavorable market factors. Solvent distillation is much more likely to be practiced on-site by both groups — 11 times more likely for the HHMI respondents and 5 times more likely for the CHAS respondents. It appears that off-site distillation of laboratory solvents has great potential for growth if impediments can be identified and overcome.

The waste management and minimization data both show that the HHMI group is able to utilize on-site distillation to a much greater extent (about 1.7 times more) than the CHAS group. This may reflect their respective waste generation rates, CHAS tending to be small or very small quantity generators while HHMI are mostly large quantity generators. The economics of on-site distillation are such that a significant volume of waste solvents is needed to have a reasonable payback on investments (staff time and equipment) in the program.

Chemical treatment — The waste management and minimization data both show that the HHMI respondents are about twice as likely to utilize on-site chemical treatment than the CHAS respondents. There might be a number of reasons for this — more restrictive policies in the private sector, differences in regulatory scrutiny, and, among academic institutions, more acceptance of treatment when conducted as part of the educational process.

It is also interesting that a high percentage of both groups are not inclined to use chemical treatment at all (33% for HHMI and 64% for CHAS). While the regulatory status of on-site chemical treatment is much more variable than for elementary neutralization, there is a growing body of literature on benchtop treatment techniques, and the practice is becoming more accepted, especially in the academic setting. Its benefits include significant cost savings and reduced handling risk as the waste is treated in the laboratory or at an on-site facility rather than being packaged, transported, and managed at an off-site location. These data indicate that — despite the benefits of chemical treatment — its uncertain regulatory status and the additional staff effort required may be somewhat discouraging to laboratories.

Redistribution, purchasing controls, and chemical inventory management — Off-site redistribution is not popular. On-site use differs between the two groups: HHMI respondents are about twice as likely to use this method than the CHAS respondents. This might be attributed to the large number and wide diversity of laboratory operations (from freshman instructional to postdoctorate research) associated with most Howard Hughes Medical Institutes. This likely increases the potential for one laboratory to use another laboratory's excess chemicals. Since the CHAS respondents represent much smaller laboratories, and some of these may have very specific functions requiring a limited variety of chemicals, it follows that the opportunities for on-site redistribution might be fewer. Considering that on-site redistribution was cited by HHMI respondents as among the most beneficial waste minimization techniques, off-site redistribution may be worth pursuing. Also, CHAS respondents and other private laboratories should fully investigate the potential of redistribution.

Another factor may be the extent to which these groups are able to use chemical inventory management and purchasing controls. For those respondents who report that they engage in waste minimization activities, CHAS respondents are over 2.5 times more likely to use purchasing controls than HHMI respondents. The CHAS respondents are also more likely to implement chemical inventory management. These findings are consistent with the experience of both authors, who have found that purchasing controls and chemical inventory management are difficult to implement centrally in large academic institutions.

Commingling — Like it or not, this practice is a significant time and money saver and should not be discounted, though it is not a true waste reduction technique. It is important that laboratories pursue the appropriate use of this practice because it will save significant dollars that can be applied toward practices that reside higher on the waste minimization hierarchy.

Waste Minimization — The good news is that over 90% of all three groups of laboratories surveyed are engaging in waste minimization. A high proportion of these practices are true pollution prevention and waste reduction techniques. A good example of this is that reducing the scale of procedures was considered the most beneficial waste minimization technique by a significant number in both the CHAS and HHMI groups.

The bad news is that laboratories in these groups are not doing a good job of documenting their programs. Many laboratories that report practicing waste minimization do not have a written plan for waste minimization and even fewer have stated goals for waste minimization. Worse yet, very few respondents have any idea of how much money is being saved by implementing waste minimization methods. However, those who are keeping track of costs are generally showing impressive savings from waste minimization. Some even report savings in the hundreds of thousands of dollars!

The reasons for documenting laboratory waste minimization programs and their associated savings are many. Documenting cost savings from waste minimization helps convince administrators that waste minimization is a worthwhile activity to expend staff resources on, even though it is not yet required by law. Also, in the event new regulations mandate specified waste reduction rates in upcoming years, having records of past waste reduction efforts may help convince regulators that your institution has already met those targets. Finally, documenting waste minimization efforts is the best way to pass on the knowledge to other laboratories, and help them prioritize their efforts.

CHAPTER 3

Investigating Waste Minimization Possibilities

Cynthia Klein-Banai

CONTENTS

Introduction 31
Informational and Compliance Programs 34
Assessing Waste Minimization 34
Planning and Organization 35
    Assessing Wastestreams 35
    Evaluating Feasibility 37
    Implementing Waste Minimization Projects 37
Prioritizing Approaches 38
Actual Projects Investigated 39
    Structure of the Study 39
    Results and Conclusions of the Study 39
        Survey Findings 39
        Wastestream Minimization Strategies 44
        Focus on Individual Labs 48
    Follow-Up 49
Conclusion 53
Endnotes 53

INTRODUCTION

    Three major parties at an institution should be involved in waste minimization in laboratories: the administration, the waste manager/environmental compliance officer, and the laboratory workers. While a waste minimization plan should have top-down support and commitment, it should also have bottom-up input and participation. However, there are many things a laboratory worker can do even if the administration has not made a policy or committed resources toward pollution prevention.

When waste minimization starts at the administrative level (top down), organization is initiated through the formation of policy and may include the establishment of a waste minimization assessment committee as discussed in Chapter 6. The committee will investigate waste minimization options for the institution. After obtaining a laundry list of options, the committee can analyze them for feasibility and implement the best ones. This assessment procedure (Figure 1) has been adopted by the U.S. Environmental Protection Agency (EPA) as the recommended approach to waste minimization assessment.[1]

Alternatively, the initiation of waste minimization may come from the side of the hazardous waste manager/environmental compliance officer. This person sees what wastes are generated and comprehends how waste minimization can ultimately reduce responsibilities. The waste manager knows what wastes are in the institution and the disposal issues, costs, management needs, and regulations related to them. From previous years of waste management, the waste manager also has the data to show where the wastestreams that need targeting are.

Many waste minimization strategies are also plain common sense. In other words, a certain level of waste minimization may be obtained by making some fairly logical, even obvious changes in the way laboratories are managed. The waste manager can provide guidelines in this area.

The best level of pollution prevention in laboratories would be achieved if this waste manager were an expert in every area of laboratory work that produces hazardous waste in her facility. A high level of technical knowledge, familiarity with the work being done, and the specific techniques used in a given lab are essential to achieve the greatest degree of pollution prevention. Therefore, the people in the laboratories need to be tapped for their knowledge as well. They know what laboratory work generates the waste, what alternatives there might be to those hazardous processes or materials, and what the advantages and disadvantages of these alternatives might be.

There needs then to be a symbiotic relationship between the waste manager and the laboratory worker. Not only does the waste manager need to be educated about what is done in the lab, the lab workers need to be educated in some basic waste management and minimization concepts. Yet, as explained in Part 4 of this book, even if there is no initiation of waste minimization from the administration or waste manager the lab workers can still implement waste minimization in their own labs.

This chapter will outline an approach to investigating pollution prevention possibilities, both from the institutional aspect and the laboratory worker's aspect. We will also discuss the results of a study that identified and implemented waste minimization strategies at a research institution.

INVESTIGATING WASTE MINIMIZATION POSSIBILITIES

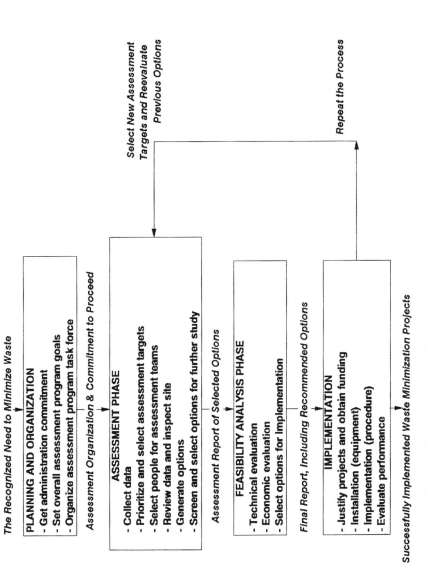

Figure 1 The waste minimization assessment procedure.

Table 1 Definitions

Generator: Any person who first creates a hazardous waste, or any person who first makes the waste subject to subtitle C regulation (e.g., imports a hazardous waste, initiates a shipment of a hazardous waste from a TSD, or mixes hazardous wastes of different DOT shipping descriptions by placing them into a single container)

Hazardous Waste: As defined in the regulations, a solid waste is hazardous if it meets one of these four conditions:

(1) Exhibits a characteristic of a hazardous waste (40 CFR Sections 261.20 through 262.240
(2) Has been listed as hazardous (40 CFR Sections 261.31 through 261.330
(3) Is a mixture containing a listed hazardous waste and a nonhazardous solid waste
(4) Is not excluded from regulation as a hazardous waste

HSWA: The Hazardous and Solid Waste Amendments of 1984 that significantly expanded both the scope and coverage of RCRA

Manifest: The shipping document used for identifying the quantity, composition, origin, routing, and designation of hazardous waste during its transportation from the point of generation to the point of treatment, storage, or disposal

Small Quantity Generator: Produces 100–1000 kg of hazardous waste at a site per month, or less than 1 kg of acutely hazardous waste per month

Large Quantity Generator: Produces 1000 kg or more of hazardous waste per month, or 1 kg or more of acutely hazardous waste per month; these generators are fully regulated

INFORMATIONAL AND COMPLIANCE PROGRAMS

While having a waste minimization program in place is of great importance, it should be clear that a waste management program must already be in place before a waste minimization program is started. Institutions with laboratories must have a compliance program in place to properly collect and dispose of hazardous waste. Researchers and laboratory workers need to know what safe and legal waste handling and disposal is. These are the roles of the waste manager, referred to above. An institution that does not have a hazardous waste management program in place cannot avoid the issue by only having a waste minimization program. Even a small quantity generator of hazardous waste is required to dispose of its waste through a hazardous waste contractor. Ideally, these programs can be initiated at the same time if neither exists already. As discussed, the waste manager's expertise is valuable to achieving waste minimization. Table 1 gives some basic definitions.[2] For more information on establishing such a program, refer to Endnotes 3 to 5.

ASSESSING WASTE MINIMIZATION

With every hazardous waste shipment, an employee of your institution signs manifests that state that your institution has "a program in place in order to reduce the volume and toxicity of waste generated" and he has "selected the practicable method of treatment, storage, or disposal currently available" to minimize "the present and future threat to human health and the environment." The EPA's "Waste Minimization Assessment Procedure" (Figure 1) can get you started on a gradual basis. First, consider the easiest implemented and most effective approaches to waste minimization. This approach will integrate the three institutional parties to investigate and implement waste minimization. By reading this book you have taken the first step of recognizing the need for waste minimization.

PLANNING AND ORGANIZATION

Administration commitment is extremely desirable. Administration or management should enact policies that encourage waste minimization. Waste minimization should be a goal of the institution. This creates an environment conducive to allowing waste managers to implement waste reduction practices. They should set goals for the amount of waste minimization desired and establish a task force or appoint someone to be in charge or implement the program. In labs where there is a commitment from administration toward safety there is more of a willingness among laboratory staff to implement changes.

The EPA recommends creating an assessment program task force that includes members of any group or department that has a significant interest in the outcome of the program. This task force would get commitment and policy from the administration, set waste minimization goals, prioritize the wastestreams, and select assessment teams, among other things. The waste minimization task force is concerned with the whole institution. The focus of an assessment team is more specific, concentrating on a particular wastestream or department. These teams should include people with direct responsibility and knowledge of the particular wastestream or laboratory work (i.e., chemistry, life sciences, engineering, operation and maintenance, etc.).[6] It may not be necessary to have both a task force and assessment teams, depending on the size of your institution.

It may not be necessary to have either of these committees. The establishment of policy might be made by an administrative unit. Policy can provide incentive and support for waste minimization activities that are carried out in laboratories, but is not a prerequisite for implementing waste minimization. There are many steps laboratories themselves can take toward reducing the waste they generate.

ASSESSING WASTESTREAMS

After organizing and getting commitment to waste minimization, you can proceed to the assessment phase. This work is best done by someone familiar with the institution's waste management program and/or someone familiar with the work done in the labs that generate waste. To know where to start, it helps to have an overall knowledge of what wastestreams there are and how they are handled.

Table 2 Labs Generating in Excess of 50 l Acid Waste Annually

Acid-waste generator	Amount	Percent of total
State analytical lab	1400	28
Chemistry teaching lab	783	16
Art teaching lab	186	3
Microelectronics research lab	172	3
Central stores[a]	118	2
Water treatment station[a]	90	2
Agriculture lab	83	2
Materials research lab	80	2
Environmental studies lab	70	1
Pesticide analysis lab	70	1
Environmental engineering lab	70	1
Engineering lab	67	1
Hazardous materials lab	50	1
Total liters (labs above)	3339	63
Total of all labs[b]	5037	100

[a] Not laboratories.
[b] 149 total laboratories.

Collect data on the wastestreams generated: what the major chemical components are, what areas they are coming from, in what quantities, and what processes generate them. These data can be obtained from more than one source. Records of hazardous waste pick-ups in your institution might be assembled similar to data shown in Table 2. This shows the lab generating the waste, the quantity, and the percentage of the total wastestream.

Alternatively, look at the purchase records. Figure 2 shows the proportion of different solvents purchased for the top five solvent generating labs within the School of Chemical Sciences at the University of Illinois. These data gives a more-detailed breakdown of the solvents labs were using than data from the waste pick-up records. However, it cannot be assumed that the same proportions of solvents purchased are disposed of as hazardous waste. In fact, data from a study conducted at the University of Illinois (UIUC) indicated that only about 50% of solvents purchased in one organic chemistry research group were disposed of as hazardous waste.[7] It was speculated that these solvents are lost through evaporation or down the drain, topics that are covered in Part 1 of this book.

Select assessment target wastestreams from information like this. Choose people for assessment teams, if necessary. Then inspect the labs, find out where these solvents are used, and how they are disposed of. In the above example, hexane, methylene chloride, and ethyl acetate are used primarily for column chromatography. Acetone is used mostly for cleaning. Once the processes generating the waste are understood by the waste management part of the team and the waste disposal issues are clear to the laboratory team member, options for waste minimization can be screened and selected. Many ideas for waste minimization will come out of discussions with people in the various labs. Suggestions of what one lab has done may be applicable to another. It may be helpful to contact various companies that produce alternative materials and equipment that result in pollution prevention. Having a "library" of these promotional materials that can be disseminated to labs is very useful. More suggestions will be explored later in this chapter.

EVALUATING FEASIBILITY

The next phase involves feasibility analysis in two areas. A technical evaluation looks at what equipment as well as facility and personnel needs there are. The economic evaluation looks at purchase costs for equipment and saved chemical purchase and disposal costs. The most feasible projects are then chosen for implementation. In the first round of waste minimization, the least costly and less technical solutions are usually the best ones to implement. In this way, results can be obtained quickly and the skeptics can be persuaded that waste minimization is worthwhile.

IMPLEMENTING WASTE MINIMIZATION PROJECTS

To implement projects you must first justify them legally and economically, and then obtain the necessary funding. By having cost analysis on payback periods, you can convince administration that this is a worthwhile investment. Some projects require little financial resources, such as improving housekeeping practices, inventory, and dissemination of information. If equipment is required, this is the stage when it is installed and the procedure is implemented. Then evaluate the performance of this procedure. Once the first waste minimization projects have been implemented, repeat the process by going back to the assessment phase and so on.

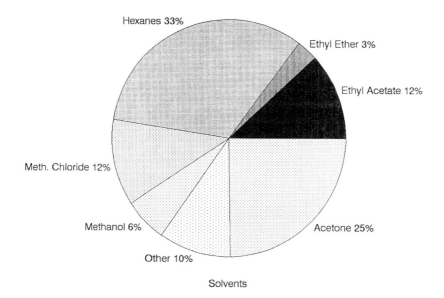

Figure 2 Annualized solvent purchases (1991) for top five organic chemistry laboratories.

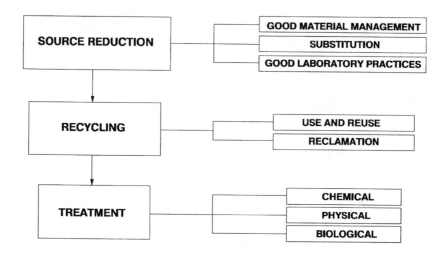

Figure 3 Waste minimization techniques for laboratories.

PRIORITIZING APPROACHES

The EPA defines the preferred order of exploration of waste minimization possibilities[6] and relative environmental desirability as shown in Figure 3. Source reduction is the most desirable level of pollution prevention. This is any activity that reduces or eliminates the generation of hazardous waste at the source. In labs this can be achieved by good material management, substitutions, and good lab practices. Figure 4 gives some more specific examples of source reduction in these categories. Many of the suggestions in this area are just common sense.

Maintaining an inventory in each lab enables workers to know exactly what chemicals they have (required by the OSHA lab standard as well), how much, and where they are located. This may be a fairly large job in some labs, but should be each lab's own responsibility. The ideal system would allow mutual access to all laboratories' inventories within a department, building, or institutions. This facilitates sharing of chemicals among labs and reduces the problem of disposing of excess chemicals.

The next area of exploration is recycling. This is where a waste material is either used for another purpose, treated and reused in the same process, or reclaimed for another process. A chemical redistribution program would be an example of use and reuse. Our experience has shown that a program that pulls unused reagent chemicals before they are disposed of and finds other users for them is cost effective and therefore an essential part of any laboratory waste minimization program. The chemical redistribution program at the University of Illinois saved approximately $30,000 in fiscal year 1991 (Table 3). We believe that this can be easily increased with minimal effort. This is enough to justify employing someone exclusively for the purpose of running such a program, even though we have found it does not require 100% time.

Another area of recycling worthy of exploration is solvent distillation (see below). If the volume of solvents used cannot be reduced, the next best alternative

SOURCE REDUCTION

Good Material Management

Better housekeeping
Regular chemical inventories
Centralized purchasing
Purchase in quantities needed
Centralized waste management
Good waste segregation

Good Laboratory Practices

Increased instrumentation
Scaling down experiments
Microscale (teaching labs)
Videodiscs (teaching labs)
Solid phase extractions (instead of
  column chromatography)
Review techniques for waste generated

Substitution

Reduce or eliminate toxic materials
Citrus based solvents for xylene
Resins for DNA preparation
Ozone treatment or soap and water
  for degreasing
Detergent or nochromix for chromerge
Red liquid thermometers for mercury
Biodegradable scintillation cocktail

RECYCLING

Use and Reuse

Chemical redistribution program
Reuse solvents after rotavaping
Use spent solvents for initial degreasing
Redistilling mercury
Solvent recycler for HPLC

Reclamation

Distillation of solvents
Off-site reclamation (especially oil and
  mercury)

TREATMENT

Biological

Phenol degradation with sludge

Chemical

Neutralization of corrosive wastes with
  no heavy metals
Oxidation/reduction
Precipitation of metals

Physical

Figure 4 Examples of source reduction, recycling, and treatment.

is redistillation. The economic viability of such a project varies with institution depending on such factors as the quantity of a given wastestream and the purchase cost of the solvents.

The last area of exploration, according to the EPA's hierarchy, is treatment. From a regulatory and waste manager's point of view, this is most easily done in the laboratory. While technically not waste minimization, this may be considered after exploring all the other alternatives. The most common treatment is elementary neutralization, which may be done either by the researchers or at the waste storage facility. Other types of treatment may be included as a final step of a

Table 3 Chemical Redistribution: Cost Savings for Fiscal Year 1991 at UIUC

	Number of containers	Kilograms redistributed	Saved purchase costs ($)	Saved disposal costs ($)
Bulk solvents	5	1000	567	1,000
Pesticides and fertilizers	151	951	5,214	11,470
Laboratory chemicals	631	709	6,732	3,685
Totals	785	2660	12,513	15,763

laboratory procedure. When considering treatment in the laboratory, the permit requirements for each state may vary.

ACTUAL PROJECTS INVESTIGATED

A study of laboratory waste minimization opportunities was carried out at the University of Illinois at Urbana-Champaign (UIUC).[7] Funding was received from the Hazardous Waste Research and Information Center (Illinois Department of Energy and Natural Resources). This supported nearly two FTE research positions for a 2-year period. The goal was to both identify waste minimization practices suitable in laboratories and to implement these practices to determine whether they could be effective.

STRUCTURE OF THE STUDY

First, it was necessary to establish what level of waste minimization already existed on campus. This was accomplished surveying 270 laboratories that generate hazardous waste. The survey (Table 4) raised different issues in the areas of waste management and minimization that also resulted in increasing the researchers' awareness of waste minimization practices. Second, a study of general wastestreams produced by labs such as solvents, acids, mercury, cleaning agents, and surplus chemicals was carried out. Third, an in-depth study of six labs working in various scientific areas was carried out. A detailed audit (Table 5) of hazardous waste practices was made for each lab at the beginning of the study period. Specific procedures used by each lab were examined. Mass balance studies were carried out by conducting detailed quarterly inventories of chemicals in stock and tracking chemical purchases and disposal.

RESULTS AND CONCLUSIONS OF THE STUDY

Survey Findings

The results of the survey indicated that 91% of the responders classified their labs as involved with research, 14% were analytical, 13% were teaching, and 8% were other types. (Some labs selected more than one category). Of the teaching labs, about one third indicated that they had adopted waste minimization techniques, such as microscale.

Table 4 Waste Minimization Survey: Division of Environmental Health and Safety, University of Illinois at Urbana-Champaign

Principal investigator: _____

Name of person completing survey: _____
                  Title: _____

Location of Lab: Room(s) _____
               Building _____

Telephone: Office _____ Lab _____

1. How would you classify your laboratory? (Circle all that apply)
 Teaching ... 1 Research ... 2 Analytical ... 3
 Other (specify) ... 4 _____
 IF YOU CIRCLED "TEACHING" ANSWER Q.2–5, OTHERWISE SKIP TO Q.6a
2. Have you reduced the scale of experiments through microscale or similar techniques? Yes ... 1 No ... 2
3. Have demonstrations or video presentations been substituted for student experiments that generate chemical wastes? Yes ... 1 No ... 2
4. Have you considered using preweighed or premeasured reagent packets for introductory laboratories where waste is high? Yes ... 1 No ... 2
5. Has waste management become part of the pre- and postlaboratory written student experience? Yes ... 1 No ... 2
6a. Does your laboratory generate nonhalogenated solvent waste (e.g., hexane, ethyl acetate, benzene, acetone, etc.)? Yes ... 1 No ... 2 (SKIP TO Q.7)
 b. What processes generate these wastes? (Circle all that apply)
 Chromatography ... 1 Reactions ... 2 Cleaning/rinsing ... 3
 Other (specify) ... 4 _____
7a. Does your laboratory generate halogenated solvent waste (e.g., chloroform, trichloroethane, methylene chloride, etc.)? Yes ... 1 No ... 2 (SKIP TO Q.8)
 b. What processes generate these wastes? (Circle all that apply)
 Chromatography ... 1 Reactions ... 2 Cleaning/rinsing ... 3
 Other (specify) ... 4 _____
8a. Does your laboratory generate acid waste? Yes ... 1 No ... 2 (SKIP TO Q.9)
 b. What processes generate these wastes? (Circle all that apply)
 Analytical ... 1 Reactions ... 2 Cleaning ... 3
 Other (specify) ... 4 _____
9a. Does your laboratory generate base waste? Yes ... 1 No ... 2 (SKIP TO Q.10)
 b. What processes generate these wastes? (Circle all that apply)
 Analytical ... 1 Reactions ... 2 Cleaning ... 3
 Other (specify) ... 4 _____
10a. Does your laboratory generate waste oil? Yes ... 1 No ... 2 (SKIP TO Q.11)
 b. What processes generate these wastes? (Circle all that apply)
 Do any processes other than vacuum pumps generate waste oils in your labs? Yes ... 1 No ... 2 (SKIP TO Q.11)
 c. What processes? _____
11a. Does your laboratory generate unused or partially used reagent chemicals?
 Yes ... 1 No ... 2 (SKIP TO Q.12)
 b. Why do you have these wastes? (Circle all that apply)
 Outdated ... 1 Overstocked ... 2 Lab cleanout ... 3
 Other (specify) ... 4 _____
12a. Does your laboratory generate mercury waste? Yes ... 1 No ... 2 (SKIP TO Q.13)
 b. Why do you have these wastes?
 Spills ... 1 Old equipment ... 2 Surplus ... 3 Mercury reagents ... 4
 Other (specify) ... 5 _____
13a. Does your laboratory generate chromic acid waste? Yes ... 1 No ... 2 (SKIP TO Q.14)
 b. Does any other process other than cleaning generate this waste? Yes ... 1
 No ... 2 (SKIP TO Q.14)
 c. What processes? _____

Table 4 Waste Minimization Survey: Division of Environmental Health and Safety, University of Illinois at Urbana-Champaign *(Continued)*

CIRCLE APPROPRIATE ANSWER FOR ALL QUESTIONS BELOW.

		Yes	No
14a.	Does your laboratory have a written waste management/reduction policy?	1	2
b.	Would you like assistance from the Division of Environmental Health & Safety in this area?	1	2
15a.	Is waste reduction included as part of student and/or employee training?	1	2
b.	Would you like assistance from the Division of Environmental Health & Safety in this area?	1	2
16a.	Are manuals such as American Chemical Society (ACS) "Less is Better" or "ACS Waste Management Manual for Lab Personnel," part of your training?	1	2
b.	If no, would you like to receive information on how to order?	1	2
17.	Is there an incentive program (awards, recognition) for waste minimization in your laboratory?	1	2
18.	Does a central purchasing system exist in your lab or department to prevent overbuying, accumulation of outdated, or off-spec chemicals?	1	2
19a.	Are chemicals inventoried at least once each year?	1	2
b.	Are solvents inventoried at least once each year?	1	2
20.	Do you ever have surplus chemicals due to buying more than needed?	1	2
21.	Is there a program of dating chemicals so that older ones are used first?	1	2
22a.	Has your laboratory been audited for waste generated (quantity, type, source, and frequency) during the past year?	1	2
b.	If yes, what is the name of the person who conducted the audit? _____		
23.	Do you have information on file about waste you generate?	1	2
24.	Has an area in your lab been established specifically for storing chemical waste?	1	2
25.	Have procedures to prevent and/or contain spills been developed?	1	2
26a.	Has someone conducted a safety inspection of your laboratory during the past year?	1	2
b.	If yes, what is the name of the person who conducted the inspection? _____		

(Continued)

Overall, laboratory workers reported that they generated the types of waste indicated in Table 6. The most commonly named source of wastes was cleaning activities. Extractions, chromatography, and reactions were other major sources of solvent waste. For waste acids and bases, reactions and analytical uses were also important. It should be noted that, although these were the most common activites, they may not be the most significant in terms of volume. Unused reagents most commonly arose from becoming outdated or from cleaning out the lab. Mercury wastes came from spills, old equipment, and mercury reagents.

Few labs had formally implemented policies relating to waste minimization; however, a large proportion demonstrated an interest in such policies by requesting assistance in developing them.

About half the labs indicated that they conducted an annual inventory of chemicals in storage and ordered chemicals through a central purchasing system. Higher proportions indicated they dated chemicals to use older ones first, had developed procedures to prevent or contain spills, and had established an area in the lab specifically for storage of chemical waste.

INVESTIGATING WASTE MINIMIZATION POSSIBILITIES

Table 4 Waste Minimization Survey: Division of Environmental Health and Safety, University of Illinois at Urbana-Champaign *(Continued)*

		Always	Usually	Sometimes	Rarely	Never	Not Applicable
27.	How frequently are the following practices being performed in your laboratory?	1	2	3	4	5	8
a.	Halogenated solvents are separated from nonhalogenated solvents.	1	2	3	4	5	8
b.	Recyclable waste/excess chemicals are separated from nonrecyclables.	1	2	3	4	5	8
c.	Organic wastes are separated from metal-containing or inorganic wastes.	1	2	3	4	5	8
d.	Nonhazardous chemical wastes are separated from hazardous wastes *and* the most toxic (cyanides, etc.) are separated further.	1	2	3	4	5	8
e.	Chemical wastes are kept separate from normal trash (paper, wood, etc).	1	2	3	4	5	8
28.	Alternates to chromic acid cleaning baths are used.	1	2	3	4	5	8
29.	Alternates to mercury thermometers are used.	1	2	3	4	5	8
30.	Experimental procedures are reviewed and evaluated to see if less hazardous or nonhazardous reagents may be substituted for highly toxic, reactive, carcinogenic, or mutagenic materials.	1	2	3	4	5	8
31.	When purchasing or designing new equipment, the type and quantity of waste generated is considered.	1	2	3	4	5	8
32.	Detoxification and/or waste neutralization steps are included in laboratory experiments.	1	2	3	4	5	8
33.	The kinds and amounts of waste products are considered when preparing a new protocol.	1	2	3	4	5	8
34.	Waste/surplus chemicals are examined to see if there are other laboratories, departments, or areas (garage, paint shop) who might wish to use them.	1	2	3	4	5	8
35.	Spent solvent is used for initial cleaning and fresh solvent is used only for the final cleaning.	1	2	3	4	5	8
36.	Corrosive wastes are neutralized at the lab bench.	1	2	3	4	5	8
37.	Highly reactive chemicals are deactivated in the hood.	1	2	3	4	5	8

Table 4 Waste Minimization Survey: Division of Environmental Health and Safety, University of Illinois at Urbana-Champaign *(Continued)*

38. Within the last year, has equipment been purchased or built that has resulted in less waste being generated?   Yes ... I   No ... 2
39. Have your procedures been reviewed within the last year to see if quantities of chemicals and/or chemical waste could be reduced?   Yes ... I   No ... 2
40a. Has the chemical recycling coordinator (CRC) from the Division of Environmental Health and Safety been contacted to see if your lab can use the surplus materials from another area?   Yes ... I (GO TO Q.36a)   No ... 2
  b. Would you like the CRC to contact you?   Yes ... I   No ... 2
41a. Has redistillation of waste solvents been evaluated?   Yes ... I (GO TO Q.36b)   No ... 2
  b. Are solvents being redistilled?   Yes ... I   No ... 2
42. Have other wastes been evaluated for reclamation in your laboratory?   Yes ... I   No ... 2
43. What other methods have you used to reduce the amount or toxicity of chemical waste generated in your laboratory?

44. Comments (if any)

45. Would you like to receive a copy of the report of this survey?   Yes ... I   No ... 2

For you convenience, our return address has been provided on the back of this survey. Please fold survey in half, staple closed and deposit in campus mail.

Close to one quarter of the labs indicated that they had pursued general waste reduction policies such as obtaining new equipment that resulted in less waste, reviewed experimental procedures to reduce waste, used surplus chemicals from the university's redistribution program, evaluated distillation of solvents, or evaluated wastes for reclamation. Responses to questions on waste reduction practices showed varying levels of activity.

Wastestream Minimization Strategies

Several processes were found related to solvent pollution prevention. Traditionally labs have used solvents for liquid/liquid extraction techniques to separate the analyte between two nonmixing solvents. An alternative to this commonly used method is solid-phase extraction, which employs small, disposable extraction columns of solid material to adsorb an analyte, which is subsequently eluted

Table 5 Waste Minimization — Laboratory Self-Audit Division of Environmental Health and Safety University of Illinois at Urbana-Champaign

1. Principal Investigator _____
2. Contact person _____ Position _____
   Telephone no. _____
3. Location(s) of laboratories in section to be monitored _____
4. Type of laboratories:
   _____ Research laboratories
   _____ Undergraduate teaching laboratories
   _____ Analytical laboratories
   _____ Other (Specify)
5. Where are the chemical storage areas in this research group? _____

   How are the chemicals segregated? _____

6. Person(s) in charge of chemical storage/stock areas: _____

7. Total number of people working in lab _____
   Number of undergraduate students _____
   Number of graduate students _____
   Number of postdoctoral fellows _____
   Number of research assistants (salaried) _____
   Other (please explain) _____
8. What proportion of chemicals are purchased from each of the "suppliers" listed below? (NOTE: This information may have to come from business office) Chemical stores _____ Other campus stockrooms _____ Direct order from vendors _____ (Specify major vendors) _____

9a. Who is authorized to purchase chemicals? _____

 b. Who in the departmental business office would keep records? _____

 c. How frequently are chemicals purchased? (e.g., daily, weekly, monthly) _____

10. Do you maintain an active inventory of chemicals in stock? _____ If yes, describe: _____

    How often do you update the inventory? _____
11a. Who is responsible for chemical waste management or is everyone responsible for own wastes? _____

 b. How are wastes currently collected for disposal within the section? _____

 c. Where are wastes stored pending disposal?
12. What cleaning agents are used for glassware and other laboratory equipment? _____

13. Are mercury, mercury compounds, mercury solutions, or equipment containing mercury used? _____ If yes, describe uses _____

14. Are radioactive materials, including uranium compounds, used?
    If yes, list _____

Table 5 Waste Minimization — Laboratory Self-Audit Division of Environmental Health and Safety University of Illinois at Urbana-Champaign *(Continued)*

15. Does the lab produce biohazardous wastes (animal carcasses, sharps, infectious agents, etc.)? If yes, describe briefly _____

16. Does the lab use compressed gases? If yes, list _____

17. List impressions of housekeeping (Give examples such as clutter, improper chemical storage, labeling) _____

18. What procedures involve the use of water? Explain. _____

19. What kinds of chemical waste does this section produce? _____

20. What are the procedures, processes, or methods that produce these wastestreams (specify which processes relate to which wastes)? (refer to hazardous waste records also) You may use additional sheets as necessary _

21. If possible, prepare a flow chart to demonstrate the destination of each type of waste described in question 20 (from cradle to grave): include chemicals used, source of chemicals, processes used, storage, disposition). Use additional sheets as necessary.

Waste Minimization — Laboratory Self-Audit Chemical Inventory

Department _____ Phone no. _____
Location _____ Room _____
Person completing inventory _____ Date _____

List all chemicals in stock including all information required below: Note: include storage areas, work areas, waste in storage, chemicals in use. Make sure solutions and mixtures are described in detail. Get copy of group's inventory (if available).

Chemical (conc. if appl.)	No. of containers	Size of cont. (kg or l)	Quanity in cont.	Location

Table 6 Percentage of Laboratories Generating Various Types of Waste

Type of waste generated	% Generating waste type (total 270)
Nonhalogenated solvents	71
Oil	64
Halogenated solvents	61
Unused reagents	60
Bases	46
Chemical contaminated solids (debris)	33
Mercury waste	30
Chromic acid waste	26

from the column for analysis. Product information on these materials are kept in UIUC's "library" of waste minimization techniques.

Waste minimization strategies that are feasible after waste solvents are produced were investigated: redistribution of surplus solvents and "useful waste" and distillation. A B/R Model 8300 spinning band still was acquired through the study grant to use in solvent redistillation studies. The feasibility of distillation on a variety of solvent wastes generated at a variety of laboratories such as chemistry, operations and maintenance, engineering, pathology lab, and histology lab was evaluated. Some of the factors that are pertinent in choosing solvent redistillation equipment are safety, temperature regulation, automation, throughput, cost, space, theoretical purity, cleaning, and maintenance.

A waste minimization option that is worthy of consideration regarding acid wastestreams is to change procedures to eliminate or reduce the use of acids. In analytical chemistry, the use of automated equipment such as gas chromatographs, atomic absorption, and other methods significantly reduce the amounts of waste generated per sample analyzed. Substitution of less-hazardous materials for cleaning, as for chromic acid discussed below, is also feasible. We found that there is a need for the development of standard test methods (e.g., EPA methods) that do not use reagents containing heavy metals.

After waste acids are produced, neutralization is the first option to examine. Neutralization is widely practiced in labs and is usually not technically difficult. However, neutralization makes the most sense when the waste is not otherwise hazardous because the resulting solution can then be sewered.

The study also looked at alternatives to chromic acid cleaning solutions. We found that 26% of the labs surveyed in the study used chromic acid. Discussions with people who use chromic acid as a cleaning agent indicate that they use it because that is what has always been used. Often one of several detergents may be used instead. We recommend that laboratories that have used chromic acid in the past replace them with Alconox, Pierce RBS-35, or similar detergents. If that is not possible they should try neutralizable/sewerable solutions such as potassium hydroxide/ethanol, dilute hydrochloric acid, aqua regia, or other oxidizing agents without chromium or other metals.

Ways to reduce or eliminate the use of mercury in laboratories was also investigated. Mercury spills (broken thermometers, manometers, etc.) were iden-

Table 7 Examples of Excess Chemicals and their Uses

Chemical	Use
Various dyes	Histological and microbiological stains
Formic acid	Decalcifier
Acetic acid	
Phosphoric acid	
Hydrochloric acid	Etching baths
General acids	Cleaning baths
Sodium hydroxide	
Potassium hydroxide	
Cupric sulfate	Spraying fruit trees
Nitrates	Fertilizer, glass-blowing oxidants
Potassium dichromate	Preserving biological samples, stains
Silver nitrate	Stains
Non-(or expired) reagent-grade solvents	Same use as bulk-grade solvents
Unopened and opened organic and inorganic solid chemicals	In organic and inorganic teaching labs as well as research experiments
Xylene waste	Cleaning street paint equipment

tified as a source of waste in half the labs generating such waste. In some teaching labs mercury thermometers were replaced with red liquid thermometers. Metal or digital thermometers can also be substituted. An instructional experiment involving the use of mercuric chloride was eliminated, reducing the generation of over 300 l of mercury waste that would otherwise be a disposal problem.

The study also found that another essential aspect of laboratory waste minimization is a chemical redistribution program. Table 7 illustrates some common uses of redistributed chemicals. This involves finding new customers for opened and unopened reagent chemicals that the original owners no longer need. This does not require much technical knowledge of laboratory techniques. However, an understanding of which laboratories can use which chemicals is an asset to this activity. This program is probably best run as if it were a sales operation. Stocking of chemicals is done as they enter the storage area for holding before disposal. Inventories should be kept and publicized. Orders, both standing and special, should be taken. Deliveries should be made promptly and cheerfully. Finally, an accounting system should be kept of the sales. For example, Table 3 shows results from redistribution efforts at UIUC for the year ending June 30, 1991. It should be noted that the saved purchase costs are rather conservative, since UIUC is able to purchase chemicals at rather low prices due to large volumes it contracts for.

Focus on Individual Labs

In selecting laboratories for this portion of the study we determined whether they generated chemical wastes on a routine basis, represented the kind of work performed in labs throughout the state, and/or represented a cross section of the types of labs within the university. The six labs chosen included a state agency analytical lab, a histology lab, a high school chemistry lab, and three research labs in the areas of life sciences, material science/engineering, and organic chemistry.

As mentioned above, a detailed audit was made of each laboratory, including a description of the common procedures used in each laboratory. Inventories of the chemical stocks in each of the study labs was taken every 3 months for 1 year. All chemicals in their original bottles and in quantities over 100 g were tracked. Solutions, mixtures, and wastes were not recorded. Chemical purchase and disposal records were also monitored.

We found that these labs purchased anywhere from 1 to 100% of the chemicals at the university's central stores. This meant tracking of chemicals only through central stores would be inadequate.

As expected, we found a substantial number of chemicals present in each lab. Initially, we found that the number of different chemicals in the labs ranged from 74 in the materials science lab to 2754 in the organic chemistry lab. After limiting the inventory to those containers with at least 100 g of material, the numbers fell to 65 and 894, respectively. We found changes in 50 to 80% of the records between the quarterly inventories. A change was considered to be a new item, a deleted item, change in storage location, or an item with a quantity difference of at least 100 g.

We examined the inventories to determine what proportion of chemicals in storage would be classified as hazardous should someone wish to dispose of these materials. In terms of weight, the organic chemistry, the histology, and analytical laboratories had the largest quantity of chemicals. Each had 300 to 500 kg (65 to 90%) of chemicals that would be classified as hazardous. However, the organic chemistry laboratory had by far the largest number of containers (1064 vs. 590 for next largest). The remaining three labs each had between 100 and 150 kg of chemicals, of which at least 40% were classified as hazardous.

Except for solvents, mass balance studies proved difficult to conduct for our study labs. In the histology lab, we found that 90% of the xylene and xylene substitute purchased ended up being disposed as hazardous waste. Similarly, we could account for 85% of the formalin purchases. The organic chemistry lab waste disposal records, on the other hand, accounted for only 51% of the solvents purchased. Waste disposal records at the anlytical laboratory accounted for 69 to 77% of solvents purchased. The reasons for this better performance over the organic chemistry lab are unclear. This could involve the use of different procedures and greater attention to solvent recovery.

In the cases of the other three labs it was difficult to perform any mass balances at all. This was either due to poor purchase records or the small amounts of waste actually disposed. These efforts illustrated the difficulty in performing mass balances on miscellaneous laboratory chemicals.

FOLLOW-UP

The waste minimization survey was followed-up by distribution of material to those who indicated interest. An anecdotal list of "101 Ways to Reduce Hazardous Waste in the Lab" (Table 8) and information on how to order pamphlets that provide information on waste management and minimization was sent out. Also, we concluded that more information about waste minimization should be provided to the laboratories. The study sponsored broadcast of a satellite seminar (produced by the American Chemical Society) on "Practical Approaches to Laboratory Waste Man-

Table 8 101 Ways to Reduce Hazardous Waste in the Lab

1. Write a waste management/reduction policy.
2. Include waste reduction as part of student/employee training.
3. Use manuals such as the American Chemical Society (ACS) "Less is Better" or "ACS Waste Management for Lab Personnel" as part of your training.
4. Create an incentive program for waste reduction.
5. Centralize purchasing of chemicals through one person in the lab.
6. Inventory chemicals at least once a year.
7. Indicate in the inventory where chemicals are located.
8. Update inventory when chemicals are purchased or used up.
9. Purchase chemicals in smallest quantities needed.
10. If trying out a new procedure, try to obtain the chemicals needed from another lab or purchase a small amount initially. After you know you will be using more of this chemical, purchase in larger quantities (unless you can use some someone else doesn't need any more).
11. Date chemical containers when received so that older-ones will be used first.
12. Audit your lab for waste generated (quantity, type, source, and frequency). Audit forms are available from DEH&S.
13. Keep MSDSs for chemicals used on file.
14. Keep information about disposal procedures for chemical waste in your lab on file.
15. If possible, establish an area for central storage of chemicals.
16. Store chemicals in storage area except when in use.
17. Establish an area for storing chemical waste.
18. Minimize the amount of waste kept in storage. Ask for a waste pick-up as often as you need.
19. Label all chemical containers as to their content.
20. Develop procedures to prevent and/or contain chemical spills — purchase spill clean-up kits, contain areas where spills are likely.
21. Keep halogenated solvents separate from nonhalogenated solvents.
22. Keep recyclable waste/excess chemicals separate from nonrecyclables.
23. Keep organic wastes separate from metal-containing or inorganic wastes.
24. Keep nonhazardous chemical wastes separate from hazardous waste.
25. Keep highly toxic wastes (cyanides, etc.) separated from above.
26. Avoid experiments that produce wastes that contain both radioactive and hazardous chemical waste.
27. Keep chemical wastes separate from normal trash (paper, wood, etc.).
28. Use the least-hazardous cleaning method for glassware. Use detergents such as Alconox, Micro, RBS35 on dirty equipment before using KOH/ethanol bath, acid bath, or No Chromix.
29. Eliminate the use of chromic acid altogether.
30. Eliminate the use of uranium and thorium compounds (naturally radioactive).
31. Substitute red liquid (spirit-filled), digital, or thermocouple thermometers for mercury thermometers where possible.
32. Use a bimetal or stainless steel thermometer instead of mercury thermometer in heating and cooling units. Stainless steel lab thermometers may be an alternative to mercury in labs as well.
33. Evaluate laboratory procedures to see if less-hazardous or nonhazardous reagents could be used.
34. Review the use of highly toxic, reactive, carcinogenic, or mutagenic materials to determine if safer alternatives are feasible.
35. Avoid the use of reagents containing arsenic, barium, cadmium, chromium, lead, mercury, selenium, and silver.
36. Consider the quantity and type of waste produced when purchasing new equipment.
37. Purchase equipment that enables the use of procedures that produce less waste.
38. Review your procedures regularly (e.g., annually) to see if quantities of chemicals and/or chemical waste could be reduced.

Table 8 101 Ways to Reduce Hazardous Waste in the Lab *(Continued)*

39. Look into the possibility of including detoxification and/or waste neutralization steps in laboratory experiments.
40. When preparing a new protocol, consider the kinds and amounts of waste products and see how they can be reduced or eliminated.
41. When researching a new or alternative procedure, include consideration of the amount of waste produced as a factor.
42. Examine your waste/excess chemicals to determine if there are other uses if your lab, neighboring labs, departments, or areas (garage, paint shop) who might be able to use them.
43. Review the list of chemicals to be recycled or contact the chemical recycling coordinator to see if chemicals needed are available before purchasing chemicals.
44. Inform the chemical recycling coordinator of the types of materials you can use from the recyclables.
45. Call the chemical recycling coordinator to discuss setting up a locker or shelf for excess chemical exchange in a lab, stockroom, or hallway in your department.
46. When solvent is used for cleaning purposes, use spent solvent for initial cleaning and fresh solvent for final cleaning.
47. Try using detergent and hot water for cleaning of parts instead of solvents.
48. Consider using ozone treatment for cleaning of parts.
49. Consider purchasing a vapor degreaser, vacuum bake, or bead blaster for cleaning of parts.
50. Reuse acid mixtures for electropolishing.
51. When cleaning substrates or other materials by dipping, process multiple items in one day.
52. Use smallest container possible for dipping or for holding photographic chemicals.
53. Use best geometry of substrate carriers to conserve chemicals.
54. Store and reuse developer in photo labs.
55. Precipitate silver out of photographic solutions for reclamation.
56. Neutralize corrosive wastes that don't contain metals at the lab bench.
57. Deactivate highly reactive chemicals in the hood.
58. Evaluate the possibility of redistillation of waste solvents in your lab.
59. Evaluate other wastes for reclamation in labs.
60. Scale down experiments producing hazardous waste wherever possible.
61. In teaching labs, consider the use of microscale experiments.
62. In teaching labs, use demonstrations or video presentations as a substitute for some student experiments that generate chemical wastes.
63. Use preweighed or premeasured reagent packets for introductory teaching labs where waste is high.
64. Include waste management as part of the pre- and postlaboratory written student experience.
65. Encourage orderly and tidy behavior in lab.
66. Polymerize epoxy waste to a safe solid.
67. Consider using solid-phase extractions for organics.
68. Put your hexane through the rotavap for reuse.

Use the following substitutions where possible:

	Original material	Substitute	Comments
69.	Acetamide	Stearic acid	In-phase change and freezing point depression
70.	Benzene	Alcohol	
71.	Benzoyl peroxide	Lauryl peroxide	When used as a polymer catalyst
72.	Chloroform	1,1,1-Trichloroethane	
73.	Carbon tetrachloride	Cyclohexane	In test for halide ions

	Original material	Substitute	Comments
74.	Carbon tetrachloride	1,1,1-Trichloroethane 1,1,2-Trichloro-trifluoroethane	
75.	Formaldehyde	Peracetic acid	In cleaning of kidney dialysis machines
76.	Formaldehyde	"Formalternate" (Flinn Scientific)	For storage of biological specimens
77.	Formaldehyde	Ethanol	For storage of biological specimens
78.	Formalin	See Formaldehyde	
79.	Halogenated solvents	Nonhalogenated solvents	In parts washers or other solvent processes
80.	Amitrole (Kepro Circuit Systems)	Mercuric chloride reagents	Circuit board etching
81.	Sodium dichromate	Sodium hypochlorite	
82.	Sulfide ion	Hydroxide ion	In analysis of heavy metals
83.	Toluene	Simple alchols and ketones	
84.	Wood's metal	Onion's Fusible alloy	
85.	Xylene	Simple alcohols and ketones	
86.	Xylene or toluene based liquid scintillation cocktails	Nonhazardous proprietary liquid scintillation cocktails	In radioactive tracer studies
87.	Mercury-free catalysts (e.g., $CuSO_4$-TiO_2-K_2SO_4)	Mercury salts	Kjeldahl digests

88. Destroy ethidium bromide using $NaNO_2$ and hypophosphorous acid.
89. Run mini SDS-PAGE 2d gels instead of full-size slabs.
90. Treat sulfur and phosphorus wastes with bleach before disposal.
91. Treat organolithium waste with water or ethanol.
92. Seek alternatives to phenol extractions (e.g., small scale plasmid prep using no phenol may be found in *Biotechnica,* Vol. 9, No. 6, pp. 676–678).
93. Use procedures to recover metallic mercury.
94. Review procedures to recover mercury from mercury-containing solutions.
95. Recover silver from silver chloride residue waste.
96. Purchase compressed gas cylinders, including lecture bottles, only from manufacturers who will accept the empty cylinders back.
97. When testing experimental products for private companies, limit donations to the amount needed for research.
98. Return excess pesticides to the distributor.
99. Be wary of donations from outside the university. Accept chemicals only if you will use them within 12 months.
100. Replace and dispose of items containing polychlorinated biphenyls (PCBs).
101. Send us other suggestions for waste reduction.

agement."[8] This gave an audio/video presentation of the materials as well as providing an opportunity to distribute information to the laboratories. In the future, a "Guide to Pollution Prevention in Laboratories" that outlines the principles of waste minimization with specific examples will be distributed. Other ways institutions can provide information include periodic newsletters, articles in their newspapers and publications, a section in the waste management guide on waste minimization, and training of new researchers and students.

CONCLUSION

Investigation of waste minimization possibilities should draw on all areas of expertise in your institution. A hazardous waste manager can provide knowledge of the regulations, wastestreams particular to an institution, an understanding of waste minimization principles, and act as a resource for information center for "technique exchange." The actual generators of the waste (the lab personnel) provide the knowledge of the processes generating the waste, ways waste minimization has been done in the past, help the waste manager to come up with new ways to work, and implement better housekeeping. The administration provides the incentive, by policy making and goal setting, for the others to work toward pollution prevention.

A central staff person or program may be necessary to initiate a laboratory pollution prevention program. However, laboratory waste minimization will succeed best when each laboratory worker makes waste minimization a priority.

ENDNOTES

1. U.S. Environmental Protection Agency, *Guides to Pollution Prevention: Research and Education Institutions*, EPA/625/7-90/010, Risk Reduction Engineering Laboratory, Center for Environmental Research Information, Cincinnati, June 1990.
2. Hill, D. C., Establishing a Waste Minimization Program, *Professional Safety*, 8, 16, 1992.
3. National Association of College and University Business Officers, *Hazardous Waste Management at Educational Institutions*, NACUBO, 1 Dupont Circle, Washington, D.C., 1987.
4. American Chemical Society, *The Waste Management Manual for Laboratories*, ACS, Dept. of Government Relations and Science Policy, 1155 16th St. NW, Washington, D.C., 1990.
5. Kaufman, J. A., Ed., *Waste Disposal in Academic Institutions*, Lewis Publishers, Chelsea, MI, 1990.
6. U.S. Environmental Protection Agency, *The EPA Manual for Waste Minimization Opportunity Assessments*, EPA600/2-88-025, Hazardous Waste Engineering Research Laboratory, Cincinnati, 1988.
7. Ashbrook, P. C., Klein-Banai, C., and Maier, C., Determination, Implementation and Evaluation of Laboratory Waste Minimization Opportunities, ENR Contract No. HWR 91085, Illinois Dept. of Energy and Natural Resources, Hazardous Waste Research and Information Center, draft report, July 1, 1992.
8. Harless, J. M. and Phifer, R. W., *Practical Approaches to Laboratory Waste Management*, Video V-5700, American Chemical Society, Washington, D.C., 1992.

CHAPTER 4

The Law on Waste Minimization in Laboratories

Jennifer L. Hernandez

CONTENTS

Legal Requirements for Waste Minimization in Laboratories 56
    Hazardous Waste Minimization Mandates 56
    Other Pollution Prevention Mandates 58
    The "Baseline" Conundrum and other Obstacles to Regulatory
    Reform in Waste Minimization 60
Legal Incentives for Waste Minimization in Laboratories 61
    Liability for Hazardous Waste Disposal Site Cleanups 61
    Liability for Hazardous Waste Handling Incidents 62
    Regulatory Economics: Fees, Taxes, Insurance, and Higher
    Transportation and Disposal Costs 62
    Private Economics: Landlords, Lenders, and Insurance Companies 62
    Enforcement Risks 62
Legal Considerations in Implementing Waste Minimization Programs
in Laboratories 63
    Waste Minimization Audits: A Lawyer's Perspective 63
    The "Treatment Trap" and other Regulatory Disincentives 64
    Cross-Media Pollution Considerations and other Regulatory Trends 64

    Laws on waste minimization in laboratories include specific statutory requirements and prohibitions as well as legal incentives relating to liability and risk management. This chapter begins with an overview of legal requirements relating specifically to waste minimization in laboratories and then addresses the larger universe of liability and risk management issues which create strong waste minimization incentives for laboratories. The chapter concludes with a discussion of the legal issues to be considered when implementing waste minimization programs in laboratories.

LEGAL REQUIREMENTS FOR WASTE MINIMIZATION IN LABORATORIES

Laws relating to waste minimization, like other types of environmental laws, occur at the federal, state, and local level. The focus here is on federal requirements with references to California requirements to illustrate how state and local agencies have expanded the laws of waste minimization.

HAZARDOUS WASTE MINIMIZATION MANDATES

One of the first, and until recently frequently ignored, waste minimization requirements was enacted in 1984 as part of the amendments to the Resource Conservation and Recovery Act ["RCRA," RCRA § 3002(b)].* Rejecting arguments for mandatory waste minimization requirements, Congress compromised by directing the U.S. Environmental Protection Agency (EPA) to obtain generator certifications regarding waste minimization efforts by including on the Uniform Hazardous Waste Manifest the following language:

> If I am a large quantity generator, I certify that I have a program in place to reduce the volume and toxicity of waste generated to the degree I have determined to be economically practicable and that I have selected the practicable method of treatment, storage or disposal currently available to me which minimizes the present and future threat to human health and the environment; OR, if I am a small quantity generator, I have made a good faith effort to minimize my waste generation and select the best waste management method that is available to me and that I can afford. (40 CFR Part 262, Appendix.)

Generators who falsely attest to having completed the required level of waste minimization efforts are subject to RCRA civil penalties of up to $25,000 per day per violation. To the extent that false certification is interpreted as "knowingly making a false material statement or representation," violators would also be subject to RCRA criminal penalties of up to $50,000 per day per violation, and up to 5 years imprisonment.

These penalties could be assessed in the same way that EPA (or the state environmental agency authorized by EPA to implement RCRA) enforces other RCRA requirements. In the course of issuing a notice of violation or investigating a potential violation, EPA may seek documents or other evidence to determine whether the generator has truthfully completed the waste minimization certification.

Despite the routine use of hazardous waste manifests and equally routine generator certification as to the existence of hazardous waste minimization efforts, most remained skeptical about the effectiveness of the certification mechanism as

---

* RCRA is codified at 42 USC § 6901 *et seq.*; herein referenced as "RCRA § 1001 *et seq.*"

a means of mandating hazardous waste minimization. This skepticism seemed particularly justified in view of the fact that there were for many years no reported violations of RCRA § 3002(b), and until recently EPA's routine RCRA inspection checklists did not even remind inspectors to review or inquire about the generator waste minimization efforts that formed the basis for the generator certification. Even today, EPA is not aggressively enforcing RCRA § 3002(b) despite the agency's policy of encouraging pollution prevention.

The federal government took a major step beyond waste minimization by creating a comprehensive statutory framework for mandatory source reduction with enactment of the Pollution Prevention Act ("PPA") in 1990. (42 USC § 13101 *et seq.*) The PPA sets forth express congressional policy supporting "source reduction" which it defines as any practice which:

1. Reduces the amount of any hazardous substance, pollutant, or contaminant entering into the environment (including fugitive emissions) prior to recycling, treatment or disposal
2. Reduces the hazards to public health and the environment associated with the release of such substances, pollutants or contaminants

The term includes equipment or technology modifications, process or procedure modifications, reformulation or redesign of products, substitution of raw materials, and improvements in housekeeping, maintenance, training or inventory control. (42 USC § 13102)

The PPA expressly endorses source reduction as an effort that is different from, and superior to, conventional "pollution prevention" efforts such as waste minimization. The PPA directs EPA to establish a multimedia Source Reduction Clearinghouse to plan and facilitate source reduction efforts, and establishes a grant program for states to develop source reduction programs.

At the generator level, the PPA is implemented in conjunction with the toxic chemical release reporting requirements established by Title III of the Superfund Amendments and Reauthorization Act ("SARA Title III") (42 USC § 11023). Generators who are required to report toxic chemical releases (those in Standard Industrial Classes 20 to 39) are required to include in their release reports specified information about pollution prevention efforts, including quantity of each chemical wastestream, estimated percentage of wastestream which is recycled or treated (including description of recycling and treatment process), other source reduction practices used (e.g., equipment/process modifications, product redesign or reformulation, materials substitution, and improved management practices such as procurement and training), actual reductions from the previous calendar year, and forecasted reductions for the next 2 calendar years (42 USC § 13106). The PPA does not yet mandate any particular source reduction requirements (e.g., an average 10% reduction per year for 5 years). However, it is anticipated that when Congress next considers RCRA authorization it will consider the data collected under the PPA and direct EPA to adopt regulations designed to attain mandatory pollution prevention targets.

The PPA requirements do not yet affect academic laboratories, but do affect laboratories that are part of facilities classified as SIC 20-39. As a practical matter, however, laboratory activities are excluded from the toxic chemical release inventory requirements and are thus excluded from the corresponding PPA requirements as long as the laboratory activities fall within the following exclusion:

> If a toxic chemical is manufactured, processed, or used in a laboratory at a covered facility under the supervision of a technically qualified individual [as defined], a person is not required to consider the quantity so manufactured, processed, or used when determining whether an applicable threshold has been met under [specified sections]. This exemption does not apply in the following cases:
>
> (1) Specialty chemical production
> (2) Manufacturing, processing, or using toxic chemicals in pilot plant scale operations
> (3) Activities conducted outside the laboratory

Also relevant to laboratories, chemicals used in routine maintenance activities, in structural building components, in articles, or at *di minimus* concentration levels, are likewise exempt from the Title III reporting requirements (40 CFR § 372.38).

States and even local agencies have also adopted hazardous and solid waste minimization requirements, some of which go beyond what is required at the federal level. The California Hazardous Waste Source Reduction Act, for example, requires all generators of over 5000 kg/year of hazardous waste to prepare and implement a source reduction plan. Laboratories are not exempt from this state law (California Health & Safety Code § 25244.15). Some local agencies in California have adopted their own local source reduction ordinances, and have also imposed source reduction requirements via land use permits and via mitigation measures included in environmental review documentation for new and expanded facilities.

OTHER POLLUTION PREVENTION MANDATES

Waste minimization efforts can also be mandated by environmental laws covering other media. For example, Title III of the Clean Air Act Amendments of 1990 established mandatory reduction requirements for the emission of 189 hazardous air pollutants (42 USC § 7412). Some of these listed hazardous air pollutants are common solvents used in laboratories. EPA is directed to establish source categories and then promulgate regulations designed to require maximum achievable control technology to reduce air toxic emissions from each source category. Research and laboratory facilities are to be considered a separate category of source [42 USC § 7412(c)(7)], and EPA recently determined that laboratories are not a high priority level for regulatory development. If and when such regulations take effect, it is probable that they will create a direct or indirect incentive to reduce use of designated hazardous materials in laboratories, and this may also reduce laboratory waste. Some local air control agencies in California are developing regulations that would assess and may require reductions in toxic air emissions from laboratories.

Wastewater discharge regulations also affect waste management in laboratories. For example, under RCRA certain types of hazardous wastes generated in laboratories are exempt from the definition of "hazardous wastes" and can be disposed of in the laboratory's wastewater stream, if the wastewater is subject to pretreatment or direct discharge effluent standards established under the Clean Water Act and the wastewater results:

> ... from laboratory operations containing toxic (T) wastes listed in subpart D of this part, provided, that the annualized average flow of laboratory wastewater does not exceed one percent of total wastewater flow into the headworks of the facility's wastewater treatment or pre-treatment system, or provided the waste's combined annualized average concentration does not exceed one part per million in the headworks of the facility's wastewater treatment or pre-treatment facility. Toxic (T) wastes used in laboratories that are demonstrated not to be discharged to wastewater are not included in this calculation. [40 CFR § 261.3(a)(2)(iv)(E)]

The RCRA regulations allow drain disposal of these types of laboratory wastes based on the rationale that laboratory wastes are produced in very small quantities which are subject to effective treatment via the normal dilution process prior to treatment at a sewage plant. It technically reduces the quantity of hazardous wastes that must be treated or disposed offsite. However, it is also a practice that is often frowned upon by local sewage agencies, which are increasingly prohibiting the sewage discharge of any hazardous materials (except household and janitorial hazardous wastes).

RCRA also allows generators of certain wastestreams to perform certain types of relatively simple hazardous waste "treatment" processes such as elementary neutralization without being required to obtain a hazardous waste treatment permit (40 CFR § 270.1(c)(2)(v)). RCRA includes exemptions from the definition of hazardous waste for certain types of laboratory and treatability study samples [40 CFR § 261.4(d) to (f)] and residues of hazardous wastes in empty containers (40 CFR § 261.7). These regulatory exemptions effectively "reduce" the quantity of hazardous wastes produced by laboratories, and also create certain incentives (and disincentives) for various types of laboratory behavior. For example, a broader exemption, authorizing solidification or other types of treatment processes for laboratory-scale quantities of waste, would further reduce laboratory hazardous waste generation. Currently laboratory facilities would be required to take the costly, controversial, and restrictive approach of obtaining a permit to operate as a treatment facility prior to engaging in such bench-scale treatment processes to reduce the quantity or toxicity of laboratory hazardous wastes.

Legal requirements that directly or indirectly encourage waste minimization can also be found in laws designed to reduce the use of specified hazardous materials, such as regulations to eliminate the use of halogenated chlorofluoroalkanes for aerosol propellants (40 CFR Part 762).

Laws limited to one media or to a particular hazardous material, however, do not necessarily have the overall effect of waste minimization in laboratories. Phasing out one type of hazardous material to reduce air emissions, for example, may require substitution of a different material that must be used in greater quantities or that is less susceptible to recycling or reuse or that is more toxic to the user or that is more toxic or dangerous to handle than hazardous waste. These concerns have led Congress and other states to adopt the type of framework sketched out by the Pollution Prevention Act, and to mandate a more comprehensive, multimedia source reduction effort.

THE "BASELINE" CONUNDRUM AND OTHER OBSTACLES TO REGULATORY REFORM IN WASTE MINIMIZATION

Before moving to other legal considerations that give rise to laboratory incentives to reduce waste, it is useful to briefly note some of the difficulties that Congress and regulatory agencies have faced in attempting to adopt mandatory waste minimization requirements. Some of the key difficulties that are most relevant to laboratories include

- **The "baseline" conundrum.** Some institutions (or parts of institutions) may have taken extraordinary steps to reduce waste; others may have done nothing. Any mandatory waste minimization requirement is likely to involve a percentage reduction from some historical use level. What should this historical "baseline" be? How can any mandatory reduction approach avoid punishing companies that have voluntarily reduced wastes? Proposed solutions tend to focus on averaging the past 5 years of waste, or on a specified baseline year of several years ago. These solutions are imperfect, however, and this remains a controversial issue in efforts to develop more stringent waste minimization laws.
- **"Actual" vs. "allowable" reductions: technology, recordkeeping, and the reality of operational flexibility.** Another key problem with any mandatory waste minimization approach is whether and how to establish "allowable" levels of hazardous waste for a category of facilities or a specific facility. This debate is very familiar to laboratories that have wrestled with how to work with concepts such as "best available control technology" and specific emission or effluent limitations for variable research activities capable of producing a literally infinite variety of chemicals. Recordkeeping associated with waste minimization efforts is also a nontrivial consideration, as small quantities of materials procured from a variety of sources are nearly impossible to accurately track even at the product stage. "Mass balance" approaches — subtracting quantities discharged to various media as waste from quantities originally purchased — are completely infeasible in a research setting. Finally, it would be impossible to adequately describe, let alone police for some level of waste minimization compliance, the range of research activities that may produce hazardous wastes.

While these difficulties may be particularly acute in a research setting, the problems also extend to manufacturers and other facilities that use hazardous materials. The difficulty of these issues has prompted Congress and many states to refrain from enacting a "command and control" regulatory approach in favor of an approach that encourages users to undertake voluntary and cost-effective waste minimization efforts. These voluntary efforts, coupled with annual reporting requirements in the case of larger users such as manufacturers, currently comprise the direct regulatory framework for waste minimization.

LEGAL INCENTIVES FOR WASTE MINIMIZATION IN LABORATORIES

There is also a very influential "indirect" regulatory framework for waste minimization: the retroactive, joint and several liability that is common for hazardous waste landfill sites now being remediated under the federal Superfund Program, the Comprehensive Environmental Response, Compensation and Liability Act ("CERCLA") (42 USC § 9601 *et seq.*). Also relevant is the liability that can result from handling or transportation spills of hazardous materials and hazardous wastes, as well as damage claims from employees or neighbors who allege they have been exposed to harmful levels of hazardous materials. While these types of legal considerations have no direct connection to waste minimization efforts, they do provide legal incentives for minimizing hazardous materials handling and hazardous waste disposal. (Of course, they also provide indirect incentives for various other activities, ranging from efforts to educate the public about the real risk of hazardous materials to advances in environmental cleanup technology.)

LIABILITY FOR HAZARDOUS WASTE DISPOSAL SITE CLEANUPS

Generators who send hazardous waste to a hazardous waste disposal facility that has released a hazardous substance to the environment can be jointly and severally liable for environmental cleanup costs. Other members of the "potentially responsible party (PRP) group sharing cleanup liability include the current and prior site owners, operators, and transporters who arranged for the waste to be delivered to the site and other generators. While there are a handful of defenses against a government-initiated cleanup order, and a larger range of claims and defenses to allocate liability among PRPs (and their insurance companies), the bottom line is that involvement in a Superfund site is generally a costly, time-consuming, and fundamentally unpleasant experience.

CERCLA liability, coupled with RCRA's ban on the land disposal of many types of hazardous wastes, has effectively diverted most laboratory wastestreams toward hazardous waste incinerators, solvent and used oil recyclers, and other alternatives to hazardous waste landfills. Because of the residual liability associated with hazardous waste landfills, waste minimization efforts that reduce the quantity of hazardous waste that must be disposed of in such landfills is, in the long term, the single most cost-effective form of waste minimization for most laboratory facilities.

LIABILITY FOR HAZARDOUS WASTE HANDLING INCIDENTS

Apart from hazardous waste landfills, CERCLA, RCRA, and other state and local laws have established increasingly stringent requirements for accidental spills of hazardous materials and hazardous wastes. Generators of hazardous wastes, in particular, are susceptible to liability for cleanup costs associated with transportation accidents and other mishaps. This is particularly common (and problematic) given the limited financial resources of many hazardous waste transporters.

Alleged harmful exposures to hazardous materials and hazardous wastes is also of concern, as it becomes increasingly common for employees and neighbors to question and object to chemical handling activities. Laboratories have historically been more successful in avoiding this type of controversy, but the specter of "toxic contamination" has proven a potent weapon in some circumstances involving disgruntled employees, opponents of certain types of research activities, and residents seeking to stop the development or expansion of research facilities near their neighborhoods.

REGULATORY ECONOMICS: FEES, TAXES, INSURANCE, AND HIGHER TRANSPORTATION AND DISPOSAL COSTS

While not directly related to liability, the fees and taxes imposed on hazardous materials used and hazardous waste generators by environmental agencies have created legal incentives to reduce waste. These fees and taxes vary substantially by jurisdiction, but in general they are increasing — and providing an increasing economic incentive to reduce the activities that give rise to the fee or tax. The regulatory system has over the long term also dramatically increased the cost of operating hazardous waste transportation, storage, treatment, and disposal operations, resulting in another economic incentive for reducing hazardous waste generation. (Explanatory note: costs are generally lower now than they were 2 years ago.)

PRIVATE ECONOMICS: LANDLORDS, LENDERS, AND INSURANCE COMPANIES

Landlords, lenders, and insurance companies also increasingly add premiums to cover the perceived additional risk of becoming involved with a type of activity that involves hazardous materials. It is increasingly common in California, for example, for ordinary commercial leases and loan documents to include detailed provisions about what types of hazardous materials and activities will be allowed at a particular location, and to require landlord (or lender) consent prior to expanding or changing the nature of such authorized uses.

ENFORCEMENT RISKS

Finally, handling larger quantities of hazardous materials and generating higher quantities of hazardous wastes generally means a higher enforcement profile for federal, state, and local environmental enforcement agencies. Over time, higher enforcement profiles translate into more inspections, more routine (and virtually unavoidable) citations, and higher fines and penalties. For example, over half of all hazardous waste manifests inspected in a random California sample contained at least one error, from a transposed waste code to a missing data entry. It was also the case that these problems were more frequently identified, and subject to an enforcement action, at larger facilities.

LEGAL CONSIDERATIONS IN IMPLEMENTING WASTE MINIMIZATION PROGRAMS IN LABORATORIES

As the foregoing discussion makes clear, if legal requirements and liability considerations were the sole factor in determining whether and what types of research would be conducted, environmental compliance costs would quadruple and/or science as we know it would be eliminated. The real message was that there are *many* legal reasons to minimize waste. The next question: are there any legal considerations in planning or implementing a hazardous waste minimization program? As with any other question involving environmental regulations, the answer is: of course.

Because we do not yet have a "command and control" system of regulatory permits and controls on waste minimization, it is also true that legal considerations play a smaller role in waste minimization than the consideration of economic and technical feasibility.

WASTE MINIMIZATION AUDITS: A LAWYER'S PERSPECTIVE

While the legal issues associated with auditing in general are beyond the scope of this chapter, it is prudent to remember a few key points. Environmental compliance audits are encouraged by EPA and by many other state and local agencies. All lawyers view environmental compliance audits as evidence, and audits are frequently conducted for or under the supervision of an attorney to protect the results under an attorney-client privilege. Compliance audits followed by prompt corrective action establish an institution's commitment to compliance and can be an extremely effective tool in defending against an enforcement action. Compliance audits followed by inaction due to resource or bureaucratic constraints can have the opposite effect, and serve as evidence of "knowing and intentional" violations in an enforcement context.

Fortunately, an audit conducted for the express purpose of identifying, prioritizing, and implementing waste minimization activities raises few of the legal considerations related to compliance audits generally. As a practical matter, and to minimize the likelihood that the waste minimization audit will have unintended legal consequences, it is prudent to conduct a waste minimization audit independent of a general compliance audit. This approach should minimize any legal risks associated with waste minimization; unless your state or local jurisdiction has a mandatory waste minimization requirement, the audit will document noncompliance, and your institution will be unable to comply even after receiving the audit report.

THE "TREATMENT TRAP" AND OTHER REGULATORY DISINCENTIVES

As noted above and in other sections of this book, there are a variety of bench-scale treatment methodologies that would effectively reduce the quantity and/or toxicity of laboratory wastes. Unfortunately, RCRA places a severe regulatory constraint on engaging in regulated "treatment" activities, and the cost and complexity of obtaining an RCRA treatment facility permit puts this option beyond the reach of most research and laboratory facilities. Similarly regulatory "traps" lie elsewhere in environmental laws, such as the local sewage agency prohibiting the discharge of laboratory hazardous wastes via the drain notwithstanding the RCRA exemption for designated types of laboratory wastewater drain disposal. These regulatory restraints on certain types of economically and technically feasible laboratory waste minimization activities should be reviewed during the waste minimization audit and planning process.

CROSS-MEDIA POLLUTION CONSIDERATIONS AND OTHER REGULATORY TRENDS

There is increasing recognition that environmental regulations that have the effect of transferring pollution from one media to another (e.g., stripping volatile organic compounds from groundwater and releasing them to air) are not acceptable. Accordingly, large capital projects that would convert waste from one media to another are likely to be of relatively limited value over time. Conversely, capital projects that create "closed loop" systems without discharges and projects that will recover enough useful material to create a "payback" in reduced purchase and disposal costs over a relatively short period are likely to have a longer term value.

In relation to other regulatory trends, there are several important opportunities to join with other research and laboratory facilities to ensure that Congress authorizes waste minimization efforts that are environmentally sound, as well as technically and environmentally feasible, without the regulatory disincentives which currently exist. While the regulatory agenda of different institutions may vary, what follows are a handful of suggestions of particular relevance to laboratory and research facilities as these issues are considered at the federal, state and local level:

RCRA: Allow treatment of laboratory-scale quantities of hazardous wastes without requiring RCRA permit.
Clean Water Act: Prevent publicly owned treatment works (POTWs) from adopting pretreatment regulations that prohibit drain disposal of laboratory chemicals.
Clean Air Act: Closely monitor EPA's eventual efforts to establish a hazardous air pollutant regulation for research and laboratory emissions. Educate EPA regarding the real risks, and difficulty of technology and operational control strategies, associated with air emissions from routine laboratory operations.

Obviously, many laboratory and research facilities share concerns with other members of the regulated community on environmental laws such as CERCLA and "command and control" waste minimization requirements. The above agenda, however, may be a starting point for focusing on those concerns warranting special attention by environmental professionals working with research and laboratory facilities.

Part 2:
Effecting Pollution Prevention and Waste Minimization

CHAPTER 5

Planning and Development of a Model Waste Minimization Program

Robert Charbonneau

CONTENTS

Introduction 70
Essential Elements 71
Hints on Program Implementation 72
Waste Minimization Program Elements 73
    Planning and Organization 73
        Executive Management Support 73
        Waste Minimization Policy 75
        Appoint a Waste Minimization Coordinator 75
        Establish Task Force 75
        Establish Specific Goals 76
        Develop a Program Plan 76
        Long-Range Strategic Planning 77
    Wastestream Assessment 78
    Feasibility Analysis 82
        Technical Analysis 82
        Institutional Analysis 83
        Regulatory Analysis 83
        Economic Analysis 83
    Implementation 85
    Program Evaluation 87
Endnotes 88

INTRODUCTION

This chapter will provide an overview of programmatic development strategies for hazardous waste minimization programs in research institutions. This approach is based on guidance from the U.S. Environmental Protection Agency (EPA) and also incorporates knowledge gained from implementation of a number of successful industrial waste minimization programs. This model also incorporates a strategic planning approach to program development in an institutional framework. The overview will begin with a review of the essential components of successful waste minimization programs, followed by a number of hints on program development and implementation. Finally, the five basic elements of an effective waste minimization program will be reviewed.

A waste minimization program is a strategically planned and systematic continuing effort to reduce or eliminate the amount of hazardous waste generated. According to the EPA's waste management hierarchy, source reduction measures should take precedence over recycling and reuse strategies. Institutional barriers, both structural and attitudinal, are the greatest impediments to program implementation, in contrast to any technical considerations. Successful organizations must effectively manage their human resources, and incorporating waste minimization into the existing administrative structure and organizational "culture" is the key to successful program development. Laboratories associated with educational institutions have the unique opportunity to instill waste minimization ethics into students that will become future leaders in their professions.

The formality of the waste minimization program will depend upon the size and complexity of the organization and its hazardous waste generation characteristics. Highly complex larger organizations will require more formal structured programs as compared to smaller organizations. Likewise, generators of large volumes of diverse wastes will need more highly organized programs than small generators due to the scale of the effort. Programs must be flexible to quickly adapt to changing waste minimization targets and priorities. Waste minimization is one of the only environmental programs that does not require stringent oversight by legal counsel because it is not governed by conventional regulatory "command and control" requirements. Refer to Chapter 4 for an overview of the waste minimization regulatory framework.

Research and educational institutions pose unique problems for developing waste minimization programs. Management structures are divided between multilayered academic research and administrative functions and are often decentralized. This makes it difficult to get clear top management support for waste minimization and also complicates policy considerations. Furthermore, there is often little control over chemical acquisition processes and hazardous waste management practices in the laboratory. All of these critical factors make implementation of waste minimization programs in research and educational institutions more complex and difficult compared to industrial settings.

ESSENTIAL ELEMENTS

All successful waste minimization programs generally have a few critical elements in common. These essential elements are:

- Top management support
- Find champions
- Recognition of successes
- Information dissemination

Without the support of executive level management, it will not be possible to overcome institutional resistance to change. Researchers will be reluctant to change their laboratory protocols and operations to incorporate waste minimization measures. Middle management and line supervisors will also be reluctant to change their policies and practices. A clear mandate in the form of a written waste minimization policy is needed to overcome institutional or bureaucratic obstacles to change. An active commitment is needed from top management to supply the human, technical, and financial resources needed to implement an effective waste minimization program. Management should be educated as to the importance and benefits of implementing a hazardous waste minimization program at the facility. Refer to Chapter 5 for additional discussion of gaining institutional commitment.

It is necessary to find one or more persons in the organization who are personally as well as professionally committed to the principles of waste minimization. These "champions" should be in a position of authority that enables them to provide the leadership necessary to overcome the power of the status quo and other institutional obstacles to implementation. These leaders should be familiar with laboratory processes as well as hazardous waste management practices. They should have a good rapport with both management and research personnel and also possess a working knowledge of waste minimization principles and techniques.

Recognition of successful projects and personal efforts is vital to positively reinforce waste minimization throughout the organization. Laboratory personnel should be provided with rewards or incentives to motivate them to succeed. These rewards do not have to be elaborate or expensive. Certificates, plaques, or other appropriate tokens are perfectly acceptable. No matter what the reward actually entails, it should be well publicized so that colleagues can acknowledge the efforts of their peers, and to provide motivation for others to succeed. Incentives can be integrated into existing award programs such as exceptional employee performance or safety awards. Incremental grassroots successes of individual laboratories should be encouraged and publicized. In the absence of a formal programmatic structure, these initiatives can motivate other labs to undertake waste minimization measures and persuade administrators to commit resources.

Finally, information must be disseminated to and among laboratory personnel through technology transfer, education, outreach, and training efforts. Technical as well as general information on waste minimization techniques and efforts must be effectively and efficiently transmitted among research personnel who will actually be implementing these measures. Good communications are essential to foster coordination and cooperation between departments and avoid duplication of efforts.

HINTS ON PROGRAM IMPLEMENTATION

- **Incorporate waste minimization into the organizational "culture."** To be truly successful, changes in attitudes and thinking processes are needed to incorporate waste minimization into the entire organization's culture. The principles of waste minimization must be instilled into the organization, so that all employees practice waste minimization as part of their daily lives, not just as another one of their job duties or responsibilities. Employees must be both involved and responsible for waste minimization. Executive policy and management support at all levels is essential to make waste minimization an integral part of standard laboratory operations and research. It helps to achieve early, highly visible success through implementation of simple reuse or recycling measures.
- **Motivation.** Laboratory personnel must be strongly motivated to implement and then maintain the changes necessary to incorporate waste minimization into their operations. If possible, waste minimization should be made a standard part of annual job performance evaluations or reviews. Financial incentives in the form of bonuses or merit pay increases can also be offered, but may prove to be counterproductive because employees may not fully cooperate with each other if they perceive that someone else will be financially rewarded for something they helped develop. Recognition of successful efforts is essential, no matter what mechanism is used.
- **Pilot program approach.** A pilot or trial approach is an efficient way of trying various waste minimization measures. Small-scale programs are more manageable and controllable, even with limited resources. This approach becomes even more valuable in times of economic hardship and competition for increasingly scarce resources. Pilot programs can be relatively easily modified or radically shifted to meet changing conditions or goals. Even unsuccessful programs have the benefit of limited losses on a small scale. Pilot programs help to build the expertise, commitment, and sustainability necessary to implement waste minimization at the larger institutional level. Successful pilot efforts can be promoted and adopted throughout the entire organization. An effective method is to integrate these waste minimization approaches into existing health and safety or other appropriate programs. Small-scale successes can also help to convince administrators to make a commitment to develop institutional waste minimization programs.
- **Provide technical support.** Technical assistance on specific waste minimization options and measures must be given directly to laboratory personnel. Educational and outreach efforts should be emphasized. Specialized training must be provided to lab personnel if necessary.
- **Provide data to lab personnel.** Laboratory personnel should be given complete data on waste generation as well as the unit costs and total costs of waste disposal for the lab. This will enable lab personnel to internally evaluate lab-specific waste minimization measures more effectively. Labs may also compare

their waste generation and disposal costs with similar research labs to identify discrepancies and opportunities for waste minimization already being implemented in other labs. Information will also facilitate brainstorming on waste minimization options within and between labs.
- **Involve lab personnel in planning and evaluation.** Supplied with technical assistance and data, laboratory personnel can become integrally involved in the planning and evaluation of waste minimization measures or systems for their own lab. This will enable researchers to gain a vested interest in waste minimization and will motivate them to succeed. Individual labs should be encouraged to try innovative approaches and techniques. Researchers know their experiments better than anyone, so avoid using outside consultants to perform wastestream assessments in laboratories. It is more efficient and cost effective to educate lab personnel on waste minimization processes and techniques.
- **Fully charge for waste disposal costs.** In terms of overall economic efficiency, laboratory generators should be charged the full economic costs of the materials used and wastes generated. Although controversial, chargeback (or surcharge/recharge) systems provide a direct financial incentive to laboratories to reduce their waste generation. However, recharges may also discourage proper waste disposal and lead to increased drain disposal or other improper practices. Proper waste disposal must be ensured by educational efforts and internal environmental monitoring programs. Front-end surcharges on chemical purchases can alleviate this potential problem, but this option requires that a centralized purchasing mechanism already be in place.
- **Use simple technologies.** If selected waste minimization measures are technical in nature, they should be as simple as possible. Low technology is best. Equipment should be simple to operate, reliable, and easy to maintain.

WASTE MINIMIZATION PROGRAM ELEMENTS

Table 1 shows the basic structure of a waste minimization program. Each of the five major program elements will be discussed in detail in the sections that follow.

PLANNING AND ORGANIZATION

Strategic planning and a high degree of organization are essential for successful development of a waste minimization program. The impetus for developing a program may come from executive management or lower-level managers or researchers. Once top management has decided to establish a waste minimization program, this commitment must be conveyed to the entire organization through a formal policy statement. A coordinator is then appointed and a task force established to set goals, develop a program plan, and direct its implementation.

Executive Management Support

It is important to reiterate the need for top management commitment to successfully implement a waste minimization program. It is the responsibility of

Table 1 Waste Minimization Program Elements

I. Planning and organization
  1. Executive management support
  2. Waste minimization policy
  3. Appoint a waste minimization coordinator
  4. Establish task force
  5. Establish specific goals
  6. Develop program plan
  7. Long-range strategic planning
II. Wastestream assessment
  1. Identify and characterize lab processes and wastestreams
  2. Target important wastestreams or labs
  3. Identify and prioritize waste minimization options
  4. Screen and select best options
  5. Long-term vs. short-term strategies
III. Feasibility analysis
  1. Technical evaluation
  2. Institutional evaluation
  3. Regulatory evaluation
  4. Economic evaluation
IV. Implementation
  1. Finalize implementation plan and schedule
  2. Resource and cost allocation
  3. Evaluate opportunities for integration
  4. Develop or modify SOPs if necessary
  5. Install or modify equipment if necessary
  6. Provide technical assistance and training
  7. Information dissemination
  8. Provide recognition and awards/incentives
V. Program evaluation
  1. Develop mechanism for evaluation
  2. Periodically review program
  3. Measure effectiveness and impacts

top management to instill formal support for waste minimization throughout the organization by clearly communicating policies and priorities to subordinate managers and all employees. Mid-level management such as department heads, laboratory managers, and other line supervisors can be major obstacles to implementation if they are not supportive and receptive to change. Educating mid-level management is just as important as education tailored to executive management. Active management commitment is essential to supply the resources needed for implementation of a waste minimization program such as personnel time allocations for waste minimization activities and funding priorities.

Educational institutions have even more complicated administrative structures than corporations. Parallel administrative and academic bureaucracies make the task of obtaining executive management support more difficult. Compounding the problem, larger institutions tend to be more hierarchical and consequently access to top level administrators is more restrictive. To obtain resources for waste minimization projects, laboratory researchers must often go through the proper channels beginning with their department chairperson up through a dean or provost, ultimately to a vice chancellor or vice president. Vice chancellors for academic affairs and administration can then coordinate commitment of resources and policy formulation for the chancellor's approval. An alternative to this bureaucracy is to investigate the institutional committee structure of the academic or faculty senate to determine if there is an appropriate committee to handle waste minimization proposals and concerns.

Waste Minimization Policy

A formal written policy statement or management directive on waste minimization must be issued by executive management and included in a formal program planning document. The policy should consider the following points:

- Statement of management commitment to protection of human health and the environment and the establishment of a waste minimization program
- Delegation of responsibility for waste minimization to every employee
- General goal to eliminate or reduce waste generation, the types of wastes targeted, reduction of risk, and any more specific goals as appropriate
- Statement of the waste management hierarchy — source reduction has the highest priority followed by recycling/reuse and treatment
- Applicability of the policy in terms of operations, sites, and wastestreams
- Authority for implementation — establish waste minimization coordinator and task force or committee structure

Appoint a Waste Minimization Coordinator

One person should be given the responsibility and authority to establish and oversee the waste minimization program to enable a coordinated, focused approach to program implementation. The coordinator works closely with a waste minimization program task force to develop and implement the program. The coordinator must also facilitate communications between all levels of the organization, so this person should be named from the highest practical level.

Establish Task Force

An administrative structure such as a task force or committee should be established to develop the waste minimization program. The task force directs implementation, secures cooperation, and builds consensus among staff and researchers to participate in the program. Units or departments most affected by the changes required to implement waste minimization should be represented. The capabilities and attitudes of task force members toward waste minimization will be major determinants of successful program implementation. Members should be carefully selected based on their substantial technical, administrative, and communications skills as well as organizational knowledge. One or more waste minimization "champions" should be appointed to serve if possible. Consensus decision making will help to overcome institutional barriers to implementation. The waste minimization coordinator works very closely with the task force to implement selected measures and projects.

The scope of the waste minimization program will determine whether a full-time commitment is needed by the coordinator or any of the participants on the task force. In cases where the coordinator or participants are only part time, management must allow them to allocate sufficient time to work effectively on the waste minimization program. In general, the task force is responsible for overseeing implementation of the program. The task force may be directly involved with implementation activities or may assume a purely supervisory or monitoring role. The task force should be intimately involved in both the planning and evaluation phases of the program and should foster good communications both within and outside the organization.

Establish Specific Goals

Explicit goals should be established for reducing waste volumes and/or toxicity to focus efforts and build consensus. These goals must be consistent with the policy established by senior management and are formulated by the task force in coordination with management. The goals are reviewed periodically during strategic planning of the program, revised to reflect changes in the program focus, and integrated into a formal program planning document.

Goals should be:

- Measurable over time
- Acceptable to laboratory personnel
- Flexible and adaptable to changing conditions
- Challenging but not unreasonable
- Useful and meaningful for every employee
- Achievable with a practical level of effort
- Incorporated into departmental goals if appropriate
- Set to some defined timeline or milestones

In dynamic laboratory environments where experiments and research changes often, chemical use and waste generation will vary considerably, making it very difficult to set quantitative waste reduction goals such as specific target levels or percentages. In this case, setting qualitative programmatic implementation goals is probably more appropriate than trying to set quantitative goals. Besides program milestones, other qualitative goals can be devised, such as requiring each lab or researcher to evaluate waste minimization options for their most significant wastestream.

Develop a Program Plan

A detailed waste minimization program plan should be developed by the task force to define program objectives, identify potential obstacles and solutions, define data collection and analysis procedures, and develop a program schedule. The task force should begin a preliminary assessment phase by reviewing data that are already available and begin defining ways to process that information. Much of the data needed for a waste minimization program may already have been collected in response to existing regulatory requirements for waste management

Table 2 Elements of a Waste Minimization Plan

Waste minimization policy statement from senior management
Description of task force makeup, authority, and responsibility
Description of how all of the groups (laboratory, environmental management, maintenance, engineering, procurement, and others) will cooperate to reduce waste generation
Plan for publicizing and gaining institutional support for the waste minimization program
Plan for communicating the successes and failures of waste minimization programs within the institution
Description of the processes that produce, use, or release hazardous or toxic materials, including clear definition of the amounts and types of substances, materials, and products under consideration
List of treatment, disposal, and recycling facilities and transporters currently used
Preliminary review of the cost of pollution controls and waste disposal
Description of current and past waste minimization activities at your institution
Evaluation of the effectiveness of past and ongoing waste minimization activities
Criteria for prioritizing candidate facilities, labs, processes, and wastestreams for waste minimization projects

Source: EPA, 1992.

and disposal. Other useful information involves laboratory processes and procedures, chemical inventory information, procurement records, accounting/budgeting information, and organizational charts.

During the preliminary assessment phase, the task force should try to identify opportunities for waste minimization and work with management to establish initial priorities. This will be the starting point for defining short and long-range objectives. Objectives are the specific tasks that will be necessary to achieve stated goals and can be defined at the facility or departmental level. Ideally, objectives should be stated in quantitative terms and have target dates.

As the task force begins to develop and implement a waste minimization program, they are likely to encounter a number of impediments. Obstacles need to be recognized, and the means for overcoming them need to be defined. Apparent barriers will be less likely to impede the process if everyone understands that there is a mechanism for addressing them at a later stage. Common obstacles fall into four broad categories: technical, institutional, regulatory, and economic. These factors are analyzed more extensively during the feasibility analysis element of the program. For a more detailed discussion of overcoming impediments to waste minimization, refer to Chapter 7.

The final aspect of program planning is to list milestones within each of the major elements from wastestream assessment through implementation and assign realistic target dates. Try to follow these milestones closely to avoid delays in program implementation. Table 2 provides a list of elements that should be included in a formal waste minimization plan.

Long-Range Strategic Planning

The model set forth in this chapter for developing an institutional waste minimization program is actually based on a strategic planning approach. In general, strategic planning provides a set of concepts, procedures, and tools that enable organizations to effectively deal with dynamic environments. A generic strategic planning process begins with broad policy and direction setting. Mission statements and mandates are developed, and consideration is given to all stakeholders that will be affected by these policies. Next, an assessment of the organization's current status and needs is performed through both internal and external analyses. Opportunities, threats, and key issues are identified. Third, goals, objectives, and strategies are formulated to deal with the key issues. Alternatives and barriers are also identified. Fourth, an implementation plan is devised to carry out selected strategic actions. Finally, continuous monitoring is conducted to evaluate and update the implementation plan.

In terms of this model program, the strategic planning emphasis is explicitly focused on the analysis of data generated from the evaluation element of the program. Waste minimization options are reprioritized and new goals or targets are selected in light of the performance of implemented techniques. This is an iterative rather than a linear process. A summary report prepared after the wastestream assessment and feasibility analysis phases can provide a basis for evaluating and maintaining the program.

WASTESTREAM ASSESSMENT

The second key component of the program is evaluation of the types and quantities of hazardous waste generated by various laboratory processes. Wastestream assessment is an information management tool designed to identify, screen, and analyze various waste minimization options in a systematic and comprehensive manner. This component is the technical focus of a waste minimization program, and generates options for further evaluation. This phase is also variably known as "waste auditing," "waste minimization opportunity assessment," and "process waste assessment." Figure 1 outlines the general wastestream assessment process.

The focus of wastestream assessments should be on lab processes and activities as sources of waste. A thorough understanding of laboratory research procedures and operations is needed to identify how wastes are generated and to formulate waste minimization options. Researchers are the best source of information on the sources, causes, and controlling factors that influence waste generation in laboratories. Educational, research, and clinical laboratories use a wide variety of diverse techniques in their daily operations, making it very difficult to generally classify the processes and operations that generate hazardous waste. However, the assessment process should be flexible enough to combine multiple activities into general laboratory processes or wastestreams to allow meaningful analysis on a manageable scale. Examples of general laboratory processes include chemical acquisition and inventory, or the collection and disposal of surplus chemicals.

The first step is to accurately and specifically identify and characterize laboratory wastes as precisely as possible. Wastestreams should be segregated and characterized as specifically as possible according to standard RCRA or state waste code classifications. Laboratory air emissions and wastewater discharges

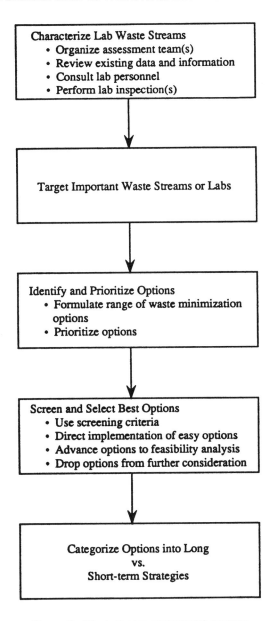

Figure 1 Wastestream assessment process.

must also be captured in this analysis. Waste information can be obtained from hazardous waste manifests and attached sheets, annual or biennial RCRA waste reports, SARA Title III submissions, wastewater and air emissions reports, and other environmental records. Mass balance analysis of lab chemical inputs vs. waste generated is often impossible to accurately quantify because of the numerous chemical transformations that occur in the laboratory. However, elemental balances can be performed for extremely hazardous substances such as certain heavy metals. After accurately characterizing the waste, significant wastestreams (in terms of volume or toxicity) can then be targeted as priorities for waste minimization efforts and for subsequent identification of potential waste minimization options.

Assessments can focus on either specific wastestreams or target individual laboratories. For example, photographic labs that generate large volumes of acidic heavy metal wastes can be targeted based on either their large waste volume or their specialized wastestream. Individual labs may be targeted based on high waste volumes, especially if many surplus chemicals are being discarded.

Information concerning laboratory design characteristics, personnel, research processes and operations, chemical inventories, procurement records, and wastestreams can all be valuable when conducting wastestream assessments. Process flow diagrams are often very helpful in identifying where and how wastes are generated. After as much data as possible have been collected and reviewed, an actual site inspection of laboratory areas should be performed to resolve any outstanding questions or inconsistencies in the data, and to gather supplemental information. The site visit also helps the team understand the data and identify factors that are not well documented. Site visits should be well planned to maximize the use of everyone's time. The site visit should be scheduled at a time when lab operations of interest are conducted. Lab personnel should be made aware of the inspection and any information needs well in advance. The site inspection should track targeted activities from acquisition of laboratory chemicals through laboratory processes to the point of waste pickup and disposal. General lab "housekeeping" practices should be observed and researchers interviewed about any concerns or processes.

Depending on the size and scope of the waste minimization program, the wastestream assessments can be performed either by individuals or organized teams. An optimal approach is to form teams tailored to specific laboratory processes or wastestreams. The size of the team depends on the complexity of the process and the time frame for completion. At a minimum, the teams must include the waste minimization coordinator and laboratory personnel with direct responsibility and knowledge of chemical acquisition, inventory, and storage, as well as the processes generating the wastes and waste management and disposal practices. At least one member of the task force should be included on each team to facilitate communications. More complex large-scale teams may also include personnel from environmental management, engineering, facilities maintenance, information systems, procurement, accounting/budgeting, and administrative staff. Multidisciplinary teams are more likely to achieve a comprehensive assessment and identify a wide range of options. Be aware of personality conflicts and inhibiting hierarchical relationships. The use of objective outside personnel or experts can be considered to bring in specialized expertise and different ideas from outside the organization. The program task force may either assist or supervise the assessment.

Computerized hazardous materials and waste tracking systems can facilitate the evaluation of wastestreams and also perform an inventory control function. A computerized database can easily identify which laboratories are using extremely toxic materials or high volumes of chemicals. These labs can then be targeted for further waste audits. Computerized systems can also promote redistribution of surplus chemicals and facilitate inventory control by ensuring proper chemical stock rotation so that the oldest chemicals are used first before their expiration dates. Centralized purchasing protocols lend themselves to bar coding identification of incoming chemicals that can then be tracked by computer through stockrooms to the laboratory. "Cradle to grave" tracking of hazardous materials may prove difficult due to the numerous chemical transformations performed in the laboratory. However, a bar coding system can be used to track secondary materials as well as hazardous wastes after they are generated in the laboratory.

Following wastestream characterization, waste minimization options for the targeted wastestreams are then identified and prioritized by the assessment team. Lab personnel and team members should independently formulate waste minimization ideas based on their own expertise and experience. Researchers and team members should then try group brainstorming about potential waste minimization measures and approaches to develop additional options. Ideas should also be sought from outside organizations through technology transfer and information exchange. Many potential options should be identified through this creative process. Prioritization of waste minimization options for further screening is necessary when waste minimization funding and/or personnel are limited. According to the waste management hierarchy, priority should be given to source reduction measures over recycling and reuse options whenever possible. Other important prioritizing criteria include quantity and toxicity of wastestreams, health and safety concerns, costs of waste disposal, regulatory compliance issues, liability concerns, and the potential for easy implementation.

After prioritizing waste minimization options, it will be necessary to screen them to identify the options that offer the best potential for waste minimization. This is necessary because the subsequent intensive feasibility analysis phase is time consuming and more costly. The screening procedure should eliminate those measures that appear to be marginal, impractical, or inferior without having to go through the in-depth feasibility analysis phase. Screening procedures can either be qualitative or quantitative and performed either by the waste minimization coordinator, assessment team, or the task force. The weighted-sum method is a commonly used quantitative technique designed for use in more complex decision-making situations. Refer to Appendix E of the Facility Pollution Prevention Guide (EPA, 1992) for a summary and example of the weighted-sum method.

Major screening criteria include the following:

- What is the principal benefit of the measure (potential impact)?
- Can the measure be implemented with existing resources and technology?
- Does the measure appear cost effective?
- Can the measure be implemented in reasonable amount of time?
- Does the measure have a dependable performance record?

- Are there regulatory constraints to implementation?
- Does the measure have a good chance for success?

Options are screened based on these major criteria and selected for further feasibility analysis or implementation. The number of options selected depends upon available resources and time constraints. Options such as procedural changes that do not involve any capital costs can be implemented with little or no additional evaluation. Measures found to be impractical or that have marginal value are dropped from further consideration. Remaining options will generally require in-depth feasibility analysis. Depending upon the scale of the program and resource availability, waste minimization efforts can also be divided into short-term implementation strategies and long-term goals to reduce waste generation.

FEASIBILITY ANALYSIS

Waste minimization options identified and screened during the wastestream assessment process are then further evaluated in a feasibility analysis phase that considers four major factors: technical, institutional, regulatory, and economic considerations. After the feasibility analysis is completed, each assessment team should prepare a summary report of its findings. The task force then compiles all of the team assessments into a summary assessment report and implementation plan.

Technical Analysis

The technical evaluation is done to determine if a proposed waste minimization option is likely to work in a specific application. This evaluation often requires the expertise of a variety of people. Technical factors include the following:

1. Potential impact on the volume or toxicity of hazardous waste generated
2. Adverse impacts or interruptions of laboratory research/teaching activities
3. Resource needs
   A. Specialized equipment
   B. Facility space for storage/handling of wastes or new equipment
   C. Personnel
4. New equipment considerations
   A. Specialized training needs
   B. Specialized maintenance requirements and warranty conditions
   C. Utility requirements
   D. Installation requirements and impacts
   E. Time requirements for delivery and installation
   F. Pilot or performance testing requirements
5. Health and safety concerns for researchers or other staff
6. Commercial availability and reliability of technological approaches
7. Database systems or other computerized support needs
8. Creation of other environmental problems (such as air emissions)

Institutional Analysis

Institutional evaluation factors include:

1. Compatibility with research or teaching environment and practices
2. Changes required in existing policies
3. Changes required in existing standard operating procedures
4. Training requirements
5. Need for interdepartmental cooperation or coordination
6. Degree of resistance from researchers or staff
7. Changes in purchasing/procurement processes
8. Shifts in authority or responsibility within or between departments

Regulatory Analysis

Regulatory evaluation factors include permitting requirements, hazardous waste storage and treatment regulations, building and fire code requirements, as well as OSHA and other occupational safety requirements. Unfortunately, opportunities to implement waste minimization measures sometimes conflict with the RCRA 90-day storage limitation for hazardous wastes and the requirements to obtain special permits for "treatment" of wastes under various regulations. Refer to Chapter 7 for a discussion of how to deal with these regulatory impediments.

Economic Analysis

Standard economic assessments typically are not adequate for evaluating waste minimization projects for several reasons. Traditional analysis methods do not account for waste minimization measures that affect multiple operations and processes. Secondly, common methods do not analyze a sufficiently long time period to fully evaluate the long-term benefits of waste minimization. Finally, they do not deal well with the probabilistic nature of waste minimization benefits, many of which can not be estimated with a high degree of certainty.

In recognition of these problems, the U.S. EPA has developed an analysis method known as Total Cost Assessment (TCA). There are four elements of TCA: expanded cost inventory, extended time horizon, use of long-term financial indicators, and direct allocation of costs. This section will only summarize the essential characteristics of TCA. For more information, refer to a report on TCA prepared in 1991 by the Tellus Institute for EPA.

TCA includes both direct and indirect costs, as well as cost factors related to liability and certain "less tangible" benefits. TCA can be used incrementally by gradually bringing each of the cost factors into the analysis, especially when it is problematic to estimate liability or less tangible costs. Unlike typical capital investments, the indirect costs of waste minimization projects are likely to represent a substantial net savings. These costs are usually hidden as overhead costs or omitted from the project financial analysis. It is necessary to estimate these costs in an economic analysis and allocate them to their source. Costs can be allocated across the entire institution, department, or operating unit level or directly to the waste generators.

Allocating future liability costs is subject to a high degree of uncertainty, but may offer significant net savings to research and educational institutions. However, there are several ways to address liability costs in project analysis. First, a calculated estimate of liability reduction can be based on penalties, fines, or claims against similar institutions involved in similar processes. Secondly, a qualitative estimate of reduced liability risk associated with a waste minimization project can be given without attaching any specific dollar value. Finally, financial performance requirements such as payback period can be loosened to account for reduced liability. Less tangible benefits, although difficult to measure, should be incorporated into the analysis whenever feasible. At a minimum, they should be highlighted after quantification of other costs and benefits.

Many of the liability and less-tangible benefits of waste minimization will occur over an extended time period. The economic evaluation should consider a long time frame, greater than the 3- to 5-year period typically used for other types of projects. Long-term financial indicators such as Net Present Value (NPV), Internal Rate of Return (IRR), or Profitability Index (PI) should be selected for economic analyses. However, increasing the time frame also increases the uncertainty of the cost factors used in the analysis. The actual length of time over which the analysis is performed is a subjective decision that ultimately must be made by management personnel evaluating the project. This decision will reflect the degree of commitment and importance that management places on waste minimization. Therefore, it is essential to educate management personnel on all the benefits of waste minimization, especially long-term reduction of liability.

Economic (cost-benefit) evaluation factors include the following:

A. Direct costs
   1. Capital expenditures
      a. New facilities and/or building renovations
      b. New equipment and materials (e.g., piping, ducts, electrical, various controls, etc.)
      c. Installation and construction (including utility hookups)
      d. Engineering and consulting
      e. Start-up and initial training
   2. Operating costs
      a. General operation and maintenance
      b. Training materials and/or programs
      c. Salary for additional personnel
   3. Expenses or revenues
      a. Hazardous waste disposal (includes fees, taxes, transportation, etc.)
      b. Raw materials inputs (laboratory chemicals)
      c. Hazardous waste generator fees and taxes
      d. Labor for waste disposal
      e. Value of recovered or recycled materials

B. Indirect costs
   1. Administrative costs
   2. Regulatory compliance costs
      a. Permitting
      b. Recordkeeping and reporting
      c. Monitoring
      d. Manifesting
   3. Insurance
   4. Worker's compensation
   5. Waste management costs
   6. Operation of pollution control equipment
C. Hazardous waste liability costs
   1. Fines and penalties
   2. Personal injury
   3. Property damage
   4. Cleanup costs for land disposal and releases or spills
D. Less tangible benefits
   1. Reduced occupational illness and injury liability
   2. Enhanced public image in the community
   3. Leadership role among research and teaching institutions
   4. Improved relationship with regulators
   5. Instilled waste minimization ethics in students and workforce

Following the feasibility analysis, each assessment team prepares a report that includes these elements: results of wastestream assessment, prioritized listing of potential options, results of option screening, results of feasibility analysis, detailed description of options proposed for implementation, and a tentative schedule. The task force then compiles each team report into a summary assessment report and prioritized implementation plan and schedule. The report can be used as a basis for evaluating and maintaining the waste minimization program and also to secure internal funding for projects that require capital investment.

Before the summary report is issued in final form, researchers and other experienced laboratory personnel affected by the proposed waste minimization measures or projects should be asked to review the report. This review will assure that the measures are well defined and feasible from their perspectives. In addition to ensuring the quality of the report and implementation plan, the review will also help secure the support of the personnel responsible for implementation.

IMPLEMENTATION

Final decisions on the implementation plan and schedule are made at this point. The assessment teams may be asked to produce additional data or to develop alternatives or modifications. An important consideration is to evaluate opportunities to integrate the selected measures into existing programs or activities. Laboratory chemical hygiene or illness and injury prevention plans and programs may prove to be efficient vehicles to facilitate implementation of waste minimization techniques. Appropriate environmental management personnel such as radiation safety officers, industrial hygienists, hazardous waste managers, and fire marshals should all be involved with waste minimization technical assistance and educational efforts within each of their own areas of expertise.

Adequate resources must be committed by senior management to implement the waste minimization program. This includes both human and technical resources such as new or modified equipment or facilities. The task force will seek funding for those measures that require capital expenditures. Grants or government financing may be available in some cases. In general, capital costs should not be significant, at least in the short-term. Waste minimization depends more often on changing people's attitudes and operating procedures, rather than on investments in new equipment or facilities.

Implementation activities may involve a wide range of waste minimization strategies depending upon which methods are selected. Key elements will always include information dissemination and technical assistance. These goals are accomplished through technology transfer, education, training, and outreach activities. Education and/or training should be provided to all management levels, laboratory researchers, procurement personnel, and staff as well as all new employees and students. Education of lab personnel on methods for more accurately estimating chemical quantities that are needed for research may prove to be especially cost effective.

Because laboratory waste minimization is a relatively new field, there are probably many opportunities to apply waste minimization techniques already proven in various industrial processes. This technology can be transferred to laboratory applications. Implementation may also entail the development or modification of standard operating procedures (SOPs) for laboratories or purchasing departments. SOPs are generally administrative or institutional measures such as material handling, inventory, and waste segregation practices that can be applied to waste minimization. SOPs can be implemented with little or no cost, and are therefore very cost effective. Administrative, operational, and procedural changes should be implemented as soon as practically feasible. An example of an effective SOP is to designate a single person in each laboratory to coordinate and process all chemical acquisition requests.

Technological measures entail modifications in processes or equipment to reduce waste. These changes may range from minor modifications that can be implemented at low cost to the replacement of processes involving large capital costs. Projects involving new or modified equipment are handled like other capital improvements through a phased process that includes planning, design, procurement, construction, and operator training. The performance of new equipment should be evaluated to determine if it meets expectations and whether any modifications are required.

Successful waste minimization efforts must be recognized and rewarded. This will provide incentives to others in the organization to implement waste minimization measures. Waste minimization is an ongoing effort. After highest priority wastestreams have been reduced as much as practical, lower priority wastestreams are assessed and the process is repeated until all wastes are reduced to the maximum extent possible.

PROGRAM EVALUATION

The final program element involves ongoing evaluation of the waste minimization program's performance. Evaluation of successes and failures will help guide future waste minimization assessment and implementation cycles. A mechanism for evaluation must be developed from the outset of the program to determine informational needs. It is easiest to work from existing data bases or structures, such as baseline hazardous waste generation data or wastestream disposal costs. Computerized database systems can effectively and efficiently facilitate the evaluation of hazardous waste management practices. No matter what mechanism is used, it must be able to quantitatively measure the impact of the waste minimization program against some baseline level. In addition, a qualitative program evaluation should be considered through mechanisms such as a survey to researchers on the acceptability and ease of implementation of waste minimization techniques.

Several quantitative measures can be considered. Usually it is best to select a combination of methods to fit your data availability, facility characteristics, and goals. The change in actual waste generation can be considered, but this is an unreliable indicator because there is no way to discern how or why waste volumes increased or decreased and, in the latter case, whether or not this was due to waste minimization practices. A better indicator is the adjusted change in waste quantity. This measure correlates with laboratory activity, so that waste generation is tied to the number of experiments or processes performed on some unit basis. This would normalize the data by taking into account increases in the level of experiments or processes run in the laboratory. Always watch for shifts of wastes to other environmental media such as water or air emissions.

Other quantitative measures include the total quantity of chemicals received, changes in total amounts of toxic materials released, changes in effluent discharge toxicity, lab research process analysis, cost-effectiveness analysis of implemented measures, and simple enumeration of the number of techniques implemented for source reduction. All of these methods have their own advantages and disadvantages depending on the characteristics of your particular organization and processes. In general, a waste minimization program is successful if source reduction measures are implemented, and there is a decrease in the adjusted quantity of waste generated.

There are many difficulties involved in measuring the impact of waste minimization in laboratories. Accurate baseline data and tracking systems often do not exist. Regulatory reporting data may not be specific enough to evaluate a targeted wastestream. In dynamic labs that change experiments or research activity levels often, new baseline data on waste generation must be formulated. Labs that conduct multiple experiments pose especially difficult evaluation problems. Determining a "unit of production" for laboratories to measure adjusted waste quantities requires innovative approaches and critical analysis. Changes in laboratory research "quality" due to implementation of waste minimization measures is also difficult to measure. These are just some of the major considerations that pose a special challenge to evaluating the success of waste minimization in laboratory environments.

Program evaluations should be conducted periodically, at least annually. The effectiveness of waste minimization techniques must be measured and compared to the explicit goals set forth in the planning phase of the program. Program goals may be modified through the strategic planning process. The summary assessment report prepared after the feasibility analysis can assist in evaluating various measures. The overall impact on hazardous waste generation or disposal costs should be assessed. This evaluation phase feeds back into the strategic planning component of the program. Waste minimization options and priorities are then re-evaluated and new options or targets are selected for further feasibility analysis or implementation. Reiteration of this process will yield maximum success in eliminating or reducing the generation of hazardous waste.

ENDNOTES

1. Beroiz, D., Director, Environmental Resource Management, Northrop Corporation, presentation given at Federal Facilities Waste Minimization conference, San Francisco, January 22, 1992.
2. Checkoway, Barry, Ed., *Strategic Perspectives on Planning Practice*, Lexington Books, Lexington, MA, 1986.
3. *Federal Register,* U.S. Environmental Protection Agency Draft Guidance on Waste Minimization Programs, Volume 54, No. 111, pp. 25056–25057, June 12, 1989.
4. Freeman, H. M., Ed., *Hazardous Waste Minimization*, McGraw-Hill, New York, 1990.
5. Higgins, T., *Hazardous Waste Minimization Handbook*, Lewis Publishers, Chelsea, MI, 1989.
6. Sorkin, D. L., Ferris, N., and Hudak, J., *Strategies for Cities and Counties: A Strategic Planning Guide,* Public Technology, Inc., Washington, D.C., 1984.
7. Tellus Institute, *Total Cost Assessment: Accelerating Industrial Pollution Prevention Through Innovative Project Financial Analysis, with Applications to the Pulp and Paper Industry*, December 1991.
8. U.S. Environmental Protection Agency, *Waste Minimization Opportunity Assessment Manual*. Hazardous Waste Engineering Research Laboratory, Publication EPA/625/7-88/003, Cincinnati, July 1988.
9. U.S. Environmental Protection Agency, *Guides to Pollution Prevention — Research and Educational Institutions*, Risk Reduction Engineering Laboratory and Center for Environmental Research Information, Publication EPA/625/7-90/010, Cincinnati, June 1990.
10. U.S. Environmental Protection Agency, *Facility Pollution Prevention Guide*, Risk Reduction Engineering Laboratory, Office of Research and Development, Publication EPA/600/R-92/088, Cincinnati, May 1992.
11. Waste Reduction Institute for Training and Applications Research, *Process Waste Assessment Implementation Guide,* Prepared for the U.S. Department of Energy, Idaho National Engineering Laboratory, Minneapolis (city of publication), 1992.

CHAPTER 6

Institutional Policy, Commitment, and Support

Fay M. Thompson

CONTENTS

Introduction 89
How Does the Process Begin? 89
Where to Find Champions 90
How to Sell Administration 91
How to Keep Progressing 94
Conclusion 95

INTRODUCTION

Pollution prevention is a state of mind. Until this new state of mind becomes the norm at all levels of the institution, pollution prevention's true potential will not be achieved. Reaching this goal requires finding and selectively educating the people who can make it happen.

HOW DOES THE PROCESS BEGIN?

An institution's foray into pollution prevention and waste minimization activities probably began simply enough, with interest from people at the hands-on level. Questions such as "Where can I recycle these cans?" and "Who's responsible for recycling?" were undoubtedly asked a number of times before the first aluminum can collection program was put in place. In the laboratory, a technician's desire for more shelf space may have meant disposal of unneeded chemicals, many of them still in sealed containers. A natural reluctance to throw out expensive chemicals that were still good led to attempts to find new users and perhaps to organized chemical recycling programs.

Other pollution prevention activities may be the result of a regulatory impetus. RCRA, Air Toxics, SARA Title III, NPDES, plus state and local regulations have caused many large quantity hazardous chemical users or waste generators to look for ways to reduce the quantity of material used.

A third (less likely) reason to begin a waste minimization program could have come from the management of an institution. To the extent that a key administrator (perhaps the chief financial officer) was convinced that there was money to be made from waste, some of the potentially profitable recycling programs might have been instituted.

The point of these examples is that most institutions already have some waste minimization/pollution prevention (WM/PP) activities in place, having been established for a variety of reasons. To move beyond these uncoordinated beginnings, the institution must make a commitment at the highest level to pursue a comprehensive pollution prevention/waste minimization program. Achieving this commitment may not come easily and will certainly require that a few well-connected champions of the program are found.

WHERE TO FIND CHAMPIONS

The support of people who believe in the value of WM/PP activities is vitally important to the development of a successful institution-wide program. The departments responsible for environmental health and safety and for waste management are certainly the first places to look for people who are committed to WM/PP activities. Other less obvious departments include:

- Purchasing, which receives much information on "environmentally safe" materials
- Storehouse, which often has helpful quantity use and inventory data
- Custodial services, which may well be concerned over inappropriate disposal of hazardous materials
- Finance, where there has undoubtedly been frustration over rapidly rising disposal costs
- Public relations, which knows that the environment is "in" and that a successful project will make good press
- Many others whose jobs are impacted in some way by the selection, purchase, storage, use, or disposal of materials

At academic institutions, two other groups can provide very valuable support. Among student populations, recent interest in environmental protection has again reached the levels of the late 1960s and early 1970s. A very strong component of this interest is conservation of resources, and there is often a significant concentration on the impact of the local institution on the environment. Students can add much enthusiasm and are often willing to commit time to special projects. A second possible source of support is the Board of Regents. As individual Regents, or preferably the entire Board, become interested in the cause and formally support it, a great deal of credibility is achieved. If the state of environmental awareness of the individual Regents is not known, it may be helpful to contact the office within administration that supports the Regents to ask for biographical information. The types of people who are most likely to have an interest in the cause are the corporate financial officers, the elected officials, the physicians, and those who list the environment among their interests. Whether direct, personal contact with individual Regents is appropriate for this topic should be checked out with the appropriate administrative individuals.

So far we have not discussed the *users* of materials that may be targeted for minimization activities. This is the group that is most directly impacted and therefore may be the most difficult to get involved. Even within this group, there will be a few hazardous material users who have taken it upon themselves to find ways to reduce the quantity or the hazard of the materials they use.

WM/PP activities are important to all of the people mentioned above, plus many others whose jobs involve the specification, use, management, or disposal of materials. A strong effort should be made to identify all of these interested individuals and to seek their support in developing an institutional WM/PP program.

HOW TO SELL TO ADMINISTRATION

The best way to use the enthusiasm and expertise of the people mentioned above is to form an ad hoc committee for the purpose of developing a comprehensive and permanent WM/PP program for the institution. This WM/PP planning committee must be broadly representative of everyone who may be impacted by decisions to change the way materials are managed. Much of the work of this committee will need to be accomplished before official recognition by administration can be expected; thus formation of the committee relies on the voluntary efforts of those who have special enthusiasm for the project and who take it upon themselves to get people together.

One of the goals of the planning committee should be to identify and evaluate a wide range of options for WM/PP projects. To fully evaluate various options, the committee must consider all the likely ramifications of each project. One of the major problems with paper recycling is the consolidation of combustible material in inappropriate locations. Similarly, bottle and can recycling often results in sanitation, vermin, and odor problems, making the program unpopular and potentially unsafe. Many attempts to begin laboratory chemical reuse programs have had insufficient storage and inventory management, resulting in creation of new "mini-superfund" sites in uncontrolled storage areas. Failure to anticipate all of the potential issues associated with WM/PP programs can lead to their early demise. The input of everyone potentially affected, particularly the users of the material being addressed, must be sought.

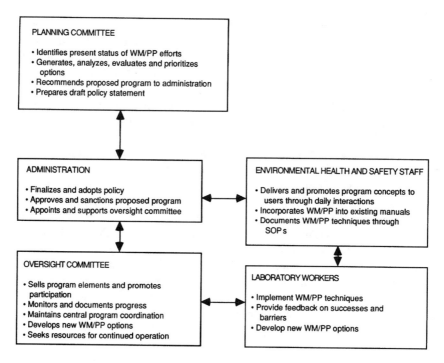

Figure 1 Elements of a waste minimization/pollution prevention program.

In conjunction with evaluation of the technical aspects of carrying out WM/PP projects, a good review of cost implications should also be done. It is not entirely obvious at first glance whether a project has a negative or positive cost impact or what the magnitude of the impact might be. Projects that appear to save money on disposal costs may increase costs at another point, and projects that appear to have large up-front costs may produce major annual savings. For each project, the capital costs and the annual maintenance and operations costs (including staffing) should be balanced against the avoided waste disposal costs and the value of recycled or recovered materials. To the extent that projects can be presented in language familiar to the institution's financial managers, administrative approval and acceptance are more likely to occur. The committee should learn about cost/benefit ratios, cost avoidance, payback periods, and other similar concepts.

For example, in 1989 the University of Minnesota Department of Environmental Health and Safety proposed building a $50,000 facility to crush scintillation vials and to sewer the solvent effluent. The ongoing operational cost of the facility was estimated at $20,000 per year. Sewer disposal had become a possible option because a new nonflammable water-miscible solvent was available to replace the commonly used flammable material. The present cost of disposal by incineration was almost $200,000 a year. Thus, the payback period for the $50,000 capital investment was 3 months and the annual operating cost ($20,000) was one tenth of the avoided disposal cost ($200,000). It was not difficult to determine that this was a good project.

The above example can also be used to bring out another significant issue in WM/PP efforts, i.e., the socio-political climate of the institution. Many of the projects that look good on paper require a significant change on the part of the waste generator or handler. The project just cited required that several hundred researchers change to a different solvent for scintillation counting. A change of this magnitude must be handled with extreme care and cannot be achieved by fiat. Early involvement of a group of the researchers' peers, demonstration by peers that the new solvent worked as well as the old, a lengthy phase-in period, and, finally, implementation of a charge for disposal of the flammable solvent all worked together to produce a 98% reduction in that particular wastestream. It is extremely important that the socio-political aspects of a project be given as much attention as the technical and financial aspects.

After a number of possible projects have been identified and evaluated, the WM/PP planning committee needs to set priorities for implementation. To help get waste minimization/pollution prevention established as a successful program, it is important to try to choose initial projects that are quite sure to work and will save money. Another high priority factor is regulatory obligation. After the easy, cost-effective, required projects have been scheduled, the going becomes more difficult, but a positive track record will help considerably in carrying out less-attractive projects.

With a group of evaluated and prioritized projects in hand, the planning committee should be ready to bring a proposal to the institution's administration for its support. The committee should carefully describe the benefits of an institution-wide WM/PP program and, most importantly, should prepare an appropriate policy or statement for administration to adopt. A well-thought-out policy accompanied by a realistic approach to implementation has a good chance of being positively received.

The University of Minnesota "Pollution Prevention and Waste Abatement" policy, passed in 1992 by the Board of Regents, is reproduced here as one example of an institutional statement that works reasonably well. This particular statement is a melding of many such policies, both internal and external, and the current authors (the University's Waste Abatement Committee) would be pleased to have this version also used by others.

Pollution Prevention and Waste Abatement

The University of Minnesota is committed to excellence and leadership in protecting the environment. In keeping with this policy, our objective is to reduce waste and emissions. We strive to minimize adverse impact on the air, water, and land through excellence in pollution prevention and waste abatement. By successfully preventing pollution at the source, we can achieve cost savings, increase operational efficiencies, improve the quality of our services, and maintain a safe and healthy work place for our students and employees. By successfully abating those wastes that cannot be eliminated at the source, we can recover useful resources and reduce the environmental burden of waste disposal.

The University of Minnesota's environmental guidelines include the following:

- Environmental protection is everyone's responsibility. It is valued and displays commitment to the University of Minnesota.

- Preventing pollution by reducing and eliminating the generation of both hazardous and nonhazardous wastes and emissions at the source is a prime consideration in teaching, research, service and operations. The University is committed to identifying and implementing pollution prevention opportunities through encouragement and involvement of all students and employees.
- Technologies and methods which substitute nonhazardous materials and utilize other source reduction approaches will be given top priority in addressing all environmental issues.
- Waste abatement programs such as recycling, reuse, and purchase of recycled materials will be vigorously pursued to reduce the need for disposal of waste that cannot be reduced at the source.
- Opportunities to encourage pollution prevention and waste abatement through changes in purchasing policies and specifications will be sought.
- The University of Minnesota seeks to demonstrate its leadership role in the State of Minnesota by aggressively adhering to all environmental regulations. We promote cooperation and coordination among higher education, industry, government, and the public toward the shared goals of preventing pollution and abating waste.

Therefore, be it resolved, that the Board of Regents directs the President to establish effective pollution prevention and waste abatement programs and to develop policies and plans to achieve that goal.

HOW TO KEEP PROGRESSING

Reading through the above policy brings up two issues. It is obvious that this policy can be applied to nonlaboratory waste and to nonhazardous waste. Although the topic of this publication is laboratory waste, and the major items of concern are the various types of hazardous waste, it may be in the best interest of the institution to coordinate all aspects into one comprehensive waste minimization/pollution prevention program. Considerable synergism can occur by bringing together people who have addressed a common goal from a number of different directions. One of the better ways to maintain an active and successful program is to keep looking for new ideas.

A successful WM/PP program needs an oversight committee, which will probably include many members who helped carry out the original effort, as well as some who represent the "institutionalization" of the program. Part of this committee's responsibility is to sell the various elements of the program to those who should be using it. Figuring out ways to increase participation of individuals and groups becomes a major activity once the basic program elements are in place. Membership on this committee should be supported by administration as a significant contribution to the institution, and appointment to the committee should be made by a high-level officer.

It is also important to monitor ongoing programs for problems that may occur. Providing adequate contacts for waste generators and encouraging them to call if questions arise will help head off undesirable incidents and negative attitudes. For example, if collection containers for a recyclable material are not emptied before they overflow, the result is at best aesthetically displeasing and at worst hazardous and in violation of regulations. An appropriate phone number on recycling containers can alleviate this problem, and adequate printed instructions will remind users both where and how to appropriately manage their waste.

Another problem that can occur after a WM/PP program has been fairly well incorporated into everyday activities is the development of counterproductive projects. At the U of M, the recycling coordinator put much effort into arranging a contract for recycling of foamed polystyrene beads. Just after the system was finally in place, an individual laboratory talked its chemical suppliers into changing their packing material to environmentally friendly foamed cornstarch beads, which look just like polystyrene at first glance. The recycling process for polystyrene involves steam cleaning of the beads before reuse; unfortunately steam cleaning of cornstarch beads is apt to lead to a very unsatisfactory result, including rejection of the entire load of material sent for recycling. Central coordination of all WM/PP activities is important to the continued success of the institution's program.

A significant source of help in promoting WM/PP ideas and in watching for problems can be found within the institution's environmental health and safety staff, whether this is all one unit or is split among several areas. Many of these people routinely survey, inspect, or collect waste from the locations where hazardous materials are used and wastes are generated. Taking time to instruct them on the basic concepts of WM/PP and identifying techniques that are appropriate to the areas in which they work will pay dividends in ideas passed on to those whom they contact. This idea can be carried even further by including WM/PP ideas and responsibilities in chemical hygiene plans, radiation manuals, and other types of operating manuals for processes where hazardous materials are handled. The implementation of good waste reduction techniques should always be documented in written standard operating procedures so that the ideas are not lost because of a change in personnel.

At some point, the institution will have established all of the money-saving programs that can be identified, and will find itself coming up with good ideas for reducing waste and improving the environment that will be costly to implement. If it is possible to establish a policy that allows some of the money saved from certain programs (i.e., costs avoided) to be used for setting up programs that will not be self-supporting, WM/PP efforts can be carried much further. Small grants for pilot projects can be very helpful in showing that specific ideas will work, or for identifying pitfalls before a large effort has been expended.

CONCLUSION

The most important factor in establishing an ongoing WM/PP program for laboratories is to identify those people within the institution who have a very strong commitment to the endeavor. Once identified, those individuals will be most effective if they take it upon themselves to design and recommend for implementation a set of carefully evaluated and prioritized projects. Prioritizing for maximum early success, based largely on cost effectiveness and ease of implementation, will help assure administrative support and continued existence of the program.

CHAPTER **7**

Overcoming Impediments to Waste Minimization

Peter A. Reinhardt

CONTENTS

Introduction 98
A Review: The Hierarchy of Waste Minimization 98
Perspectives on Laboratory Waste Minimization 101
Why is Waste Minimization Occurring in Laboratories? 101
    Waste Minimization Variables from Laboratory-Initiated
    Practices 102
        Microscale Experiments in Teaching Laboratories 102
        Distillation of Waste Organic Solvents 104
    Waste Minimization Variables from Institution-Facilitated
    Practices 104
        Mercury and Lead Recycling 105
        Neutralization of Waste Mineral Acids and Bases 105
        Redistribution of Surplus Chemicals 105
        Bulking or Commingling of Organic Solvents 108
        Product Substitution 108
        In-Laboratory Chemical Treatment 108
    Comparing Key Waste Minimization Variables 109
Encouraging Waste Minimization in Laboratories 109
    Safety Department Plans to Facilitate Waste Minimization 110
        Communication to Overcome Informational Barriers 111
        Methods of Communicating with Laboratory Personnel 111
    Institutional Options Not Chosen 112
        Barrier of Economic Isolation 113
            The Economist's View of Waste Minimization 113
        Barriers to Redistribution of Surplus Chemicals 114

Waste Minimization Initiatives 114
    Chemical Stockroom on a Computer Network Database 115
    Exchanging Waste Minimization Ideas 115
    Making a Solvent Still Available for Loan 115
Overcoming Regulatory Barriers 116
    Laboratory Equity and Waste Minimization Act................................. 116
        Provisions of the Act 117
            Group Permit for Laboratory Treatment............................ 117
Conclusions 118
Endnotes 118

INTRODUCTION

Overcoming impediments to minimizing hazardous waste in laboratories can be explored by asking the questions:

- For those laboratories where waste minimization is now practiced, why is it occurring?
- For those laboratories where there is no effort to minimize waste, why is there not?
- Can the answers to those questions provide insights to overcome impediments to waste minimization?
- What can a hazardous waste manager or an organization do to encourage waste minimization?

This chapter will probe these questions in the context of the author's experience at the University of Wisconsin–Madison and his knowledge of waste management practices of other laboratories, after 15 years of experience in this field. In the author's opinion, the University of Wisconsin–Madison does not have a model program for minimizing hazardous waste. There is much to do. Although the University probably devotes more than average resources, staff, and facilities to environmental management for an institution of its size, and the people who manage the University's hazardous waste to minimize waste, there exist significant barriers to implementing and improving waste minimization practices. Laboratory personnel at the University do not face unique barriers to waste minimization. In the author's view, the problems and solutions in overcoming impediments to waste minimization at the University are shared by many laboratories and organizations.

Before examining the above questions it is useful to review the hierarchy of waste minimization and to gain a historical perspective on this subject.

A REVIEW: THE HIERARCHY OF WASTE MINIMIZATION

Several sources have devised a hierarchy of waste minimization. The hierarchy ranks waste minimization practices in order of preference, based on their

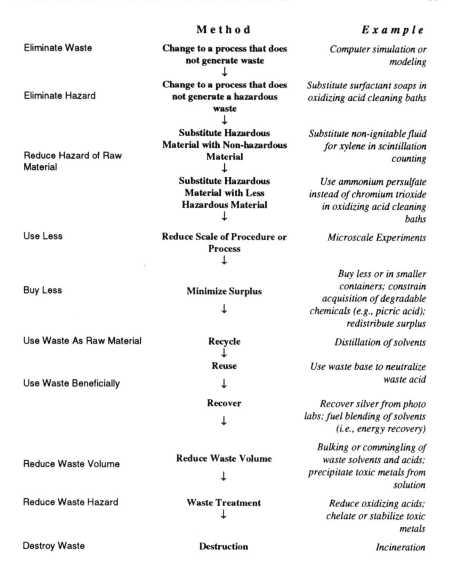

Figure 1 Hierarchy of waste minimization, ranked by increasing environmental, occupational, and accidental risks. (Adapted from *Laboratory Waste Management: A Guidebook,* The American Chemical Society, Washington, D.C., 1994. Reprinted with permission.)

overall efficiency, effectiveness, and environmental impact. When planning or allocating resources for waste minimization, it is best to focus on those methods at the top of the hierarchy.

Figure 1 is the author's view of such a hierarchy. Figure 1 contains more information than will be explained in this chapter.

Waste minimization methods in Figure 1 are ranked by approximately increasing environmental, occupational, and accidental risks. For example, the most effective minimization method is to eliminate waste altogether by changing a process to one that does not generate waste. Some laboratories have replaced or augmented wet-lab experiments that generate waste with computer simulation, modeling, or demonstrations. This example also illustrates the limitations of the waste minimization hierarchy. The hierarchy is based on the goal of preventing pollution, but is not necessarily consistent with the purpose of laboratories: teaching, testing, and research. Because chemistry students must learn to handle chemicals and manipulate laboratory equipment, computer demonstrations of experiments cannot meet many of the objectives of teaching and graduate research laboratories. Further, quantitative testing laboratories must use real samples, and development of a material requires research using them.

Another highly effective minimization method is substitution of a hazardous material with a less hazardous or nonhazardous material. This results in the generation of a less hazardous or nonhazardous waste. Two examples are given in Figure 1. The most successful example of waste minimization at the University of Wisconsin–Madison is substitution of toluene-based liquid scintillation fluid with nonignitable liquid scintillation fluid. Liquid scintillation fluid is a solution, usually of toluene, that is used in research and medicine to measure the quantity of radiolabeled chemicals. Nonignitable liquid scintillation fluids are advertised as being "safe," "biodegradable," and "sewer disposable," but their hazards differ according to their formulations, biodegradability depends upon the capabilities of the wastewater treatment system (which varies greatly by locale), and laboratories must receive approval from their treatment works prior to disposing of nonignitable fluid in the sanitary sewer system.

After evaluating the efficacy, benefits, and acceptance by the local sewerage treatment works of nonignitable liquid scintillation fluid, the University's Radiation Safety Committee adopted a policy to substitute toluene-based liquid scintillation fluid with the nonignitable variety in all but a few uses. This policy eliminated the generation of thousands of gallons per year of waste toluene solutions from liquid scintillation.

Material substitution often has many benefits. When compared to the toluene-based material, the nonignitable liquid scintillation fluid has the benefits of reducing the risk of exposing workers to a toluene, reducing the risk of fire, and reducing the environmental hazard of toluene if spilled or released. Because the selected substitute is nonignitable and can be safely disposed in Madison's sanitary sewer, it reduces the cost of disposing of the toluene fluid by incineration or fuel blending.

Figure 1 differs from waste minimization hierarchies that have been proposed by the U.S. Environmental Protection Agency (EPA) and others; it includes some methods that are not universally recognized as waste minimization. Some sources do not consider waste treatment or destruction to be a minimization method. Other sources do not recognize volume reduction or off-site practices to be a minimization method, such as fuel blending and incineration of waste laboratory solvents in cement kilns.

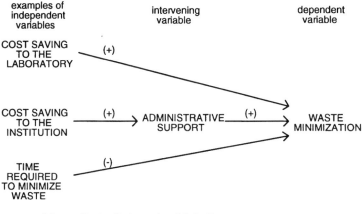

Figure 2 Waste minimization model. This simple model does not include all variables and relationships.

PERSPECTIVES ON LABORATORY WASTE MINIMIZATION

At the 1983 Washington, D.C., National Meeting of the American Chemical Society, the author spoke on reducing laboratory hazardous waste.[2] Many of the methods discussed then are important today and are shown in Figure 1. These older methods include substitutes for chromic acid, controlling acquisition of laboratory chemicals, distillation of organic solvents, neutralization of mineral acids, and recovering silver in photography laboratories. Since 1983, several promising laboratory waste minimization methods have been developed or have been more widely implemented. Among these new methods are computer simulations and substitution with nonignitable fluid, which have been mentioned. New developments include microscale laboratory methods and a great proliferation of published methods to chemically treat laboratory waste, both of which are noted in Figure 1.

WHY IS WASTE MINIMIZATION OCCURRING IN LABORATORIES?

To probe the first three questions posed at the beginning of this chapter, consider a simple conceptual model of laboratory waste minimization (see Figure 2). Such a model has independent variables that encourage or inhibit waste minimization. Those variables can be further classified as variables that work at the level of an individual laboratory or laboratory worker and variables that work at an institutional or organizational level. Institutional variables also affect an intervening variable, administrative support for waste minimization. Institutional

Table 1 Waste Minimization Practiced by Laboratories at the University of Wisconsin–Madison

Method	Why
Silver recovery from photo fixer	Initiated because of financial incentives
	Continues today to meet sewer limits
Product substitution	Improved safety
(e.g., chromic acid substitutes)	Desire to reduce environmental impact
	Encouragement from the safety department
Computer simulation	Enhanced teaching and research
	Ease
	Monetary savings compared to the cost of running experiments
	Improved safety
Microscale experiments in	Technological trend
teaching laboratories	New laboratory manuals include microscale methods
	Improved safety
	Reduced cost of materials
Solvent distillation	Monetary savings
	Environmental ethic

Note: Waste minimization methods initiated by laboratory personnel in the absence of institutional encouragement.

variables are typically reflected in activities of the organization's hazardous waste manager or environmental health and safety office.

Tables 1, 2, and 3 list and classify these variables. In large part, this information is anecdotal and based on the author's experience. Since these tables are fairly complete, only some information in them will be explained further in the text.

WASTE MINIMIZATION VARIABLES FROM LABORATORY-INITIATED PRACTICES

Table 1 lists five waste minimization methods that are now being used in laboratories and notes why, in the author's opinion, these activities are occurring. At the University of Wisconsin–Madison, most of the methods in Table 1 have generally been initiated and implemented at the laboratory level with little or no support from the institution.

Financial incentives and greater safety appear to be the two most important variables that positively affect the use of minimization methods.

Microscale Experiments in Teaching Laboratories

Much progress has been made in reducing the scale of experiments in undergraduate chemistry teaching laboratories, which results in smaller amounts of chemicals being used, released to the environment, and disposed of. This technique seems to be less important for research laboratories, of which there are many at the University. To preserve valuable samples, researchers have used microscale techniques for many years. Physical techniques for evaluating chemicals require ever smaller samples, which complements the traditional small scale of the research laboratory. Still, microscale glassware and apparatus developed for the undergraduate

Table 2 Laboratory Waste Minimization Practices Facilitated by the University of Wisconsin-Madison Safety Department

Method	Why	How
Recycling of mercury and lead	Financial incentives (initially)	Staff available for transport Source separation
In-lab elementary neutralization of mineral acids	Lower on-site management and offsite disposal costs Personal responsibility ethic	Safety Department provides instruction and the laboratory provides staff, facilities, and materials
Redistribution of surplus chemicals within campus	Environmental ethic Lower disposal and purchasing costs	Staff available to source separate, list, advertise in newsletter, and distribute storage space
Acquisition constraints (e.g., to minimize picric acid storage prior to use)	Improved safety difficulty in disposing of dry picric acid	UW policy Cooperation of purchasing and stores departments
Bulking/commingling of organic solvents prior to shipment	Lower disposal costs compared to lab packing Volume reduction	High waste volume Staff available and capable to handle bulk hazardous materials Facilities for temporary storage and personal protection Cooperation of laboratories to identify constituents and separate incompatible substances
Selection of incineration as a preferred waste disposal method for off-site disposal	Early 1980s: liability concerns Now required by land ban	Decision to pay more for lower liability
Selection of fuel blending for off-site disposal of waste organic solvents	Lower disposal costs Fewer concerns regarding legitimacy and liability of fuel blending than in 1980s	Staff available to handle bulk hazardous materials and capable of chemical analysis Source separation by laboratories of unacceptable components Provision of two waste containers by safety department and instructions to keep halogenated solvents separate Mixture composition quality assurance program and gas chromatograph Facilities for bulking high waste volume
Product substitution through acquisition constraints: nonignitable liquid scintillation fluid	Difficulty in disposal of mixed radioactive and chemicals waste Availability of substitute product Lower disposal costs Decreased risk of fire and personal exposure	University of Wisconsin-Madison policy Cooperation of purchasing and stores departments
In-laboratory chemical treatment	For wastes that Are difficult or costly to dispose of Are in small amounts Easy treatment methods exist	Safety department provides instruction; in most cases the lab provides staff facilities, and materials

Table 3 Key Waste Minimization Variables

Variables that Work at Both the Laboratory and Institutional Level	Variables that Principally Work at the Institutional Level
Financial incentives (i.e., revenue or lower disposal or purchasing costs)	Institutional policy that promotes waste minimization
	Sufficient staff with ability to manage hazardous materials
Ease of employing waste minimization method (i.e., method is easier than disposal or method is widely accepted)	Institutional communication with laboratory personnel
Safety	A facility for safe storage and handling of hazardous materials and waste
Cooperation of laboratory personnel to keep wastes separate (i.e., source separation)	Sufficient waste volume to use industrial waste disposal and recycling methods, and their economies of scale
Environmental ethic or ethic of personal responsibility	Desire to avoid liability of off-site disposal
	Manifest certification requirement

teaching laboratories have facilitated the down-scaling of experiments in research laboratories, and thus lower costs for chemical purchases and disposal.

Distillation of Waste Organic Solvents

The distillation and reuse of waste xylene is growing in popularity nationally and at the University of Wisconsin–Madison. Without encouragement from the Safety Department, and with no direct charge to laboratories for the disposal of waste xylene, histopathology laboratories have purchased stills for recycling. Payback periods have been calculated to be 2 to 3 years, based only on the savings from avoiding the purchase of xylene. Owners of stills express the belief that recycling is the right thing to do. Also facilitating use of solvent distillation is the relatively clean waste generated by histopathology laboratories and the commercial availability of safer stills. Stills can now be purchased that operate automatically and have safety switches that turn off the unit in case of a dry still bottom or overheating.

It should be noted that organic solvent distillation occurs in many other laboratories on campus, to repurify solvents and for other reasons — which may include waste solvents recycling.

WASTE MINIMIZATION VARIABLES FROM INSTITUTION-FACILITATED PRACTICES

At the University of Wisconsin–Madison, how is the Safety Department helping to minimize laboratory hazardous waste? Table 2 lists waste minimization practices facilitated by the Safety Department. Table 2 also reveals the motivation that institutional support staff (such as a Safety Department) has in helping scientists prevent pollution, and notes how these minimization methods are implemented at the institutional level.

It is important to explain the institutional role of the Safety Department at the University. The Safety Department's mission is to serve campus personnel who teach, perform research, or serve the public. Services provided by the Safety Department include removal of chemical waste at no direct cost to laboratories or departments, chemical and environmental safety consultation, and, by a variety of means, *facilitating* campus safety. Policy making is available through a campus Chemical Safety Committee. Whenever possible, the Safety Department wishes to avoid being a barrier to the University's primary mission of teaching, research, and public service.

Table 2 reveals that financial incentives (i.e., providing income or avoiding costs) is also an important variable for minimization that is facilitated at the institutional level.

Mercury and Lead Recycling

Table 2 begins with mercury and lead recycling — activities that laboratories have been undertaken for as many decades as there have been markets that pay for these scrap materials. At one time the Chemistry Department operated a mercury still (a hazardous venture due to risk of personal exposure) to repurify dirty mercury to avoid the high cost of purchasing mercury. Today, the Safety Department collects waste mercury and lead pigs (used to shield radioactive materials) from many laboratories on campus and recycles them to a local scrap dealer.

Neutralization of Waste Mineral Acids and Bases

Included in Table 2 is neutralization of waste mineral acids (e.g., hydrochloric acid, sulfuric acid, nitric acid), although some sources do not consider chemical treatment to be a minimization method. The University has a policy that laboratory personnel who generate waste mineral acids and bases are responsible for neutralizing them and disposing of them in the sanitary sewer. This policy reduces the amount of waste that is sent off site and managed by the Safety Department and lowers commercial disposal costs. The Safety Department assists laboratory personnel by providing advice, written guidance, and sometimes in-laboratory assistance, as necessary. Occasionally laboratory personnel are excused from neutralizing acids when their facilities are unsatisfactory (e.g., poor ventilation or sanitary sewer access), when there is a large quantity of waste acid or base, or when untrained personnel make in-lab neutralization too hazardous. In these cases the waste acids and bases are moved to a laboratory with satisfactory facilities and neutralized by Safety Department personnel.

Redistribution of Surplus Chemicals

The Safety Department began a campus-wide redistribution program for surplus chemicals in 1980. A study of the University's chemical wastestreams was prompted at that time by the U.S. Environmental Protection Agency's original hazardous waste regulations. This study found that one wastestream consisted of a large number of usable, but unwanted, stock chemicals, still in their original containers. (*Stock chemicals* are those fine chemicals and reagents that are purchased from suppliers who meet certain quality standards. Laboratory personnel typically buy stock chemicals in quantities sufficient for several uses and experiments, to be used over 1 or more years, and for use by other laboratory workers.) These surplus chemicals remain today as a significant laboratory wastestream. Some of these chemical containers still have their factory seals on them, with no appearance of degradation. The conjecture is that these surpluses result from laboratory purchases that misjudge the future needs of researchers, in both amount and chemical. When researchers retire from or leave the University, they often leave a large inventory of chemicals they had planned to use someday. Volume price discounts for laboratory chemicals encourage overpurchases, as do budgeting policies in which funds expire at the end of a fiscal year if they are not used.

In 1980 and before, most of these surplus chemicals were disposed of in a commercial landfill as hazardous waste. This practice was deemed to be an unwise waste of usable resources by University policy makers, especially when new regulations increased the cost and liability of off-site disposal. Instead of disposing of reusable chemicals, the Safety Department began to try to redistribute them to other laboratories on campus, thereby reducing the amount of waste generated and chemicals purchased.

In 1994, the redistribution program for surplus chemicals appears to be one of the University's most successful waste minimization efforts. When an unwanted chemical is received by the Safety Department, a chemist examines the container to judge if it can be used by another campus laboratory. Factors that contribute to a chemical's likelihood of being redistributed are the following:

- The chemical must be in its original container
- No visible deterioration or decoloration
- Powders are flowable and not caked
- For hydroscopic chemicals, no moisture is present
- The presence of a factory seal
- Adherence to storage conditions and expiration dates specified on the container
- Knowledge by the Safety Department chemist of campus laboratories that can use a certain chemical

Not all redistributed chemicals meet each of the above criterion; opened containers are often successfully redistributed to other laboratories, as are larger quantities of slightly degraded mineral acids and organic solvents.

Eligible surplus chemicals are listed on a Safety Department database of waste and surplus chemicals, which is then printed and distributed about three times a year to campus laboratories in a safety newsletter, *Lab Scan* (see Figure 3). Laboratory personnel call in orders, and the chemicals are labeled as a "redistributed chemical" and delivered free. About 30 to 50% of each list is redistributed. Not counting Safety Department labor, this saves the University about $10,000 to 20,000 per newsletter in the cost of purchasing new chemicals (which is the largest share of the savings) and avoidance of disposal costs.

OVERCOMING IMPEDIMENTS TO WASTE MINIMIZATION

L a b S C A N

Laboratory Safety Comments And News ♦ July 1994

Chemical & Environmental Safety Program
University Of Wisconsin–Madison Safety Department
Diane L. Drinkman, Editor (262-9644 or diane.drinkman@mail.admin.wisc.edu)

Safety Relocated in May

In early May, the Safety Department moved its offices to 103 North Lake Street. This is near SERF and the Southeast Dorms. All telephone numbers remain the same.

As a result of this move, all the programs of the Safety Department- Chemical and Environmental, Biosafety, CORD, General and Radioactive, have been consolidated under one roof. We hope this will improve services offered to the campus.

Redistriution List Now Available on WiscINFO

In past years, the only access the campus had to the list of chemicals for redistribution was through the publication of LabSCAN. Through the efforts of a group of students enrolled in IES 600 this past semester, a listing of chemicals currently available is on WiscINFO. To access this information follow these simple steps:

1. Login to WiscINFO through any Internet connection.
2. Select the menu option *General Campus and Community Information*.
3. Select *Chemical and Environmental Safety*.
4. Select *Listing of Chemicals for Redistribution*. This list is separated into alphabetical groups. The name, amount, manufacturer and purity are contained in the listing.
5. If a chemical is available that you would like delivered to your laboratory call the Safety Department.

A current listing can be found on pages 5-7 of this newsletter. If there are materials that are not listed and you would like us to watch for, call or e-mail us.

Diane Drinkman can help you with your chemical redistribution needs. She can be contacted at 262-9644 or diane.drinkman@mail.admin.wisc.edu.

Safety's Electronic Mail Address

Do you have questions or comments about Chemical or Environmental Safety issues on campus? Or General Safety? Or Radiation Safety? The Safety Department recently opened an electronic mailbox. We can be reached at:

safety.department@mail.admin.wisc.edu

Please Help to Prudently Manage the University`s Hazardous Wastes

It costs the University $10 to dispose of each 5 gallon carboy of organic solvents; an average of 40 carboys are generated each week on campus. This does not include the cost of removing it from your laboratory, transporting it on campus, storing it at our storage facility, and complying with the myriad of environmental and safety laws and their requisite paperwork. The disposal of one lecture bottle of gas can cost $2,500.

For many types of low-level radioactive waste, there is no facility in the nation that

---

University of Wisconsin-Madison Safety Department
608/262-8769 or safety.department@mail.admin.wisc.edu

The following chemicals are currently available for redistribution by the Safety Department. Contact Diane Drinkman (voice 262-9644, e-mail: diane.drinkman@mail.admin.wisc.edu) to have your selections delivered to your laboratory.

Chemical Name (Quantity)	Amount	Manufacturer	Purity
Adenosine 5'-Triphosphate, Magnesium Salt	1 g	Sigma	
Allolpurinol	5 g	Sigma	
Allyl Alcohol	100 mL	Aldrich	99%
Allyl Cyanide	100 mL	Aldrich	

Figure 3 Labscan newsletter lists surplus chemicals available for redistribution.

Bulking or Commingling of Organic Solvents

Bulking pertains to the combining (or commingling) of several small containers of similar wastes into a larger, bulk container. Although bulking of organic solvents is not considered a waste minimization method by some, it certainly represents a significant gain in the efficient management of hazardous waste. It is perhaps the best way to reduce waste volume and disposal costs of laboratories that generate organic solvents. Without bulking, solvents are commonly disposed of in 1-gal bottles, 15 to a 55-gal lab pack drum, which fills the drum to one fourth of its capacity. Further, disposal of lab packs is usually limited to incineration, whereas bulk solvent drums can be disposed of by fuel blending, a disposal method that is less expensive and recovers the waste's energy. For example, in 1993 it would have cost the University $30 a gallon to incinerate lab-packed solvents, whereas fuel blending bulk solvent drums cost $2 a gallon. As noted in Table 2, however, bulking requires sufficient waste volume to substantially fill a bulk drum, and staff and facilities that are able to safety transfer the solvents to a drum and store them.

Product Substitution

Difficulty in disposing of a waste is a powerful motivation for waste minimization. An excellent example of this is the substitution of nonignitable liquid scintillation fluid for xylene-based fluid, as was described previously in this chapter. In 1993 only three sites in the nation were available for the disposal of low-level radioactive waste. Because all three sites are landfills, none are appropriate for the disposal of flammable liquids. A mixed radioactive-chemical waste, such as xylene-based liquid scintillation fluid, has few commercial disposal options. The decision by the Nuclear Regulatory Commission to allow the disposal of most fluid without regard to radioactivity has eased the problem, but not solved it.

Unfortunately, there are few other product substitution success stories. Laboratory personnel have been cautioned to find less-toxic substitutes for benzene, carbon tetrachloride, and other carcinogens. There are many commercially available substitutes for chromic acid cleaning solutions. The risks of mercury use and accidents have prompted laboratory personnel to investigate replacing mercury thermometers with red liquid or electronic versions.

In-Laboratory Chemical Treatment

Lastly in Table 2, chemical treatment is encouraged by the Safety Department through the dissemination of information, consultation with researchers who generate treatable waste, and occasional in-lab demonstrations. Many treatment methods promoted by the Safety Department are more advanced variations of elementary neutralization of simple mineral acids, such as the neutralization, reduction, precipitation, and filtration of chromo-sulfuric acid (i.e., chromic acid cleaning solution). Chemical treatment is also used for wastes that are difficult to dispose of, such as gases. It is the author's impression, from discussing laboratory disposal problems with researchers, that there is great interest in chemical treatment methods. Methods for the detoxification of ethidium bromide and osmium tetroxide have been disseminated among researchers who use those materials. Further, it is often expressed by laboratory personnel that people should take responsibility for the wastes that they generate and laboratories should lessen their impact on the environment.

COMPARING KEY WASTE MINIMIZATION VARIABLES

The above discussion reveals several key variables that influence waste minimization. These are listed in Table 3 as those variables that work in a laboratory, and those that influence institutional decisionmaking for waste management. Variables listed for those in the laboratory (i.e., financial incentives, ease, safety, source separation, and ethic) are also very important at the institutional level.

Several variables seem to be most important at the institutional level, but less so in the laboratory. Laboratory personnel have little knowledge of the extent of institutional liabilities for off-site disposal, such as those liabilities posed by Superfund (or the Comprehensive Environmental Response, Compensation and Liability Act of 1980). Under this law an institution can be held liable for cleanup of a waste disposal site to which it sent waste. Liability concerns are more important to institutional risk managers, administrators, and trustees. The desirability of avoiding off-site disposal by on-site waste minimization can be an important motivation to allocate institutional resources.

One institutional variable that was not obvious in the review of existing minimization practices is the EPA requirement that all generators of hazardous waste must certify their waste minimization program. The transport manifest requires a certification that large quantity generators have a program in place to reduce the volume and toxicity of the waste they generate and small quantity generators have made a good faith effort to minimize their hazardous waste. This is listed as an additional variable in Table 3. The certification allows generators to consider the practicality and cost effectiveness of waste minimization methods. In most EPA regions and states this requirement has not been an enforcement priority. However, this federal requirement (which is more stringent in some states) motivates many laboratory managers to consider minimization methods.

The key variables reveal that there are many things an organization can do to promote waste minimization. If an organization's managers, administrators, and trustees desire to minimize hazardous waste, they can provide institutional support (such as policy, staff, or facilities) or can act upon those variables that work in the laboratory.

ENCOURAGING WASTE MINIMIZATION IN LABORATORIES

Organizations can encourage waste minimization in their laboratories. The benefits of waste minimization appear in Table 3. They include financial incentives, improved safety, ethics, avoidance of liability, and compliance with legal requirements.

SAFETY DEPARTMENT PLANS TO FACILITATE WASTE MINIMIZATION

At the University of Wisconsin-Madison, the Safety Department and the University Chemical Safety Committee have discussed what can be done to encourage minimization of hazardous waste. Four strategies are being pursued:

- **Continue doing what we're doing, but do it better.** Decision makers at the University are pleased with our minimization efforts, but acknowledge room for improvement. For example, laboratory personnel would have more confidence in our surplus chemical redistribution program if we improved our grading and quality assurance when selecting candidate chemicals. Higher confidence would likely result in greater use of the program.
- **Expand our off-site recycling options.** The State of Wisconsin has a single contract, with a multiple-year option, for hazardous waste disposal that is used by all state agencies, including the University. As a result, the contract is large enough to attract hazardous waste disposal vendors that are eager to meet the needs of state agencies. The vendor is selected on the basis of services provided, disposal site quality, and cost. In future contracts, the State intends to favor vendors who can provide recycling and waste minimization services, such as recycling of organic solvents and batteries. The potential of this strategy, however, may be limited. Compared to industry, state agencies generate many types of hazardous waste from many sources in relatively small volumes and often cannot use the recycled materials themselves (e.g., high-purity solvents are required for research). This makes it difficult to utilize existing recycling services. Even when a chemical engineering project was willing to reuse the distilled waste, the University was unable to recycle a relatively clean waste mixture of isopropanol and water because 750-gal batches were deemed too small.
- **Source separate wastes.** Separation of wastestreams at their source (i.e., source separation) is key to maintaining minimization options. As is described in Table 2, the Safety Department asks scientists to keep waste halogenated organic solvent mixtures (which are incinerated) from nonhalogenated waste solvents (which are blended for fuel use) by providing instructions and two types of containers. It is anticipated that, if recycling services become available for some organic solvents, campus waste generators will be asked to keep additional waste types separate. Source separation is especially important to the success of solvent distillation, where mixtures can result in azeotropes, which can be a difficult technical barrier to waste minimization.
- **Communicate with laboratory personnel and investigators at other institutions to facilitate technology transfer.** For an institution to encourage waste minimization, the Safety Department needs to nurture a dialog with laboratory personnel. Such as dialog would solicit pollution prevention ideas from laboratory personnel and result in a better understanding of how the University can support minimization efforts.

Communication to Overcome Informational Barriers

Communication between the UW–Madison Safety Department and laboratory personnel is important to improving chemical and environmental safety. Laboratory personnel need to know why waste minimization is important for reducing health and environmental risks. Even in the case where laboratories are not charged directly for waste disposal, laboratory personnel often change how they manage hazardous waste when they are told of the high and increasing costs of hazardous waste disposal and how this indirectly impacts them.

Overcoming informational barriers through communication facilitates the transfer of pollution prevention technology to laboratories not practicing waste minimization. Laboratory personnel should be regularly informed of newly available substitute products, distillation equipment, chemical treatment procedures, and other minimization methods.

Laboratory personnel must also be made aware of how their laboratory practices affect waste minimization. Source separation is often a requisite to making waste minimization possible, but its success requires that researchers and students be informed to keep dissimilar waste separate, and to keep waste that can be safely disposed of in the normal trash and sanitary sewer out of the hazardous waste collection container. These practices have become more important under EPA's 1990 toxic characteristic rule, which requires that some wastes be regulated as a hazardous waste, at very low levels of contamination.

Methods of Communicating with Laboratory Personnel

The Safety Department uses several means of communicating with scientists, including the publication of a *Chemical Safety and Disposal Guide*, the *Lab Scan* newsletter and, in a recent development, by holding laboratory safety symposiums at the beginning of each semester. The *Chemical Safety and Disposal Guide* includes information on the chemical and environmental risks of laboratory chemicals and explains why waste minimization is important. The new version, published in 1993, also has detailed procedures for a variety of larger-scale neutralization and chemical treatment procedures.

Another effort to build a partnership with laboratory personnel is a service that helps them manage their hazardous materials on site. This is an expansion of the Safety Department's waste pick-up service that will include:

- Informing chemical users of the risks and costs of using chemicals
- Disseminating information on federal and state laws pertaining to chemical use and disposal
- Assistance in sorting chemicals for the normal trash, sanitary sewer, neutralization, and chemical treatment
- Disseminating information on pollution prevention and waste minimization methods

Table 4 Waste Minimization Strategies Considered But Not Chosen

Waste minimization method	Why not
Solvent still operated by the institution	Staffing and facilities
	Technical limitations (e.g., azeotropes, difficulty of distilling solvent mixtures, limited number of beneficial applications)
	Fear of quality assurance problems
	Lack of markets for product
	Permits and compliance with legal requirements
Institutional chemical purchase fee	Cost and time of implementation
	Encourage purchases by nonapproved, illegal, or improper means
	Because fees are paid up front, they would not encourage in-lab waste minimization (e.g., in-lab distillation or chemical treatment)
	Not all chemical use creates hazardous waste
Institutional disposal fee	Cost and time of implementation
	Encourage storage
	Encourage improper disposal
Expand surplus chemical redistribution program beyond campus	Adverse selection
	Transportation costs
	Liability concerns

INSTITUTIONAL OPTIONS NOT CHOSEN

Table 4 lists waste minimization strategies that were considered at the University, but have not been pursued. Some institutions have supported laboratory distillation and reuse of organic solvents by operating a solvent still that serves many laboratories. Facilities, equipment, and staff are provided as part of institutional support services. The advantages of this strategy are many. For many laboratories, the organic solvent wastestream is their largest, so even a modest increase in solvent recycling can significantly improve an organization's waste minimization results. An institutional solvent still can take advantage of the economies of scale. With experience, still operators gain expertise necessary for safe, routine operation that can produce a product of consistently high quality. Such an operation, however, requires additional equipment and facilities, both of which have not been readily available at the University. The cost of permitting (when applicable), compliance, and assuring product quality are additional concerns.

Many organizations have considered paying for hazardous waste disposal through the imposition of purchase or disposal fees. Such fees would likely promote waste minimization. Demand for chemicals decrease when their prices are increased. In today's era of tight research budgets, scientists would be more careful to not order more chemicals than they need, which would reduce the amount of surplus chemicals that need to be disposed of as hazardous waste. Direct fees for waste disposal follows the ethic that the "polluter pays" for their pollution.

The University has chosen not to add purchase or disposal fees at this time, for the reasons given in Table 4. However, because financial incentives are so influential to waste minimization (it was found to be one of the key variables earlier), further discussion of the barrier of economic isolation and the economic basis for fees is merited.

Barrier of Economic Isolation

As discussed above, laboratory personnel are typically economically isolated from the institution's cost and liability of hazardous waste disposal. In the author's experience, most organizations pay for hazardous waste disposal through their budget for overhead, maintenance, or expenses. Administrators and hazardous waste managers, who are responsible for those budgets, feel pressure to reduce these costs through waste minimization, but laboratory managers, scientists, and students do not. Laboratory personnel, who are often in the best position to minimize waste, are isolated from the institutional problem of sacrificing new equipment or salaries to pay hazardous waste disposal. Economically isolated researchers may not be motivated to minimize waste and may not be fully aware of the risks of hazardous waste. Chemical purchase or waste disposal fees, imposed by the institution to pay for hazardous waste management, can overcome this barrier by directly impacting laboratory budgets.

The Economist's View of Waste Minimization

Like laboratory personnel, institutions have historically been isolated from the true cost of hazardous waste management. Prior to the existence of laws and special practices for waste disposal, the cost that institutions paid for hazardous waste management was less than the social cost of that waste. Social costs of hazardous waste management include these external costs (economists call them *externalities*):

- The cost to the public health due to the risk of chemical exposure from contaminated air, water, soil, and food, such as the cost of excess cancers
- The cost to worker health due to risk of chemical exposure from using chemicals and handling waste
- The cost to property and welfare due to risk of fire, explosion, and contamination, such as the cost of cleaning up contaminated property
- The cost to the environment due to risks to flora, fauna, and aesthetic degradation

Legal requirements for hazardous waste management impose some of these costs on hazardous waste generators, but do not account for all of hazardous waste's costs to society. One economic solution to this problem is to add a raw material or waste tax so that purchase or disposal cost is equal to the social cost. With a purchase tax, the cost to society of risks to health and the environment would be included with every purchase. Recent proposals to increase the national gas tax, a BTU tax, or add a carbon tax are similar attempts to include external environmental costs in purchase costs. Alternatively, manufacturers can include social costs in their price of goods and services by accounting for their true environmental costs (including production, use, recycling, and disposal), a practice known as "full-cost accounting."[3]

The discrepancy in cost and social cost is exacerbated at institutions where laboratories are not directly charged for hazardous waste generation. To laboratory personnel, the cost of waste generation may appear to be negligible.

Note that, despite the logical economic argument for purchase or disposal fees, Table 4 lists several important problems with their implementation.

Barriers to Redistribution of Surplus Chemicals

Redistribution of surplus chemicals is discussed above as a successful endeavor, and later in this chapter as a method that has more potential for waste minimization. There are, however, several limitations to chemical sharing and redistribution that deserve mention. First, surplus management is subject to adverse selection. People tend to store valuable items speculatively, and offer items of little value as surplus. This human trait lowers the value of any surplus management program. In chemical redistribution, laboratory personnel are continually wary of receiving someone else's waste instead of a valuable chemical. When valuable items *are* offered as surplus, they may be sought for speculation and hoarded. The University's redistribution program has experienced all of these problems.

Second, people tend to value shared items less than owned items. Laboratory personnel may be less inclined to maintain the quality of a nonowned chemical. When a stock chemical is shared, purity and degradation are always a concern.

A proposal to expand University's successful redistribution program to the other 25 University of Wisconsin System campuses has met many questions.[4] In addition to the problems mentioned above, the expense of hazardous material transportation, packaging, labeling, and recordkeeping is significant, especially for laboratory personnel who are not trained in U.S. Department of Transportation regulations.

Finally, exchange of surplus chemicals between organizations poses serious liability questions. If a chemical is used or disposed of improperly by the recipient, there is a fear that the donor organization may share some liability for damages that may result.

In practice, the University of Wisconsin-Madison has successfully redistributed surplus chemicals to other University of Wisconsin System campuses on a few occasions, and one time to an out-of-state college. In another instance, the University Chemical Safety Committee rejected a large shipment of surplus chemicals from a private research laboratory because many of the chemicals appeared to be in poor condition.

WASTE MINIMIZATION INITIATIVES

If the resources become available, the author wishes to pursue three waste minimization initiatives at the University of Wisconsin-Madison: a chemical stockroom on the campus computer network, creating a forum for minimization ideas and making a solvent still available to laboratories on loan.

Chemical Stockroom on a Computer Network Database

In the past, chemical stockrooms played an important role in minimizing surplus chemicals and waste by controlling purchases, monitoring inventories and quality, and encouraging shared use of chemicals. Chemical stockrooms typically are used to serve several laboratories for ordering, receiving, and storing of both new *and* opened containers of stock chemicals. Today, however, many investigators keep their opened containers in their laboratories, and rarely share stock chemicals with other laboratories. Poor chemical inventory habits contribute to this.

Although an organization could attempt to reestablish the redistribution of surplus chemicals from their stockroom, an alternative available today is to create an "electronic" stockroom on the campus computer network. A database of surplus chemicals on the network could be accessed by University chemical users, and perhaps other campuses of the University of Wisconsin System. The database could include surplus chemicals received by the Safety Department as well as surplus chemicals located in various campus laboratories that are available for loan.

An up-to-date surplus chemical database on the network would facilitate the exchange of surplus chemicals from nearby labs without any involvement of the Safety Department. Campus purchasing agents could check the database prior to ordering new chemicals. The Safety Department could provide transfer services for exchanges between laboratories, ideally on an as-needed basis rather than sporadically. These steps would encourage redistribution of surplus chemicals and minimize this wastestream.

Exchanging Waste Minimization Ideas

Although this chapter expresses the concern that some laboratory personnel are not fully aware of the problems of hazardous waste, many individuals who work in laboratories are aware of the problem, are seeking solutions, and have discovered successful minimization methods. There probably are many little-known, simple, and easy methods. A researcher recently explained to the author how they reuse ethidium bromide or eliminate its use altogether.

A forum to exchange waste minimization ideas would be beneficial. The American Chemical Society has discussed the possibility of a laboratory waste minimization contest, in which ideas would be solicited, awards would be given for the best methods, and many useful minimization methods would be published so that all laboratories could benefit from them.

Making a Solvent Still Available for Loan

Table 4 and above describe the benefits and difficulties of a solvent still operated by the institution. Another way for an institution to encourage solvent distillation would be to provide a still, on loan, to prospective laboratories. Laboratory personnel could then determine if solvent distillation and reuse would work for their operations, and thereby overcome the risk of an expensive purchase that may not be beneficial. The free loan of a still would delay start-up costs. If successful, their experience with a loaned still would allow laboratory personnel to better justify the purchase of their own still. Since the still would be operated by product users rather than by institutional personnel, the burden for quality control would shift to the people who had the greatest interest in it.

OVERCOMING REGULATORY BARRIERS

A survey by the Wisconsin Department of Natural Resources found that EPA regulations and OSHA regulations for worker safety were important incentives for hazardous waste management and reduction.[5] As discussed above, fear of Superfund liability is another incentive. For laboratories, however, regulations themselves can create barriers to waste minimization. For laboratories, accumulation limits often do not allow accumulation of sufficient quantity of a waste type to cost effectively utilize recycling or waste minimization methods.

Regional EPA offices and state and local hazardous waste authorities differ in their regulation of chemical waste treatment. Under certain conditions, regulators may allow treatment without a permit or may temporarily waive the permit requirements. Most regulators allow treatment of chemical waste in its collection container or as part of a process without an RCRA permit.[6] Some state and local authorities, however, consider treatment of chemical waste at the benchtop to be no different than treatment at a hazardous waste facility. Regulatory staff have insisted that a laboratory using benchtop treatment methods must obtain a permit and adhere to the stringent requirements applicable to facilities that treat large commercial volumes of hazardous waste. Obtaining a hazardous waste permit and compliance with these additional requirements, however, are prohibitively expensive for most laboratories.

The American Chemical Societys (ACS) Department of Government Relations and Science Policy (GRASP) and Task Force on Laboratory Waste Management has discussed these problems with policy makers at EPA.[7] An EPA representative expressed an understanding of barriers to laboratory waste minimization, but that staff shortages and Congressional mandates prevent EPA from issuing alternative rules or guidance to facilitate waste minimization in laboratories.

LABORATORY EQUITY AND WASTE MINIMIZATION ACT

As a result, the American Chemical Society drafted language for a bill, The Laboratory Equity and Waste Minimization Act, and assembled a coalition of over 60 colleges, universities, organizations, and firms with interest to promote waste minimization in laboratories. In 1990, ACS and members of the Task Force began discussing the need for regulatory changes with Congress in the hope of finding support.

Several members of Congress, both in the House and the Senate, agreed to sponsor the bill in the 1992 session. The Act, however, was not included in the 1992 legislation that partially reauthorized the Resource Conservation and Recovery Act of 1986.

In 1993, ACS asked members of Congress to encourage EPA to issue rules that included provisions of the act, on the grounds that there was general agreement as to its merits. Congressmen sent letters to the EPA Administrator, noting bipartisan support for the provisions, and asked EPA to investigate its legal authority to implement them. EPA replied that the provisions of the bill were "unnecessary as they essentially duplicate existing regulations. There is already regulatory authority for generators to treat their hazardous waste on-site in appropriate tanks or containers."[8] This interpretation from EPA headquarters, however, did not alter the position and inconsistency of EPA regions and states in addressing these issues.

In 1994 the ACS GRASP Department worked with Congress to add language to the report accompanying the Independent Agencies Appropriations bill, requesting that EPA modify its regulations to address these concerns.

Provisions of the Act

In summary, the provisions of the Laboratory Equity and Waste Minimization Act[9] are

- It pertains to laboratories
- So that laboratories can accumulate sufficient quantities of waste for cost-effective recycling and other waste minimization methods, the Act would allow a longer on-site accumulation period. Generators of more than 1000 kg/month would be able to store hazardous waste up to 180 days and 1500 kg. Generators of 100 to 1000 kg/month would be able to store hazardous waste up to 270 days regardless of shipping distance
- It allows the campus of an educational institution to have only one hazardous waste generator identification number
- It allows laboratory treatment of hazardous waste under a group permit

Group Permit for Laboratory Treatment

A group permit, under the provisions of the Act, would be far easier to obtain than permits issued to hazardous waste treatment, storage, and disposal facilities. A group permit for laboratory treatment of hazardous waste could be utilized by any laboratory that adhered to the following conditions:

- They must notify the Regional EPA Administrator
- Treatment of hazardous waste can only be done by nonthermal treatment methods. For example, a laboratory could not incinerate waste in a fume hood
- Treatment activities could only be used for wastes generated on site; a laboratory could not accept waste for treatment from another hazardous waste generator

- Laboratory treatment must be limited to 100 kg of hazardous waste per week and 25 kg/day per laboratory. This is consistent with EPA's limits for research and development permits
- Laboratory must annually report their treatment activities

CONCLUSION

Two additional impediments to waste minimization in laboratories are worth mentioning. First, as noted previously, the University Safety Department has the responsibility to serve the faculty, staff, and students in their mission to teach, conduct research, practice medicine, and serve the public. At the same time, the University is subject to public and regulatory pressure to responsibly manage hazardous waste. Some faculty and staff may view the additional effort to management and minimize waste as an interference in their (and the University's) mission. However, it would be unwise for the Safety Department to serve laboratory personnel in a way that completely insulates them from the incentives for waste minimization and disincentives for pollution felt by the institution.

Another impediment is that, as much as the Safety Department staff and the University administration wish to minimize waste, the University is subject to increasing environmental regulations, inspections, threats of fines, environmental liability, and a limited budget. As a result, Safety Department staff must often attend to crises and deadlines, rather than preventing them, as waste minimization would do.

These are both frustrations and challenges in achieving waste minimization.

ENDNOTES

1. Klein, R. C. and Gershey, E. L., "Biodegradable" liquid scintillation counting cocktail, *Health Physics*, 59(4), 461–470, 1990.
2. Reinhardt, P. A., In: Abstracts of Papers, 186th National Meeting of the American Chemical Society, Washington, D.C., 1983.
3. Popoff, F. P. and Buzzelli, D. T., Full-cost accounting, *Chem. Eng. News,* January 11, 8–10, 1993.
4. Ma, J. C. K., Leonard, K. L., and Kandziora, P., Chemical Conservation Program Proposal, University of Wisconsin System Administration, Spring 1992.
5. DNR Bureau of Research, Reducing Hazardous Waste in Wisconsin, Report V: Barriers and Incentives to Hazardous Waste Reduction, Wisconsin Department of Natural Resources, Madison, PUBL-MB-007(91), August 1992.
6. Correspondence from Marcia Williams, Director of EPA Office of Solid Waste (Washington, D.C.) to Bernard E. Cox, Jr., Chief, Hazardous Waste Branch, Alabama Dept. of Environmental Management (July 1, 1987) and Robert F. Greaves, Acting Chief of EPA Waste Management Branch, Requested Re-Interpretation of On-Site Treatment Exemption (December 15, 1987); Also see Title 40, Code of Federal Regulations, Section 264.1(g)(3).

7. This Task Force is sponsored and funded by the American Chemical Society's Committee on Chemical Safety and Committee on Environmental Improvement.
8. Correspondence from Senators D. P. Moynihan, F. R. Lautenberg, and J. H. Chaffee to Administrator C. M. Browner (20 April 1993). Reply from T. C. Roberts to F. R. Lautenberg (June 10, 1993).
9. A copy of the Act is available from the American Chemical Society's Department of Government Relations and Science Policy.

Part 3:
Approaches by Media, Source, and Waste Type

CHAPTER 8

Management of Laboratory Air Emissions

Ralph Stuart and Milly Archer

CONTENTS

Introduction 124
Laboratory Air Emissions: Characterization and Control 125
    The Source — Laboratories 125
        Some Terms 125
            Laboratory Use of Chemicals 125
            Fugitive and Exhausted Emissions 126
            Emissions Targets 126
        Sources of Emission within the Laboratory 126
        Sources of Laboratory Emissions to the Environment 127
    Estimating Laboratory Emissions 127
        Purpose of Estimating Laboratory Emissions 127
        Identifying Laboratory Emissions 128
        Measuring Laboratory Emissions 128
        Measuring Fugitive Emissions 129
    Controlling Laboratory Emissions 129
        Elimination and Substitution 129
        Administrative Measures 129
        Engineering Controls 130
        Operational Controls 131
    Regulatory Considerations 131
        Occupational Health Regulation 131
        Environmental Regulations 132
            Some Terms 132
                Hazardous Air Pollutants 132
                Source 132
                Ambient Air Standards 133

        Emission-Based Standards 133
        The Clean Act Amendments 133
        State Programs 133
        Regulatory Implications 134
            Scenario #1 — Establishment of a Federal MACT for
            Laboratory Emissions 134
            Scenario #2 — State Imposed Ambient Air Standards 135
Developing a Laboratory Emissions Management Program 135
    Within the Laboratory 135
        Planning Chemical Use 135
        Control Equipment 136
        Chemical Storage Practices 136
    Institutional Policies 136
        Laboratory User Education 136
        Administrative Infrastructure 137
        Regulatory Relations 137
    Research Needs 138
        Alternatives for Common Laboratory Practices 138
        Measurement of Emissions 138
Conclusions 138

INTRODUCTION

Emissions of air pollutants from laboratories have the potential of creating occupational and environmental health effects. Until recently, it has been assumed that these effects can be resolved satisfactorily by capturing contaminants in the laboratory and exhausting them outside to be diluted in the general environment. This assumption has changed recently, for several reasons. First, the occupational health status of laboratories is undergoing new scrutiny as a result of the OSHA laboratory standard. Second, laboratory buildings have become more dense, with many buildings having hundreds of labs interspersed with public and office spaces. This makes the labs' effects on the building's air quality a concern. At the same time, air pollution regulators have begun to include chemicals beyond combustion products as part of their sphere of concern. Finally, public concern about the possible health effects of exposure to low levels of chemicals has increased.

These factors have led to a need to assess the environmental impact of laboratory air emissions. These assessments will be compared to emissions goals that can be based on two different criteria. The first of these criteria is that emissions be as low as reasonably achievable (ALARA, borrowed from radiation safety). This principle means that, even if there is no immediate evidence of a hazard, emissions should be reduced as much as possible because it is the safest thing to do. The alternative approach to establishing emissions limits is based on target-based health models. These models tend to be very conservative, assuming long-term, continuous exposure of a fragile population to the worst-case emissions scenario. These assumptions generate very low emissions goals, usually unmeasureable and often unachievable without the elimination of the use of many chemicals.

Because "dilution has been the solution" to laboratory air emissions, their characteristics have not been carefully studied and their management has been neglected until recently. However, unless management of air emissions becomes a priority of laboratories as they plan and conduct their work, regulatory schemes based on the risk assessment goal described above will likely be imposed by government agencies and the result is likely to be much more costly and intrusive into the laboratory process than is necessary. A proactive approach to this issue by laboratories should provide significant long-term benefits.

This chapter has three sections. The first describes some of the physical aspects of laboratory airborne emissions. The second discusses some of the regulatory concepts that affect the management of these emissions. The third section makes recommendations for controlling emissions and their physical and regulatory impact. It also describes some of the research necessary to help clarify some of the murkier issues surrounding this issue.

LABORATORY AIR EMISSIONS: CHARACTERIZATION AND CONTROL

This section follows the traditional industrial hygiene approach to management of airborne contaminants — identification, measurement, and control of the chemicals used by processes in the workplace — in order to describe some of the physical aspects of laboratory emissions.

THE SOURCE — LABORATORIES

Laboratories are general-purpose workplaces designed to house a wide variety of physical, chemical, and biological processes. By its nature, laboratory work is sporadic — work is done for short periods of time as a particular project progresses and then stops. Different stages of a project may involve completely different processes. These factors make the characterization and estimation of laboratory emissions difficult. To simplify this discussion, some terms characterizing these emissions will be defined.

Some Terms

Laboratory Use of Chemicals

Laboratory is used in this chapter as it is used in the OSHA laboratory standard (29 CFR 1910.1450). Labs are defined there as workplaces where a variety of chemicals are used and one person can handle the quantities used. This definition applies especially well to research and educational laboratories where work is on projects of short duration, which are often set up and finished in the same day.

Fugitive and Exhausted Emissions

Fugitive emissions are releases of chemicals that are not captured by any specific piece of equipment. Fugitive emissions are controlled by the lab's general ventilation system, through dilution of the contaminant. Exhausted emissions are chemical releases that take place under the control of a local ventilation system, usually a fume hood.

Emissions Targets

Emissions targets are the receptors of the air pollutants. There are two types of targets that are important for regulatory considerations. One is the laboratory workers and their co-workers inside the building, who present an occupational health concern. The other target is the environment outside the laboratory building. The effect on these targets present an environmental health regulatory issue.

Sources of Emission within the Laboratory

Contamination of the air in the laboratory results from four primary sources: use of chemicals in areas that are served by the general ventilation system; use of local ventilation devices in such a way that they are ineffective at capturing the products of the process; storage of volatile chemicals in a way that the containers emit the chemical, and puddles resulting from chemical spills. Because many different chemicals are used in the laboratory, the cumulative effect of all these sources must be considered in assessing laboratory air quality.

Probably the strongest source of emissions in the laboratory is the benchtop use of volatile chemicals. While this practice may be acceptable when planned and carried out carefully, it can also create hazardous situations when engaged in casually. An example of this is when flammable chemicals are allowed to disperse throughout a laboratory uncontrolled. Fires resulting from solvent vapors accumulating near a spark are common laboratory accidents.

Perhaps a higher risk concern is the improper use of fume hoods or other local ventilation devices for the control of highly toxic chemicals. This can result in a significant amount of the chemical not being captured and escaping to the laboratory. This situation can result in exposures large enough to create acute health effects, such as eye and nose irritation, headaches, and nausea. Proper selection and use of local ventilation devices is important in controlling this type of emission.

Leakage from chemical storage containers is a source of fugitive emissions in the laboratory which creates unknown risks. The corrosion commonly observed in chemical storage areas testifies to the long-term effect of these chemical vapors on their containers and environment. Adequate ventilation of these areas can control this problem satisfactorily, particularly if inventory control prevents long periods of storage of volatile chemicals.

While chemical spills large enough to create laboratory emission concerns may be unusual events, they can be of sufficient magnitude to be a significant risk. Attention to proper planning for and clean-up of hazardous materials spills is important in controlling this source. Further consideration of this issue will be left to emergency response discussions.

Sources of Laboratory Emissions to the Environment

Laboratory emissions to the environment come from two sources: the exhaust of the local ventilation devices and fugitive emissions. Local ventilation devices usually emit the material they capture directly to the environment without treatment, relying on dilution to mitigate the hazard associated with the chemical. Fugitive emissions reach the environment through the exhaust of the building's ventilation system and from seepage from the building. The magnitude of these fugitive emissions is unclear, but if there is a nearby receptor of these emissions (e.g., the air intake of a building) they may present an issue to be considered.

ESTIMATING LABORATORY EMISSIONS

There are two major issues involved in characterizing air emissions: identifying the chemicals emitted and measuring either the amount of chemicals emitted or the amount that reach a particular emissions target. The nature of laboratory work makes both of these goals problematic. However, the increasing interest in laboratory emissions makes some attempt at developing an estimate necessary.

Purpose of Estimating Laboratory Emissions

There are two purposes for estimating the types and quantity of chemicals emitted in laboratories. The first is to provide information about how the ventilation system serving the laboratory needs to work. Design practices used in developing ventilation systems for laboratory buildings tend to rely on a large amount of dilution air to control chemical vapors. This is an expensive method due to energy costs of conditioning this excess air. For the ventilation system to achieve more effective control of laboratory air contaminants, more specific information about laboratory emissions must be developed.

The second purpose for estimating laboratory emissions is to be able to provide the basis for analysis of the hazards these emissions present to various populations. The populations of concern vary, but include laboratory workers, their neighbors (both inside and outside the building), the maintenance workers who service the infrastructure of the building, and the environment. While the quantities of the chemicals used in a particular laboratory are often small, the cumulative effect of many small emissions from the many labs in a single building is unknown. This problem can only be resolved by further research.

Identifying Laboratory Emissions

Identifying the chemicals airborne in the laboratory is a significant problem for a number of reasons. Chemical use in laboratories is often unplanned and recordkeeping of the use of utility chemicals (e.g., solvents and acids) is sporadic. In addition, there is often little knowledge available about the identities of intermediates or products of particular chemical reactions. Chemicals can also change phase in the course of the laboratory work so that a liquid reactant can become a solid waste or a solid reactant can become a volatile product. These factors make identification of airborne chemicals by examination of the processes difficult.

Methods of identifying the chemicals of air emissions concern have included use of purchasing records, interviews with laboratory workers, and some air sampling. These have not been particularly satisfactory, because they do not provide physical or theoretical corroboration of the results. Whatever method for completely identifying the chemicals of concern is developed will require a significant amount of resources to implement and the cooperation of laboratory occupants to be successful.

Measuring Laboratory Emissions

Measuring the concentration of air contaminants is a complex matter. To do so, you need to know the identity and approximate level of the chemical to be measured, where and when the concentration of interest will be, and have the appropriate equipment available. In particular, sampling of air contaminants in the laboratory is a significant technical problem. This is because it is difficult to predict the chemical to be sampled and the time of occurrence of the emission of interest. Given this, there are several approaches to estimating the concentration of chemicals in laboratory air. Unfortunately, these methods consider only a single chemical within the laboratory and the combined effect of multiple chemicals is difficult to extrapolate from the information collected.

The most direct method is to physically sample the atmosphere of interest. A discussion of the various methods of conducting air sampling is beyond the scope of this chapter. They range from relatively cheap systems for specific chemicals (e.g., colorimetric tubes) to expensive microprocessor-controlled analytical equipment which can identify and measure several chemicals simultaneously. This method is limited in practice by the factors noted above.

A new feature of laboratory management arose with the promulgation of the OSHA laboratory standard. This standard requires that the various operating procedures in the laboratory be evaluated by a chemical hygiene officer for potentially hazardous chemical exposures and that air sampling be done if any are suspected. If the results of these evaluations, both in terms of the chemicals considered to be of possible concern and the results of any monitoring, are collected, estimation of chemical emissions in labs will improve.

It is sometimes possible to use the odor threshold of a chemical to estimate the concentration of that chemical in the laboratory. Information about the odor threshold of chemicals is available on material safety data sheets or through other chemical safety reference sources. It must be remembered that this is an unreliable guide in situations where there are many different chemicals and where olfactory fatigue is likely. However, this may be the only method available in many laboratory situations. For this reason, laboratory workers should be familiar with the odor properties of the chemicals they work with.

Measuring Fugitive Emissions

While fugitive emissions are likely to be much less than established occupational health limits for the chemicals involved, they can be a cause for concern, for a couple of reasons. First, it is unpredictable what the health effects of the combination of chemicals that result from the fugitive emissions of many labs may be. Many laboratory buildings have a consistent smell to them which is unidentifiable as being a particular chemical. This odor creates concern for occupants of the building. The second reason that fugitive emissions may present a risk is that there can be local pockets of accumulation within the building, depending on the operations within the building and the operation of the ventilation system. These pockets can create both physical hazards (corrosion of fixtures) and potential health hazards.

CONTROLLING LABORATORY EMISSIONS

After laboratory emissions that present unacceptable risks have been identified, an appropriate control strategy must be developed to enable the level of the emissions to be brought within acceptable levels. These controls fall into categories, borrowed from industrial hygiene.

Elimination and Substitution

Controlling airborne laboratory emissions is most effectively accomplished by eliminating the chemicals that can become airborne. While this option is not always available, it should always be the first considered. As laboratory pollution prevention efforts proceed, more alternatives for standard laboratory uses of hazardous chemicals will become established. Examples of successful substitutions in laboratory procedures that are important in eliminating laboratory emissions include the substitution of acetone for benzene when washing glassware, using organic dishwashing solutions instead of chromic-sulfuric acid mixtures, and replacing mercury thermometers with alcohol thermometers.

Administrative Measures

The second level of a laboratory emissions control program is the development of administrative controls to minimize airborne release of chemicals. These controls are likely to be the most important in laboratory situations because they are the most flexible in application to individual situations.

Education and training of laboratory workers is probably the most important factor in controlling laboratory emissions. Because it is impossible to predict all the various operations and processes that will take place in the laboratory workplace, it is necessary to provide the technical background for laboratory workers to enable them to identify and control unnecessary laboratory emissions. Issues that should be included in this education include proper use of fume hoods, chemical storage practices, and development of alternatives to use of problematic chemicals. Of course, part of this education process is to provide the continuing motivation for this effort. Tying emission prevention efforts into laboratory worker's health and safety can be an important element of this program.

A second administrative measure is planning the work to be done with emission control in mind. Individual experiments need to be considered in light of the facilities available to control the processes involved. Some experiments may lend themselves to the use of filters (for particulates) or carbon beds and cold traps (for vapor emissions). Local ventilation needs should be evaluated to see whether devices other than fume hoods can be used. This can often result in significant energy savings and provide more effective capture of process emissions than general purpose devices such as fume hoods. Long-term planning of the types of research and the related ventilation and equipment needs are elements that facilitate the use of substitution and elimination of hazardous chemical usage.

Engineering Controls

Engineering controls are devices that are designed with a specific purpose in mind. Their effectiveness is limited when the assumptions behind this design are not met. Because of the general purpose nature of laboratories, engineering controls are difficult to apply to a laboratory setting. General-purpose engineering controls provide limited control over most operations. There are three engineering devices commonly found in laboratories that control laboratory emissions: the general ventilation system, local ventilation devices, and chemical storage cabinets.

The general ventilation system deals with emissions by changing the air in the room often enough to prevent vapor accumulation. Unfortunately, this strategy can be defeated by strong emission sources and incomplete mixing of the laboratory air. For these reasons, the general ventilation system should not be expected to control airborne chemicals effectively.

Fume hoods are the most common local ventilation devices in laboratories. However, they are not simple pieces of equipment to operate effectively and their capture mechanisms (the sash and the air flow) can be easily defeated by improper design, environment, or use. Education of hood users and regular monitoring of hood performance are essential to assure their effectiveness.

Ventilated chemical storage cabinets are the engineering element of a chemical storage program that has an important role in limiting air emissions. It should be combined with administrative controls such as an inventory control system which limits the amounts of chemicals stored for long periods of time to effectively control a lab's emissions.

Operational Controls

The last level of controlling emissions is relying on the laboratory operator to control emissions. This involves the careful conduct of the work at hand to minimize chemical volatilization. An example of such operational controls is closing chemical containers whenever possible to prevent unnecessary evaporation. While operational controls can be important parts of an emission control program, they are subject to operator error and accidental failures and thus may not work as hoped. In addition, for operational controls to be effective, an aggressive training and enforcement policy is demanded of laboratory management. This is particularly the case in academic labs, where personnel are often inexperienced and turnover is high.

REGULATORY CONSIDERATIONS

Regulation of air pollutants falls into two separate spheres — occupational health and environmental health. The assumptions are fundamentally different in these spheres. Occupational health regulations have traditionally focused on preventing exposures that create immediate symptoms or specifically identified irreversible diseases (e.g., cancer or birth defects). Labs have been set aside as a special case by occupational health regulations. Environmental health focuses primarily on long-term chronic health effects resulting from low-level chemical exposures. Labs are being treated similarly to other emitters by many environmental regulations. This section describes the style of both types of regulation and their implications for managing airborne emissions both for the individual laboratory and at the institutional level.

OCCUPATIONAL HEALTH REGULATION

Regulation of occupational exposure to airborne contaminants in laboratories is very difficult. Permissible exposure limits (PELs) are based on long-term exposures to time-weighted averages of the chemical of concern. It is the minority of laboratories that use the same chemicals consistently enough to meet the assumptions behind the PELs. For some chemicals, OSHA also sets a short-term exposure limit (STEL), which applies to a 15-min exposure, which may be more appropriate for laboratory settings.

It is unlikely that any 8-h time-weighted average levels of chemicals in labs are near regulatory limits in most laboratory settings. However, this does not mean that laboratory exposures are always insignificant with respect to health effects, particularly acute symptoms. In 1991, OSHA recognized the unique nature of laboratories and established a separate standard for them that requires that each laboratory develop a chemical hygiene plan which includes an assessment of the potential personnel exposure to hazardous chemicals. This assessment must include a consideration of possible exposures to airborne contaminants and must be conducted by a qualified individual. The weakness of this system is that it continues to focus on an individual chemical or process. The cumulative effect of exposure to many different chemicals is an important concern in laboratory settings.

ENVIRONMENTAL REGULATIONS

Regulating laboratory air emissions as an environmental health concern is a relatively new concept. Prior to the Clean Air Act Amendments of 1990, only seven toxic air pollutants were regulated, none of which posed compliance problems for laboratory operations. The air toxics provisions of these amendments list 189 chemicals and compounds that are defined as hazardous air pollutants, many of which are commonly emitted from laboratory fume hoods. Examples of common laboratory chemicals included on this list are benzene, formaldehyde, and methylene chloride.

Some Terms

Anyone who attempts to read, let alone understand, environmental regulations will soon find out that the regulations are filled with jargon. Commonly used words placed together can mean something obscure to the novice. The following list is limited and serves only to explain some of the terms used in this section.

Hazardous Air Pollutants

The 1990 Clean Air Act Amendments list 189 chemicals and compounds that are defined as hazardous air pollutants. The EPA is required to periodically review and, where appropriate, revise the list. Third parties such as environmental groups and industry can petition to add or remove chemicals from the list. State regulations can be more stringent, and may have additional chemicals listed as hazardous air pollutants.

Source

A source of air pollutants includes all emission points that are located on contiguous or adjacent properties. For example, all of the buildings on a college campus that are located on contiguous grounds are considered to be one source. A source category is a group of industry facilities classified by their service or product. For example, research or laboratory facilities are considered to be a source category, as are dry-cleaning plants. A facility that has the potential to emit 10 tons or more per year of any single hazardous air pollutant or 25 tons or more per year of any combination of hazardous air pollutants is considered a major source. (Editor's note: in other chapters of this book "source" may be used to refer to individual discharge points, rather than a group of discharge points at one location.)

Ambient Air Standards

Ambient air is the air around us, the air we breath. Ambient air standards are set based on impacts to human health and the environment. When determining whether a source meets an ambient air standard, its emissions are added to the hazardous air pollutants that already exist in the ambient air.

Emission-Based Standards

Emission-based standards are established based on the best technology available to control a specific pollutant emitted from a specific process or source. These standards apply to the strength of a single source, not a single source plus the pollution in the ambient air. The maximum achievable control technology (MACT) is defined by the Federal EPA for each source category that emits hazardous air pollutants. This level of pollution control specifies the maximum degree of reduction in emissions of hazardous air pollutants that EPA determines is achievable based on technology. This determination takes into account the cost of reducing emissions as well as health and environmental impacts. EPA is authorized to impose MACT standards on sources below the major source threshold on a category-by-category basis.

The Clean Act Amendments

The federal Clean Air Act Amendments of 1990 began the process of regulating hazardous air pollutants. They require source categories that emit more than specified quantities of certain air pollutants to meet emission standards to obtain an operating permit. For example, dry cleaners as a source category will be required to meet a particular standard for perchloroethylene emissions. Emission standards are based on the best-demonstrated technologies for air pollution control for a specific type of industry or activity.

The Clean Air Act recognizes research and laboratory facilities whose primary purpose is to conduct research or engage in teaching activities are a source of air emissions which is different and more difficult to control than other industries. The law mandates that the EPA establish a separate category of emission standards for research and laboratory facilities. This proviso, which is found in Title III Section 112(c)(7) of the Clean Air Act Amendments, is stated to be "necessary to assure the equitable treatment of such facilities." Whether or not this will serve to exempt research laboratories from the federal law has yet to be seen.

State Programs

The federal law does not replace existing or new state laws, as long as those state programs are at least as stringent as the federal program. For example, the federal program requires that any major source of emissions of one or more of the 189 hazardous air pollutants be required to meet a new emission standard and obtain an operating permit. To be considered a major source of hazardous air pollutants under the federal law, a facility must emit approximately 25 tons of hazardous air emissions each year. Very few, if any, educational institutions have hazardous laboratory emissions in quantities that would cause them to fall under the jurisdiction of the federal air toxics program.

However, if a state requires all sources that emit hazardous air pollutants to comply with a public health-based ambient air standard, then laboratory air emissions will likely fall under the state's jurisdiction. Unlike the federal program, which is based on best-demonstrated technology and addresses emission standards applied to categories of industries, several states are moving toward programs based on ambient air standards. Ambient air standards include background concentrations of hazardous air pollutants in determining if a single source is in compliance with the air quality standard. Compliance problems arise if a single source's emissions, when added to ambient concentrations, exceed the health-based standards established in the regulations.

The 1990 Clean Air Act Amendments require states to develop and manage an operating permit program, subject to EPA approval and supervision. The federal law also mandates that state programs be fully funded by the sources that emit air pollution in that state. Sources of air pollutants are also required to register their emissions annually with the state in which they reside. In most cases there is a fee schedule associated with the quantity and type of air pollutant emitted. Traditionally, research and teaching institutions have limited their registered sources to flue gas emissions from boiler plants. As toxic air pollutants become included in the universe of regulated emissions, new challenges arise with regard to quantifying and qualifying (and putting a price tag on) the types and amounts of air toxics emitted from laboratory fume hoods.

Regulatory Implications

To help clarify the above discussion, this section looks at two possible scenarios: one in which federally mandated pollution control technologies are imposed on laboratory emissions, the second when a state ambient air standard is exceeded. These examples are conceptual and not intended to be a guide to compliance. Compliance strategies will differ on a case-by-case basis depending on many variables including EPA-imposed emission standards for laboratories, state regulations, the types and amounts of hazardous air pollutants emitted, as well as the method used to identify and measure laboratory emissions.

Scenario #I — Establishment of a Federal MACT for Laboratory Emissions

As stated earlier in this section, the EPA has been mandated to establish emission standards for research and laboratory facilities. Furthermore, EPA must define an appropriate pollution control technology for all major source categories, and is authorized to impose those technology based standards on sources below the major source threshold. Thus, EPA could require laboratories emitting certain air toxics, e.g., benzene, to use activated charcoal filters to absorb the emissions, or they could determine that laboratories using methylene chloride must employ an MACT to recover methylene chloride vapors.

Scenario #2 — State Imposed Ambient Air Standards

Theoretically, a state-established ambient air standard for hazardous air constituents could be lower than the background concentration of that chemical or compound that already exists in the ambient air. Compliance problems can arise if a single source's emissions added to ambient concentrations exceed the emission's specified standard in the regulations. The source would then either attempt to control the emission by using pollution control equipment, (e.g., activated charcoal or vapor recovery) or cease emitting the chemical that exceeds the ambient standard.

It must be remembered that, from a regulatory standpoint, a source is defined as all emission points on contiguous properties. Therefore, the management of laboratory emissions requires campus-wide controls. If benzene and methylene chloride emissions are a compliance concern, it may be necessary to require that these chemicals only be used in fume hoods equipped with the proper pollution control devices. The ultimate solution is to eliminate, where possible, the use of hazardous chemicals that can become airborne.

DEVELOPING A LABORATORY EMISSIONS MANAGEMENT PROGRAM

To manage the safety and health concerns identified above, and to cope with increasing regulatory pressure, laboratory management needs to take an active role in managing emissions of airborne chemicals. This section describes some physical, administrative, and research activities that can be put together to develop a laboratory emissions management program, both for a specific laboratory and on an institutional level. It must be recognized at the outset that neither an institutional administration or individual laboratories can make significant progress in dealing with this issue alone. Cooperation and teamwork on both sides are necessary to deal with an issue as complex as this.

WITHIN THE LABORATORY

The most effective steps for controlling air emissions are only available within the laboratory, where the emissions originate. These steps include both administrative practices and physical operations.

Planning Chemical Use

Within the laboratory, probably the most important step that can be taken in improving the management of emissions is adding a formal "environmental impact statement" to the planning process for laboratory work. This does not have to be an extensive document. It should identify all chemicals to be used in the work being considered, including utility chemicals such as solvents for washing glassware, and identifying their expected disposal fate. Intermediates and products of chemical reactions should be identified when possible and capture of these materials should be planned to prevent their uncontrolled release. Consideration of scaling down the amount of chemicals to be used should be included in this process. This environmental impact statement would fit directly into the standard operating procedure and exposure evaluation system required by the OSHA laboratory standard. As part of this planning effort, the practice of evaporating waste chemicals in a fume hood should be eliminated. This is both to prevent unnecessary emissions and to promote worker safety. Leaving waste chemicals open in a fume hood provides a source of fuel to become involved in any accident that occurs in the hood. The practice is illegal in many jurisdictions.

Control Equipment

The second part of the emissions control process is identifying the equipment required to successfully capture the products of the work identified above. Fume hoods are limited in their effectiveness for many operations and do not provide any air-cleaning capability. In situations where static set-ups occur, specific ventilation controls should be developed, perhaps using the fume hood as an overall enclosure. Such set-ups also present the opportunity for developing systems for cleaning the air of chemical emissions.

Chemical Storage Practices

Chemical inventories should be managed to avoid storage of any materials for more than 1 year. The seals of chemical containers are potentially permeable to the chemicals they contain after periods longer than this. Also, storage cabinets containing volatile chemicals should be ventilated to the outside (possibly using the fume hood's ductwork) to provide for gradual release of the accumulating vapors over time, rather than in large doses when the cabinet door is opened. This minimizes laboratory workers' exposure and decreases the probability that the chemicals will be contaminated during storage.

INSTITUTIONAL POLICIES

A laboratory pollution prevention program cannot succeed solely on the basis of efforts within individual laboratories. Institutions housing laboratories must provide an administrative and physical infrastructure to enable the labs to take advantage of emissions control opportunities.

Laboratory User Education

An education program for laboratory users about the need for emissions control, incorporated within the laboratory health and safety program as a whole, is critical to show institution-wide decreases in laboratory emissions. Elements specific to emissions control that should be included in such a training program include the proper use of fume hoods and the determination of when alternative forms of local ventilation are more appropriate (e.g., biosafety cabinets, glove boxes, or dedicated exhaust for equipment). Also, an understanding of the design expectations of the general ventilation system and how it actually operates will enable laboratory workers to anticipate the effects of their work on others in the building and neighborhood. This is particularly important in the newer microprocessor-controlled ventilation systems, where the controls are often not intuitively obvious.

Administrative Infrastructure

The administrative infrastructure necessary to develop an effective emissions control program includes the development of a chemical delivery system which precludes the need for storage of large quantities or varieties of chemicals within the laboratory. The establishment of a chemical warehouse on campus can provide purchasing advantages that enhance significantly the health and safety of laboratory operations. For example, maximum-size chemical containers can be instituted to minimize the likelihood of a large chemical accident and assist in inventory control in the laboratory. If purchasing is organized centrally, pricing may not have to suffer from this policy.

Regulatory Relations

The regulatory requirements pertaining to laboratory air emissions are unclear and therefore not set in stone. Most regulators are not knowledgeable about the types of air toxics emitted from laboratory fume hoods, or the problems associated with identifying and quantifying those emissions. Since regulations are written with industrial sources and processes in mind, they are often difficult to apply to the institutional research and teaching laboratory setting. Understanding existing laws, monitoring changes in those laws, and anticipating the implications of new laws requires that institutions establish a cooperative relationship with their regulators.

Environmental inspectors are the first line of interaction, and a good working relationship with them is crucial. Explain to them how the institution is unique in the community of regulated sources and why specific requirements are perhaps inappropriate. Discuss difficulties such as quantifying and qualifying laboratory emissions. The complexity of certain issues is not necessarily intuitive. If you don't have an open line of communication with inspectors, they won't be able to bring your concerns to their supervisors, who generally decide how a particular set of facts will be interpreted.

Understanding how laws and regulations are changing are important factors in establishing a working relationship with the governing agencies. A pertinent example of this is that the 1990 Clean Air Act Amendments require states to develop and manage a permit program will be fully funded by the sources that emit air pollution in that state. For this reason, institutions are likely to be included in the regulated community where fees apply. These fees may be based on inventories of air emissions, including fume hood emissions. In these cases, determining the amount owed will involve cooperation and compromise between the regulators and institutions.

RESEARCH NEEDS

As the public and regulatory interest in the issue of toxic air emissions grow and as awareness of indoor air quality issues increase, the need for information about laboratory air emissions will increase. This section identifies two of these needs.

Alternatives for Common Laboratory Practices

The most important information need is for a critical examination of laboratory practices and the development of alternatives to those practices that create air contamination. The most famous example of such an alternative is microscale chemistry, whose introduction into educational laboratories was spurred by the desire to avoid the expense of rebuilding a laboratory building's ventilation system. Even without going to a completely microscale environment, institutions have been able to eliminate the odor commonly associated with chemistry laboratory buildings by raising awareness of laboratory occupants and preparing experiments with an eye to controlling the chemicals from the original container to final disposal.

Measurement of Emissions

For risk assessment and regulatory purposes, reliable measurements of laboratory emissions are necessary. These measurements will have to cover the many different forms laboratory emissions take. They should be based on physical measurements of both worst case and typical laboratory situations. Questions about what, when, and how much is emitted from fume hood exhausts have both occupational and environmental policy implications. Without an established, reliable protocol for monitoring emissions, it will be difficult to fulfill regulatory requirements. Such a monitoring protocol could become an expensive effort that significantly impacts laboratory operations if it is imposed from outside by government agencies. Research into minimizing both the cost and effect on labwork is a critical concern if such sampling protocols are to be successful.

CONCLUSIONS

While it has been traditionally accepted that dilution of laboratory exhausts will control the problem of airborne contaminants from labs satisfactorily, this idea has been recently challenged. This is a result of both physical and social factors. The increasing density of laboratory buildings and their comingling with nonlaboratory facilities increases the likelihood that laboratory emissions will create a physical or health problem. Regulators have begun the process of assessing the risks of the emissions of all chemicals to the general environment. While the federal government has been reluctant to begin this process with laboratories, state and local governments are being more aggressive. In addition, neighbors of laboratories, both within the laboratory building itself and next door to the labs, are concerned of their effect on air quality, sometimes with specific reason.

Based on the limited work done so far, it is unlikely that laboratory emissions will be found to present a chronic health risk. However, controlling emissions is an important part of a laboratory safety program. In addition, laboratory management will need to assess and control emissions of chemicals to resolve specific incidents of environmental contamination and to avoid involvement in controversy. Laboratory management should be proactive in this issue for two reasons: (1) it ties in with the requirements of a laboratory safety program and (2) establishment of emissions management priorities by outside agencies is likely to be both unnecessarily expensive and obstructive to laboratory work.

CHAPTER 9

Management of Laboratory Effluents to the Sanitary Sewer

Lloyd Wundrock and Jeff Christensen

Editor's Note: Many laboratory workers dispose of liquid wastes down the drain. When disposed in accordance with the requirements of the local sewer authority, this disposal method may not only reduce disposal costs, but also be the most desirable disposal method. However, laboratory workers should be aware that discharge limits vary widely by locale. Further, currently acceptable discharges may not be acceptable under future regulatory changes. The following article describes the experience of the University of Arizona, which must contend with some of the most stringent discharge limits in the country.—*PCA*

CONTENTS

Introduction 142
Regulatory Background 142
Regulations Applicable to the University of Arizona 143
Implementation and Operation of a Wastewater Management Program 148
    Identification of Industrial Sources of Wastewater 148
    Training for Compliance 148
    Sampling/Analytical 149
    Pitfalls and Potential Solutions 150
    Spill Protection 151
Conclusions 151
Endnotes 152

INTRODUCTION

For years pouring chemicals down the drain was considered an effective method of disposal. This attitude was best described in an 1899 report by the New Orleans Board of Health:

> To dump the garbage of a large city into a running stream from which is also derived the water supply of the city, might seem, at first glance, a rather crude and imperfect, as well as unsanitary method of getting rid of the city's waste; but when it is remembered that the Mississippi River is at this point a half mile wide, from 50 to 100 feet deep, with an average current of 3 miles per hour, as much as 1.5 million cubic feet of water passing a given point during every second at the stage of high water, we readily imagine how little influence a boat-load or two of garbage per day can have upon such an immense body of water constantly in motion.[1]

However, as time passed, public concern about trash littered beaches, bacterial contamination of drinking water, and increased knowledge of eutrophication highlighted the need to regulate the discharge of pollutants to surface waters.

This chapter discusses how the University of Arizona has been affected by wastewater discharge restrictions. In the past a significant portion of laboratory-generated wastes were disposed of via the sanitary sewer. Due to the Resource Conservation and Recovery Act (RCRA) Amendments of 1980 and Pima County Industrial Wastewater Ordinance (PCIWO) of 1984, materials suitable for disposal to the sewer have been severely restricted. All UA wastewater discharges are to the publicly owned treatment works (POTW) as there is no storm water sewer system. We offer some suggestions on how to comply with the discharge limitations, monitoring, and reporting requirements.

REGULATORY BACKGROUND

The Federal Water Pollution Control Act (FWPCA) of 1948 was the first attempt to regulate water quality. One of the FWPCA's amendments, the Water Quality Act of 1965, established the first parameters for pollutant concentrations for dischargers. The states were to establish a discharge-permitting system but there were difficulties in establishing individual discharge parameters and enforcement. Events, such as the Cuyahoga River fire, showed that a better form of regulation of discharge was needed.

In 1972 the Federal Water Pollution Control Act, also known as the Clean Water Act (CWA), was amended to update the existing regulatory scheme for preventing degradation of the surface waters of the U.S. by uncontrolled discharge of pollutants. Surface waters include, but are not limited to, rivers, streams, lakes, oceans, and wetlands. The ultimate goal of the CWA is to eliminate the discharge of toxic and nontoxic pollutants into surface waters. The interim goal is to make U.S. waters fit for fishing and swimming. The CWA focuses on two types of control to achieve these goals. The first type are water quality-based regulations. These regulations stipulate the permissible amounts of pollutants allowed in a specific body of water. Water quality-based regulations concentrate on the ability of the receiving water to absorb or dilute a given pollutant. The second type of water pollution control is by technology-based standards. Technology-based standards use pollution control technology to reduce a pollutant's ability to degrade the quality of the receiving water.

The mechanism for implementation and enforcement of the CWA is the National Pollutant Discharge Elimination System (NPDES). As defined in 40 CFR 122.2, NPDES is "the national program for issuing, modifying, revoking and reissuing, terminating, monitoring and enforcing permits, and imposing and enforcing pretreatment requirements...." To discharge any pollutant to the surface waters of the U.S., the discharger must first obtain an NPDES permit from EPA or a state that has been granted authorization by EPA to administer the NPDES program.

One kind of discharger is the POTW. These wastewater treatment facilities have received the necessary NPDES permit for discharge of their treated wastewater to a local surface water body. To comply with the parameters of the NPDES permit, the governing body of the POTW in turn establishes parameters for those entities that will discharge wastewater to the POTW. In essence, the POTW issues a "mini-NPDES permit." In some cases, a discharger is issued a permit by the POTW. In other cases, a discharger abides by certain conditions and is assumed to have a "permit by rule." In most cases, dischargers are required to adhere to the POTW's ordinances or rules. The holder of the industrial wastewater discharge permit must comply with the discharge limitations, monitoring, and reporting requirements specified in the permit.

A small sampling of universities in the western U.S. showed that the industrial wastewater discharge from their facilities has limitations on the levels of pollutants. As seen in Table 1 the levels of pollutants allowed varies widely. This difference is based on the quality of the receiving water and the unique water pollution problems for the respective area.

The implications of the discharge limits affect all levels of decision making. A recent occurrence in Tucson, located in Pima County, Arizona, illustrates this point. In an attempt to reduce the discoloration of drinking water due to old pipes, Tucson Water, operated by the City of Tucson, proposed to introduce a zinc-bearing anticorrosion chemical into the water system. Before this could be carried out, the Pima County Wastewater Management Department warned the City that such action would lead to violating the zinc levels of the NPDES permit issued to the County by EPA.

REGULATIONS APPLICABLE TO THE UNIVERSITY OF ARIZONA

The following information applies specifically to Pima County, Arizona, where the University of Arizona and Tucson are located. It must be understood

Table 1 Industrial Wastewater Discharge Limit Comparison Western U.S.

Total	U. of Arizona Tucson, AZ	Stanford Palo Alto, CA	Arizona State U. Tempe, AZ	U. of Utah Salt Lake City
Arsenic	0.4	0.1	0.1	NL
Barium	10.0	5.0	NL	NL
Boron	5.0	1.0	10.0	NL
Cadmium	0.1	0.1	0.047	0.11
Chromium	1.2	2.0	1.4	2.77
Copper	1.2	2.0	10.0	3.38
Cyanide	0.6	1.0	2.0	1.2
Lead	0.5	0.5	0.5	0.69
Mercury	0.05	0.05	0.05	NL
Nickel	3.98	1.0	5.0	3.98
Selenium	0.5	2.0	0.1	NL
Silver	5.0	0.25	0.5	1.43
Sulfides	2.0	NL	10.0	NL
Zinc	2.6	2.0	5.4	2.61

Note: NL — denotes not listed or no local limit at present; all limits are in units of milligrams per liter.

that the provisions of other local wastewater discharge ordinances may differ significantly from that of Pima County. The management and control strategies outlined in this document have been effective at the University of Arizona and are only intended as a guideline for other groups or institutions dealing with wastewater compliance issues.

The University of Arizona (UA) has been regulated by the Pima County Industrial Wastewater Ordinance (PCIWO) since its adoption in 1984. This ordinance applies to all industrial users, which includes the UA, discharging wastewater to the Pima County POTW. The term "industrial" is used to describe all wastewater generation other than from domestic sources. The purpose of this ordinance is to provide for the protection of Pima County's sanitary sewer system and the treatment processes utilized, groundwater resources, effluent and wastewater sludge disposal methods, and operating personnel through adequate regulation of industrial wastewater discharges.

Because the dry Santa Cruz river bed is the discharge point for two POTW treatment plants the resultant waterway is considered effluent dominated. This means the treatment plants must meet certain effluent standards based upon their NPDES permit. To assist in meeting these standards the PCIWO limits many industrial wastewater generator contaminant levels. Private households in Pima County are exempt from the ordinance.

The UA currently manages and maintains 11 separate industrial wastewater discharge permits and monitors wastewater at 18 specific discharge locations. Each permit is an individual control mechanism, authorization, or contract issued by the Director of the Pima County Wastewater Management Department which allows the discharge to the POTW of industrial wastewater.

The PCIWO was last amended on December 10, 1991, and included significant reduction in discharge limitations. Many portions of the ordinance can cause compliance difficulty for a laboratory facility or industry. In particular, some definitions in the ordinance, such as "Significant Industrial User,"

"Prohibited Wastes," and "Dilution" are open to wide interpretation. These terms and provisions are commonly found in the ordinances of all POTWs. However, note that the ordinances of other POTWs may have significantly different provisions.

Pima County may issue one of two types of discharge permits:

1. Categorical Discharge Permit. This permit is issued to facilities that fall under pollutant discharge limits promulgated by the EPA in accordance with Section 307(b) and (c) of the Clean Water Act, which applies to specific categories of industrial users [40 CFR 403.6 and Parts 405 to 471]. In other words, the very nature of the industrial user's business will dictate the type of permit issued. Activities that may require a categorical permit can range from meat packing to metallic plating. The UA presently holds one categorical discharge permit at the Electrical and Computer Engineering Building for small-scale wafer processing.
2. Significant Industrial User (SIU) Permit. This permit will apply to:
   a. Any industrial user subject to categorical pretreatment standards.
   b. Any industrial user that discharges an average of 25,000 gal/day or more of process wastewaters.
   c. Any industrial user that is designated as such by the control authority (director) on the basis that the industrial user has a *potential* for adversely affecting the POTW's operation or for violating any pretreatment standard or requirement.

2.c above is called the Director's Prerogative and is subject to wide interpretation and application. Based upon frequent visits by county inspection teams, it has been Pima County's assumption that any material of a chemical nature can find its way to the sewer. Therefore, if chemicals are stored and used in a laboratory setting it can be assumed that the permit issued will be considered that for a Significant Industrial User regardless of flow volume. Generally, all SIUs will require a discharge permit.

Prohibited Wastes (at point of entry to the POTW) include industrial wastewater that may be adverse or harmful to the POTW, POTW personnel, POTW equipment, or POTW effluent quality, including but not limited to:

1. Any wastestream with a closed cup flash point of less than 140°F (ignitability test).
2. Any wastes containing a concentration in excess of the discharge limitations for the following parameters: arsenic, barium, boron, cadmium, chlorine, chromium, copper, cyanide, lead, manganese, mercury, oil and grease, nickel, phenol species, selenium, silver, sulfides, or zinc.
3. Any waste having a pH lower than 6.0 or greater than 9.0 standard units.
4. Any excessive quantities of radioactive material wastes (no quantification).
5. Dilution is defined as the prohibition of industrial wastewater that *may* cause dilution or POTW hydraulic loading problems, including, but not limited to:
   a. Any water added for the purpose of diluting wastes that would otherwise exceed maximum concentration limits.
   b. Any rainwater, stormwater runoff, groundwater, street drainage, or roof drainage.

c. Any blowdown or bleed water from heating, ventilating, air conditioning, or other evaporative systems exceeding one third of the makeup water in a 24-h period.
d. Any single-pass cooling or heating water.

POTW hydraulic loading is the condition where tap water or other water that normally would not require wastewater treatment is introduced into the POTW. This water could be generated by any of the four conditions listed above. The commingling of domestic wastewaters with those considered industrial due to building or sewer design is not prohibited. However, this commingling does have an effect on the discharge permit parameters.

Tables 2 and 3 outline the discharge limits for selected pollutants under the Pima County Industrial Wastewater Ordinance.

The limits listed in the tables may not seem too difficult to comply with, but in many cases across the country, universities own and operate buildings that are decades old. In these buildings there can be any combination of mixed floorspace allocation. Lecture halls, teaching laboratories, and research lab space may all be located within the same structure. In addition, many buildings tie into the sewer

Table 2 Maximum Allowable Daily Discharge Limits (mg/l, based upon composite sample)

Substance	Limit
Arsenic — total	0.4
Barium — total	10.0
Boron — total	5.0
Cadmium — total	0.10
Chromium — total	1.20
Copper — total	1.2
Lead — total	0.5
Manganese — total	83.0
Mercury — total	0.05
Nickel — total	3.98
Silver — total	5.0
Zinc — total	2.6
Phenols — total	0.05
Cyanide — total	0.6
Selenium — total	0.5
Oil and grease	200.0[a]
Sulfide — total	2.0
Sulfide — dissolved	0.5[a]
Chlorine — total	10.0[a]

Note: A composite sample is a combination of no fewer than eight individual samples obtained at equal time intervals for 24 h for the duration of the discharge, whichever is shorter. In the case of a batch discharge with a flow duration of less than 15 min a single grab sample will meet the intent of a composite sample.

[a] Based on grab sample. A grab sample is any individual sample collected over a period of time not to exceed 15 min.

Table 3 Discharge Limits Based Upon Fume Toxicity (mg/l, Based Upon Grab Sample)

Compound	Limit
Acrylonitrile	1.24
Benzene	0.13
Bromomethane	0.002
Carbon disulfide	0.06
Carbon tetrachloride	0.03
Chlorobenzene	2.35
Chloroethane	0.42
Chloroform	0.42
Chloromethane	0.007
1,2-Dichlorobenzene	3.74
1,4-Dichlorobenzene	3.54
Dichlorodifluoromethane	0.04
1,1-Dichloroethane	4.58
trans-1,2-Dichloroethylene	0.28
1,2-Dichloropropene	3.65
1,3-Dichloropropene	0.09
Ethyl benzene	1.59
Ethylene dichloride	1.05
Heptachlor	0.003
Hexachloro-1,3-butadiene	0.0002
Hexachloroethane	0.96
Methyl ethyl ketone	249.00
Methylene chloride	4.15
Tetrachloroethylene	0.53
Toluene	1.35
1,2,4-Trichlorobenzene	0.43
1,1,1-Trichloroethane	1.55
Trichloroethylene	0.71
Trichlorofluoromethane	1.22
Vinyl chloride	0.003
Vinylidene chloride	0.003
Aroclor 1242	0.01
Aroclor 1254	0.005

line before the designated wastewater sampling/monitoring location. In situations such as this, the Combined Wastewater Formula applies. The formula is calculated when industrial and domestic discharges are combined before a permitted location; the individual discharge limits are then *decreased* by the mathematical percentage of the domestic or nonindustrial portion of that wastestream. For example, the ordinance limit for total lead is 0.5 mg/l. If the monitored wastestream is 20% industrial and 80% domestic in nature, then the final limit for lead in the combined wastewater is *0.1 mg/l*. Sometimes this situation forces a normally compliant wastestream into an ongoing status of violation. It must be understood that other POTWs may not be this strict.

Finally, the PCIWO provides for the protection of the POTW from unusual or hazardous discharges. This portion of the ordinance reads: "All USERS shall provide protection from the accidental discharge or spill into the POTW of prohibited, hazardous or other waste materials that are regulated through this Ordinance. Such protection shall be provided and maintained at the USER's expense. No USER shall commence discharge to the POTW without accidental discharge protection facilities or procedures."[2] Usually if specific spill protection measures, such as those listed below, are not in place in individual laboratories this will be noted in every site inspection report conducted by wastewater inspectors.

IMPLEMENTATION AND OPERATION OF A WASTEWATER MANAGEMENT PROGRAM

Any wastewater management program will require several key components to function independently and as part of the total plan. Some of these components and suggestions for application are included below.

IDENTIFICATION OF INDUSTRIAL SOURCES OF WASTEWATER

The identification of potential industrial sources is the first step necessary to set up further management strategies. This must be done prior to formal application for a discharge permit. Industrial sources can be itemized by individual laboratory, process wastestream, or in many cases by building in the university setting. Those buildings that generate only domestic wastewaters should be removed from consideration as far as effluent management is concerned. Current and accurate "as-built" drawings of your facility and associated wastewater or sewer systems are essential. Potential sources need to be identified and should be "grouped" to determine the most effective monitoring/sampling locations.

TRAINING FOR COMPLIANCE

Adequately training the generators of laboratory effluents is the most effective means of achieving discharge compliance. All the monitoring, sampling, and analytical work that the group or institution can afford will do nothing toward meeting regulatory limits; these activities can only prove compliance at best. Those individuals who may improperly pour something down the drain must be notified of the local regulations.

The UA experienced great difficulty in changing the mind set of laboratory personnel who assumed that regulated constituents in solution were in "trace" quantity. In our experience, "trace contamination" we frequently find that is above any regulatory limit set in the single-digit ppm range. Researchers, principal investigators, lab managers, and employees must know the level of regulated contamination of any waste that will be sewer disposed. If not, it should be assumed that the waste is not a candidate for sewer disposal and that material should be turned over to the hazardous waste program for proper handling. Laboratory personnel should never flush spilled material into the laboratory floor drain.

One segment of UA's "mandatory" laboratory safety training presents the information necessary to determine if a material can be discharged to the sewer. In addition, the following warning is posted on all sinks in laboratory areas:

DO NOT POUR THESE CHEMICALS DOWN THE DRAIN IN ANY QUANTITY

1. HALOGENATED SOLVENTS — Chloroform, Dichloromethane, TCE, PCE, Carbon Tetrachloride
2. FLAMMABLE SOLVENTS — Benzene, Toluene, Xylene, Hexane, Acetone
3. PHENOLIC COMPOUNDS — Phenol, Trichlorophenol, Pentachlorophenol, Hydroquinone
4. CORROSIVE MATERIALS — of pH less than 6 or greater than 9
5. AQUEOUS SOLUTIONS — of Arsenic, Barium, Boron, Cadmium, Chromium, Copper, Lead, Mercury, Nickel, Selenium, Silver, Zinc
6. CYANIDE AND SULFIDE SOLUTIONS

   If you have any questions about sink disposal of any material, contact Risk Management (1-1790) for assistance. Violations of the Pima County Wastewater Ordinance may result in building closure and interruption of laboratory activities.

SAMPLING/ANALYTICAL

You should attempt to monitor as many regulated discharges as possible from one location. This will reduce the general personnel requirement and hold down analytical costs. Further, verify that the proposed monitoring location will be of the proper configuration to meet the surveillance requirements. Use of dye testing may be required at this stage to confirm the flow assumption based upon "as builts." Consideration must be given to grab and composite sampling capability, pH recording, and flow monitoring.

The ideal sampling location is a control manhole. This installation, a retrofit of an existing sanitary sewer system, can be located to limit the traffic problems encountered by sampling personnel at most manholes. A control manhole is usually constructed solely for wastewater monitoring access to an existing sewer line. The design should incorporate a flat or bench area to allow placement of the sampling equipment and locked access to guard against theft. Sewer access should be of "open channel" design to allow for composite and grab sampling capability. This configuration will also accommodate the insertion of a "mouse and ring" flow monitoring device and pH recorder if required. Integral steps or a ladder must be installed if the depth of the manhole warrants it. Be advised that many, if not all, entries into sewer manholes will fall under OSHA confined space entry regulation.

If in-house staff will be conducting the sampling, they must be trained. Manhole location, sampling frequency, sample type, composite sampler operation, container preservation, traffic control, and personal protective equipment are just a few of the things that must be considered. Sometimes an outside vendor can provide these services. While this option may appear more expensive, it can free valuable resources to meet other pressing needs. It is vital to spend the time and effort to choose the highest quality analytical laboratory that will not overrun your budget. Audit the laboratory to see if analytical work is done following EPA-outlined analytical methods. Erroneous analyses or lengthy turnaround times for methods with relatively short holding life increases frustration and stress levels.

PITFALLS AND POTENTIAL SOLUTIONS

The segregation of domestic and industrial discharge plumbing in new construction is very important for facilities that will potentially be permitted and require regular monitoring. Domestic sources include bathroom facilities and handwashing sinks. Industrial fixtures will include janitorial sinks, lab sinks, and laboratory floor drains. This will allow for the highest individual discharge parameter limits because the Combined Wastewater Formula will not apply. All necessary action should be taken to assure only industrial discharges are monitored. This segregation also simplifies composite sampling activities due to the absence of domestic wastewater solids. Solid materials clog and restrict a composite sampler which leads to sampler failure.

Neutralization tanks are another source of problems. These units are usually constructed of clay, brick, iron, steel, or plastic and are a wastewater reception vessel. Historically, these vessels have been charged with limestone chips and are expected to neutralize acidic laboratory discharges. UA experience has been that these installations only function under very low flow conditions and for small quantities of acid. Many were installed improperly and allowed inlet wastewater to flow over the rock bed and be discharged without sufficient limestone contact time. It is critical to note that these installations cannot guarantee compliance with strict pH discharge limitations. These neutralization tanks also turn into collection basins for solvents and heavy metals.

A specific case at UA illustrates the failure of neutralization tanks. Over several years laboratory occupants of a chemistry building discharged chlorinated and aromatic hydrocarbons into laboratory sinks. These materials collected over time in a large neutralization tank and then bled off slowly into the designated sampling location and ultimately the POTW. These contaminants were detected through wastewater monitoring and caused the closure of eight laboratory buildings on campus for as long as 6 weeks. Extensive sampling and research was undertaken to discover the source of contamination. Upon determination that the neutralization tank was the source of contamination, the tank was decontaminated by an environmental response contractor at a cost in excess of $2000. Due to this incident, the UA has aggressively deactivated or bypassed many of these units. These units are now omitted from all new UA construction.

Many laboratories use a vacuum trap apparatus where the vacuum is generated by attaching a t-hose to a laboratory water line. This system has the potential to introduce excessive quantities of prohibited organic solvents to the POTW. Roto-vap systems with a primary liquid nitrogen trap have been found to "leak" organics into the laboratory sinks. For these reasons both system types have been banned from laboratory use at the UA.

SPILL PROTECTION

The need for spill protection applies to all areas where materials potentially damaging to the POTW or materials prohibited by local ordinance are stored without secondary containment. These areas include but are not limited to teaching and research laboratories, cooling/power plants, paint shops, mechanical shops, and vehicle maintenance yards.

In the laboratory environment the required spill protection can be provided in many ways. First, liquids and metal salts should never be stored on a shelf above a sink. If possible, liquids should be located below the lab bench or counter level. If this is not feasible due to frequency of use of small liquid containers, secondary containment must be provided. This is easily accomplished by using polyethylene wash basins for containment.

Floor drains should be eliminated from new construction. In many cases floor drains have been installed to support emergency shower/eyewash units. To prevent discharge to the sewer from larger chemical containers, existing floor drains must also be protected. This can be done in three ways: (1) installation of a "guarded" floor drain. This fixture incorporates a short lip that will not allow chemical spills to enter the drain, offers some level of containment, and allows for proper spill cleanup. This drain will function to remove excessive quantities of water should a water pipe leak or fire sprinkler system be activated. (2) A friction-fitted "test plug" may be inserted. This device is inserted into a floor drain and manually compressed to form a watertight seal. This option is cheaper but offers less flexibility. The plug prevents release of a catastrophic spill to the POTW but must be removed manually to discharge flood accumulation. (3) The floor drain may be permanently sealed with concrete.

Finally, fume hood cup sinks should be guarded or closed off. This is accomplished by sealing a cover over the cup sink if the sink is not needed. If the cup sink must remain in service, a lip of gasket material can be glued around the perimeter of the sink to protect it from chemical spills in the fume hood. A perimeter guard ring of this type can also be installed around larger laboratory sinks to protect them as well.

CONCLUSIONS

Limitations on pollutant levels in discharged wastewater forces universities to evaluate all materials that have the potential to be disposed of via the sewer, either intentionally, by mistake or by accidental release. Materials that do not meet the criteria for hazardous waste may be prohibited from sewer disposal due to pH limitations or metal concentrations. The hazardous waste management team must then make a decision as to how this type of material will be disposed of. Further evaluation may reveal that some custodial products must be banned and replaced with a product that will not violate wastewater discharge permit limits.

Wastewater discharge monitoring is very time and resource consuming, but these efforts are well spent when compared with the cost of fines and other enforcement actions. University personnel must continually be educated about what can and what cannot go down the drain. The number of permits and their complexity can be reduced if critical discharge information is determined and verified prior to permit application and issuance.

Building a working relationship with the wastewater regulators is vital to the success of a monitoring program. Any problems concerning monitoring frequency or sampling parameters can be solved through professional discussion rather than from an adversarial confrontation.

ENDNOTES

1. Cited in Melosi, M. V., *Garbage in the Cities: Refuse, Reform, and the Environment, 1888–1980,* Chicago, The Dorsey Press, 1981.
2. Industrial Wastewater Ordinance, Pima County, Arizona Wastewater Management Department, December 12, 1991.

CHAPTER 10

Pollution Prevention in Clinical Laboratories

Richard J. Vetter, John F. O'Brien, and Gregory D. Smith

CONTENTS

Introduction 153
Waste Volume Reduction through Treatment 154
    Chemical Treatment 154
        Neutralization 155
        Precipitation and Evaporation 155
        Fuel Blending 155
    Storage for Radioactive Decay 156
        Decay of RIA Waste 156
        Elimination of RIA Tests 156
        Management of Radioactive Microspheres 157
Waste Volume Reduction through Minimization and Source Reduction 157
    Updating the Hazardous Waste Management Process 157
    Hazardous Waste Minimization and Source Reduction Task Force 159
        Actions to Address Specific Hazardous Waste 159
        Development of a Long-Range (Cradle to Grave)
        Approach to Tracking Hazardous Waste 160
Conclusion 161
Endnotes 162

INTRODUCTION

Clinical laboratories generate hazardous chemicals, infectious materials, and radioactive waste as well as regular trash. The increased rigor of disposal standards requires clinical laboratories to invest in well-designed facilities and careful procedures to handle and process wastes. Likewise, increased environmental standards have resulted in the development of high-integrity disposal sites. The U.S. Department of Transportation also has increased its requirements in an attempt to assure the integrity of hazardous waste shipments. These changes have resulted in significant increases in waste disposal costs. In response to increased standards and costs, clinical laboratories must take steps to control the generation of hazardous and low-level radioactive wastes through source reduction and waste minimization.[1]

While the ultimate goal of every clinical laboratory is to significantly reduce the production of these wastes, the multitude of processes makes it impossible to completely eliminate their production at this time. Therefore, it behooves clinical laboratories to set small, achievable goals targeted to specific processes. The purpose of this chapter is to describe efforts of one clinical laboratory enterprise to understand the entire laboratory process that results in the production of hazardous chemical and low-level radioactive wastes to minimize the volume of some wastes and eliminate the generation of others.

Since their beginnings, Mayo laboratories have used chemicals. Their use results in the generation of nearly 22,000 lbs and 86,000 gal of chemical waste and the shipment of 22,000 lbs of hazardous waste each year. With the advent of the radioactive tracer technique, clinical and research laboratories at Mayo produce low-level radioactive waste in the form of dry solids, aqueous waste, liquid scintillation fluids, biological tissues, or mixed radioactive hazardous waste. Cost of disposal and regulatory and public pressures to reduce production of these wastes have increased considerably in the last 2 decades, and both have resulted in efforts to reduce waste volumes.

Obviously, energies to reduce the volume of waste disposed off-site can be directed toward reduction of the volume of waste generated (waste minimization). Alternatively, efforts can be directed toward elimination of the use of products that become hazardous wastes (source reduction). Several treatments have been incorporated into the waste disposal program at Mayo to reduce volume of waste shipped. These processes have contributed to the goal of reducing the cost of disposal, but they have no impact on the production of waste. More recently, efforts have been made to minimize the volume of waste produced or eliminate their generation. Waste treatment and minimization and source reduction both contribute to pollution prevention and reduce the costs associated with disposal of wastes. Both will be described in the context of reducing hazardous chemical wastes and low-level radioactive wastes produced by laboratories at Mayo.

WASTE VOLUME REDUCTION THROUGH TREATMENT

CHEMICAL TREATMENT

Hazardous chemical wastes are collected from each laboratory and transported to a chemical treatment laboratory. Acids and bases are neutralized and disposed in the laboratory sink. Organic liquids are collected in safety cans and transferred to a permitted waste-holding facility on site and pooled into 55-gal drums to await transportation to a hazardous waste processor. The volume of hazardous nonhalogenated organic solvents is minimized by burning the nonhalogenated organic liquids in an approved manner in the Mayo power plant boiler.

Neutralization

Acids and bases in aqueous solution are collected in plastic carboys and transported on laboratory carts to the treatment laboratory. Hazardous waste technicians check the pH and slowly mix waste acids and bases to a nominal pH of 7 (range of 6 to 8). When there is a shortage of waste bases, sodium sesquicarbonate powder is slowly dissolved in the surplus waste acid. The neutralized solution is then emptied into the sewer through a laboratory sink and flushed with several equal volumes of water.

Precipitation and Evaporation

Silver nitrate waste solutions are treated by addition of hydrochloric acid. The supernatant is sewered through the laboratory sink and the silver chloride precipitate is air dried in a chemical fume hood and placed in plastic containers for disposal.

Fuel Blending

Nonhalogenated organic solvents such as xylene, alcohols, and acetone have a high BTU content and burn well when injected into an operating boiler. These wastes are poured into a 400-gal storage tank located within the laboratory building, which is connected to a boiler in the campus power plant as permitted by the U.S. Environmental Protection Agency (EPA) under Boiler-Industrial-Furnace rules.[2] Approximately 15,000 gal of this waste are burned annually as a source of fuel for the boiler. In addition to reducing environmental liability associated with disposal of these solvents, estimated annual cost avoidance exceeds $50,000. Solvent from the storage tank is sampled semiannually and analyzed by a third-party laboratory to assure no halogenated solvents or mercury or other heavy metals are disposed of by this route.

Off-site, commercial fuel blending (cement kiln) is used for disposal of all other organic solvents including paint and paint thinners and outdated pharmaceuticals. This method of disposal is considered a better environmental alternative to disposal in a permitted hazardous waste incinerator, since it provides a beneficial reuse of the waste in the form of fuel. In addition, fuel-blended wastes are exempt from Minnesota taxes, which avoids a cost of approximately $2000 per year. Fuel-blending facilities are also more readily accessible than hazardous waste incinerators which generally have a waiting list.

STORAGE FOR RADIOACTIVE DECAY

Decay of RIA Waste

Radioimmunoassays (RIA) are used by many laboratories to conduct a number of clinical tests. Though most RIA procedures are being replaced, those for smaller molecules which use the competitive binding principle remain popular. While radionuclides used in clinical laboratories include ^{3}H, ^{14}C, ^{57}Co, ^{125}I, and ^{131}I, the radionuclide most commonly used in RIA is ^{125}I. The half-life of ^{125}I is 60 days; therefore, it is feasible to store the waste for radioactive decay. Regulations and license conditions permit decay in storage of radioactive waste for ten half-lives or longer for radionuclides with half-lives of 65 days or less in order that the remaining activity is not measurable above background, so that it can be disposed of without regard to its original radioactivity.[3] RIA wastes not only contain radioactivity, but they might also be potentially infectious since they are often mixed with body fluid samples for assaying. When this is the case, many laboratories choose to incinerate the mixture after its radioactivity has declined to zero. Laboratories that possess insufficient space to store this waste for decay must transfer it to a licensed radioactive waste facility for storage or disposal, a procedure that is becoming increasingly expensive.

In 1983 the Mayo clinical laboratories and Mayo Radiation Safety Office conducted a study to determine the feasibility of storing ^{125}I waste for ten half-lives with subsequent incineration of the waste. The volume for several RIA procedures was high, e.g., over 100,000 thyroxine tests per year with 2 or 3 tubes per test, which generated a large volume of low-level radioactive waste. An institutional incinerator was in existence, and it was determined that the ^{125}I assay tubes and other dry, radioactive waste could be added to the wastestream which included office paper and similar materials from nonradioactive assays. Results of the study showed that a steel storage building could be constructed near the incinerator and that the cost of construction would be offset in 15 months by reduced low-level radioactive waste disposal fees. Since then commercial disposal fees have increased from $16 to approximately $350 per cubic foot which includes a federal surcharge for development of new disposal sites. After 5 years of successfully storing clinical laboratory low-level radioactive waste for decay, the program was expanded to include research laboratories. It is estimated that the current annual cost for commercial disposal of these wastes would be approximately $2 million.

Elimination of RIA Tests

In addition to reduction in volume of low-level radioactive waste shipped to disposal sites, the feasibility study discussed above recommended that clinical laboratories consider alternatives to radioimmunoassay tests (RIA). Several laboratories have now developed enzyme fluorescence and chemoluminescence assays for thyroxine, thyroid-stimulating hormone, luteinizing hormone, follicle-

Table 1 Radionuclides Available as Labeled Microspheres

Nuclide	$T_{1/2}$ (days)	keV (%)
^{51}Cr	27.7	320 (9.8)
^{141}Ce	32.5	145 (48)
^{95}Nb	35	765 (100)
^{103}Ru	39	497 (88)
^{85}Sr	64	514 (100)
^{46}Sc	83.9	889 (100), 1120 (1000)
^{113}Sn	115	393 (64)
^{57}Co	270	122 (87), 136 (11)

stimulating hormone, and thyroglobulin. Wastes from these procedures can be disposed of directly in red bag waste as is usual for infectious materials. Development is underway to replace the use of ^{57}Co and ^{125}I in tests for ferritin, vitamin B12, folic acid, human chorionic gonadal protein, and digoxin. While RIA will continue to be important for clinical laboratories, acceptable alternatives to some assays exist for those who choose to avoid the issues surrounding disposal of low-level radioactive waste.

Management of Radioactive Microspheres

Radiolabeled microspheres are an excellent tool for measuring the effect of an agent on blood flow in tissues of experimental animals. Specifically, this technique is used to study blood flow and vascularization of soft tissues, tumors, and bone. To conserve the number of experimental animals used, the effects of various agents can be observed in a single animal when the investigator uses microspheres labeled with several different radionuclides whose radiation is distinguishable by gamma spectroscopy. For this to be practical in terms of radioactive waste disposal, by first storage and then incineration, one must choose tracers with half-lives less than 65 days. The selection of radionuclide is generally limited to those with half-lives between 27.7 and 270 days, but a sufficient variety exists to permit the use of several microspheres with half-lives less than 65 days (Table 1). At Mayo, investigators wishing to comply with the storage/incinerator/waste disposal method agreed to eliminate the use of ^{51}Cr and to limit, whenever feasible, their use of labeled microspheres with an extended half-life.

WASTE VOLUME REDUCTION THROUGH MINIMIZATION AND SOURCE REDUCTION

UPDATING THE HAZARDOUS WASTE MANAGEMENT PROCESS

In today's world minimization of an organization's adverse environmental impact is the right thing to do from both the ethical and the risk management points of view and is becoming required by a plethora of federal, state, and local

Table 2 Mercury-Containing Compounds

B-5 Mercuric chloride waste
Zenker's mercuric chloride waste
Mercuric bichloride/acetic acid waste
Mercuric chloride (unused chemical)
Mercuric oxide (unused chemical)
Mercuric bromide (unused chemical)
Hematoxylin stain waste
Thimerosal waste
Mercurochrome (merbromin)

regulations. The EPA defines hazardous waste as a solid waste (including liquids) that is not excluded from regulation, exhibits a characteristic of hazardous waste (ignitability, corrosivity, reactivity, toxicity), or is listed as a hazardous waste.[4] Minnesota adds the characteristic of lethality to the definition of hazardous waste. Examples include most organic solvents such as toluene, xylene, heptane, and hexane; most alcohols such as ethanol, methanol, and isopropanol; many laboratory chemicals and their solutions such as sodium azide, picric acid, and cyanide; many acids and bases; mercury and mercury-bearing compounds such as those in Table 2; heavy metals such as chromium and chromic-sulfuric acid; and in Minnesota most prescription drugs. While this discussion is directed toward hazardous waste generated in the laboratory, other activities that support the laboratory such as building and grounds maintenance also generate hazardous waste including paint and paint thinner.

Generators of hazardous waste are required to register with the state or federal EPA program and annually report the amount of hazardous waste generated. Most generators are required to have a program in place to reduce the volume and toxicity of waste to a degree that has been determined to be economically practicable. Compliance with this requirement often has been achieved through individual efforts of the hazardous waste manager.

Waste management by reduction at the source is often more difficult than reducing waste volumes once generated. It requires waste managers to convince technical directors that there is a need for change. The best-case scenario involves laboratorians who are environmentally motivated *de novo* and eagerly accept opportunity to reduce their impact on the environment. Second best is when the laboratory result is not affected by a change in method. The most difficult case is when the user is sufficiently more committed to his routine and the answers obtained then to the means by which they are obtained. This is particularly true when the laboratory must spend much effort to make the change. The evolution of the replacement of ^{131}I by ^{125}I on the basis of safety and more convenient half-life (60 days rather than 8) is now being followed by replacement of radionuclides altogether with fluorescence or chemiluminescent endpoints. Actually, manufacturers have led the way in these developments and have used the convenience of nonisotopic methods as a marketing tool. As a final resort for change where it has been difficult, specific assignments of waste removal costs to the specific laboratories (generators as opposed to the institution) may be an impetus for change.

HAZARDOUS WASTE MINIMIZATION AND SOURCE REDUCTION TASK FORCE

To initiate a more vigorous program at Mayo, the Safety Committee appointed a Hazardous Waste Minimization and Source Reduction Task Force chaired by a member of the clinical staff with representation from laboratory supervisory management, pharmacy management, research staff, purchasing, and safety. The Task Force was charged with identification of departments and laboratories that utilize hazardous chemicals that comprise the largest one half of hazardous wastes by volume; identification of chemicals, regardless of volume of hazardous waste produced, that could easily be replaced by other chemicals that would not be classified as hazardous waste; obtain suggestions from laboratory generators of hazardous wastes on how to further reduce or eliminate generation of these wastes; and examine the feasibility of source reduction through the dispensing of popular chemicals from a central chemical supply. The Task Force made specific recommendations on the need to update the hazardous waste management process, took specific actions, and made further recommendations to address specific sources of hazardous waste, and recommended a long-range approach to the tracking of hazardous (cradle to grave) waste. The third recommendation was to carry the process beyond a piecemeal approach to hazardous waste minimization and source reduction to a general approach of inventory management that would also provide the efficiency of electronic monitoring of wastes. The objective of this recommendation was to provide tracking of chemicals from order to disposal and to provide an inventory of potentially usable hazardous materials as a shared resource. To date, laboratories must call an inventory clerk to learn the location of a specific chemical, but the objective is to make an electronic inventory available to all laboratories.

The Task Force recommended a prospective, long-range view of hazardous waste management be developed at Mayo. This process should include discussions with laboratory directors and managers, encouragement to develop tests that utilize water-based rather than organic solvents or environmentally benign vs. catalytic toxic metals, development of an awareness of the continuing change in the perception of society and knowledge of the potential adverse effects of chemicals, optimization of collection and disposal procedures to minimize potential adverse effects, and establishment of a mechanism to control acquisition of chemicals and minimize the production of hazardous waste.

Actions to Address Specific Hazardous Waste

To address some potential short-term gains, the Task Force developed a "most wanted" list of chemicals and gathered data on the site and quantity of production. Specific laboratories were asked to consider substitutes for the chemicals on this list which consisted of compounds that contained mercury or chromium, two elements which are the most difficult to dispose of at this time. Laboratories that use thimerosal or merbromin were asked whether they were willing to look at alternative, less hazardous chemicals, whether they were willing to dispose of

Table 3 Contrad 70 Evaluation

	Emission Intensity			
	Carbon	Phosphorus	Copper	Calcium
Blank (uncoated tube)	12212	284	129	1608
H_2SO_4 dichromic	20339	388	140	3604
Unwashed	24113	523	181	5456
Contrad 70	12334	278	117	1780

Note: Average of three samples. The Contrad 70 was used as directed on the package.

present supplies and discontinue future use, or whether they were unable to change at this time. Likewise, laboratories (primarily cytology) that generated mercury-containing hematoxylin were informed that this material could no longer be disposed of in the U.S. As an initial response, laboratories separated non-mercury hematoxylin from mercury-containing hematoxylin, which reduced the hazardous waste that contained mercury by 85%. Thereafter, it was found that mercury was used in hematoxylin because of convention and it did not add significantly to slide quality. Laboratories responsible for generation of hematoxylin waste subsequently agreed to discontinue the use of mercuric oxide in the staining procedure. Laboratory methods references such as the Armed Forces Institute of Pathology *Laboratory Methods in Histotechnology*[5] and *Histopathologic Methods and Color Atlas of Special Stains and Tissue Artifacts*[6] provide formulations for various types of hematoxylin stains, some of which contain mercuric oxide. Many of these staining solutions are available commercially without mercury. We suggest that future editions of these and similar references encourage the use of non-mercury hematoxylin.

No consistent use of other mercury-containing compounds was observed. However, the current system for purchasing chemicals does not specifically forbid the purchase of these compounds. This provides further justification for adding an element of control to the purchase of these chemicals.

Laboratories that use sulfuric dichromate were identified and asked to evaluate their use of these chemicals, specifically as a glassware-cleaning agent. Preliminary results with a cleaner that contains no chromate are shown in Table 3. Since each laboratory must approve the cleaner used in their procedures, this information was shared with laboratories, and they each explored other alternatives that would meet their needs. All laboratories found a suitable product to replace their use of dichromic acid.

Development of a Long-Range (Cradle to Grave) Approach to Tracking Hazardous Waste

In a large laboratory with high throughput, it is difficult to develop a singular focus on policies to minimize waste or eliminate sources of waste. Even though current waste management efforts are supported by laboratories, heightened social and economic pressures, increasingly stringent disposal requirements, and environmental risk management strongly suggest the need for a "cradle to grave" waste management program. To attain this, the Task Force recommended an integration of purchasing and inventory management with environmental waste management. Discussions are currently ongoing and have developed an awareness among purchasing and laboratory staff.

It is the prevailing norm that hazardous materials management is a parallel but nonintegrated component of materials management. Purchasing, inventory, and ordering information is held separately from environmental data. This results in ineffective duplication of effort and places hazardous materials management in a reactive rather than proactive stance relative to control of chemical acquisitions. Hazardous waste minimization and source reduction efforts would be most effective when considered initially rather than as an afterthought. This would also provide a first-hand opportunity to encourage laboratories to seek nonhazardous materials in place of chemicals that result in the generation of hazardous waste. The Task Force specifically recommended that components of the current materials management system be integrated and expanded in scope to permit identification of hazardous materials at the time of purchase, recommendation of nonhazardous alternatives, and implementation of a bar code system for real time inventory of laboratory stock, which could be utilized to develop an "electronic warehouse" to encourage user exchange. While several components of the suggested system already exist and would simply need integration into the proposed program, additional software development is necessary to handle the burden of the large hazardous materials inventory and to permit networking of the component systems. A network that links computer databases will provide the opportunity for tighter management of chemicals, minimization of the production of hazardous wastes, and opportunities for source reduction. This proposal is currently under study at Mayo. One of the aims of the proposal is to develop an "electronic warehouse" without the introduction of additional bureaucratic levels but to provide a system that encourages laboratories to share resources, minimize the generation of hazardous wastes, eliminate sources of hazardous waste, and provide the hazardous waste manager with additional opportunities to monitor the process.

CONCLUSION

Clinical laboratories at Mayo have made significant progress in the source reduction (elimination) and minimization of hazardous and radioactive wastes. Key to the success of a laboratory pollution prevention program are communication of issues to laboratory personnel and active involvement of laboratory directors and managers in seeking and implementing alternatives to tests that utilize hazardous chemical and radioactive substances. Where alternatives do not exist, efforts to develop a real time (electronic) inventory of hazardous chemicals and integrate the purchasing process with waste disposal will significantly reduce volume of hazardous wastes, disposal costs, and environmental impact.

ENDNOTES

1. National Committee for Clinical Laboratory Standards, *Clinical Laboratory Waste Management; Approved Guideline,* NCCLS document GP5-A (ISBN 1-56238-218-7), NCCLS, 771 East Lancaster Avenue, Villanova, Pennsylvania 19085, 1993.
2. U.S. Environmental Protection Agency, *Hazardous Waste Burned in Boilers and Industrial Furnaces,* Title 40, CFR, Part 266, Subpart H. U.S. Government Printing Office, 1992.
3. U.S. Nuclear Regulatory Commission, *Waste Disposal,* Title 10, CFR Part 20, Subpart K, U.S. Government Printing Office, 1994.
4. U.S. Environmental Protection Agency, *Lists of Hazardous Wastes,* Title 40, CFR, Part 261, Subpart D, U.S. Government Printing Office, 1992.
5. Armed Forces Institute of Pathology, *Laboratory Methods in Histotechnology,* Armed Forces Institute of Pathology, Washington, D.C., 1992.
6. Luna, L. G., *Histopathologic Methods and Color Atlas of Special Stains and Tissue Artifacts,* American Histolabs, Inc., Gaitheresburg, MD, 1992.

CHAPTER 11

Minimization of Waste Generation in Medical Laboratories

Judith G. Gordon and Gerald A. Denys

CONTENTS

Introduction 164
    Generators of Medical Waste 164
    Types of Medical Waste 165
    Concept and Scope 166
Minimization of Waste Generation through Management Practices 166
    Strict Definition of Waste Types 166
    Source Separation of the Different Wastestreams 168
        Implementation of Source Separation 168
            Infectious Wastes 169
            Chemical Wastes 169
            Radioactive Wastes 170
            Multihazardous Wastes 170
            Wastewater 170
            Ordinary Waste Amenable to Reuse or Recycling 170
        Benefits of Source Separation 171
    Product Substitution 172
Minimization of Waste Generation through New Technology 173
    Microbiology 174
    Immunology/Serology 175
    Chemistry/Hematology 175
    Product Substitution 176
    Use of Nonradioactive Methods 176
    The Paperless Laboratory 177
Treatment of Wastes On-Site to Minimize Special Disposal
Requirements 177

        Infectious Wastes 177
            Treatment within the Laboratory 179
            Treatment within the Facility 179
        Chemical Wastes 183
        Radioactive Wastes 183
        Multihazardous Wastes 184
        Wastewater 185
    Reuse of Wastes 185
        Reuse of Disposable Items 185
        Solvents 186
            Redistillation and Reuse 186
            Incineration with Heat Recovery 186
        Zero Discharge through Recirculation of Treatment Effluent 186
        Packaging Materials 187
    Recycling of Wastes 187
        Paper 187
        Solvent Redistillation 187
        Packaging Materials 187
    Implementing a Waste Reduction Program 188
        Management Program 188
        Purchasing Strategies 189
        Employee Awareness Programs 189
        Employee Training 190
        Cost Savings: Using the Bottom Line to Obtain the Support
        of the Administration 190
    Endnotes 191

INTRODUCTION

GENERATORS OF MEDICAL WASTE

In the context of this chapter, the term *medical laboratory* refers to every laboratory that handles human or animal specimens or biological products related to the health care of people or animals. This term therefore includes clinical laboratories as well as research and industrial laboratories.

Clinical laboratories are usually situated in hospitals or other health care facilities, but some are free standing. A large variety of tests on specimens of tissue and body fluids from ill and also healthy individuals are performed in clinical laboratories. See Table 1 for a listing of laboratory specialties and types of testing that are performed in clinical laboratories.[1]

Laboratories that engage in biological and medical research and development are also generators of medical waste. These laboratories are usually situated at industrial facilities, academic institutions, and research institutions. In some of

Table 1 Types of Clinical Laboratories

Specialty	Subspecialty
Microbiology	Bacteriology
	Mycobacteriology
	Mycology
	Parasitology
	Virology
Diagnostic immunology	Syphilis serology
	General immunology
Chemistry	Routine chemistry
	Endocrinology
	Toxicology
Hematology	Routine hematology
	Flow cytology
Urinalysis	
Pathology	Cytology
	Histology
	Electron microscopy
Immunohematology	ABO and D(Rho) testing
	Compatibility testing
	Unexpected antibody detection
	Antibody identification
Reproduction and transplantation	Transplant immunology
	Experimental pathology
	Reproductive biology
Research and development	Animal testing

Adapted from Table 4 (Specialties and subspecialties for proficiency testing, as specified in the regulations for implementing the Clinical Laboratory Improvement Amendments of 1988), in Regulations for Implementing the Clinical Laboratory Improvement Amendments of 1988: A Summary, *Morbid. Mortal. Weekly Rep.*, 41(RR-2), 9, 1992.

these laboratories, basic research is performed whereas others are used for the research, development, and testing of biologicals and other products.

Some industrial laboratories are generators of medical waste. This occurs in laboratories at pharmaceutical and biotechnology companies, but medical wastes can also be generated at other industrial laboratories as well.

TYPES OF MEDICAL WASTE

Medical laboratories are unique in the types of wastes that are generated. Infectious wastes are always generated, but there may also be other wastestreams such as hazardous chemicals and radioactive materials. Multihazardous wastes (i.e., wastes that are infectious and radioactive, or infectious and hazardous, or infectious and radioactive and hazardous) are frequently generated because of the nature of the tests or experiments that are performed in medical laboratories.

Some of the wastes that are generated in medical laboratories can be managed in accordance with the commonly accepted practices that are typically used for such wastes (e.g., see, Chapters 15 and 20 on solvent recycling). However, many of the wastes generated in medical laboratories — especially the infectious and the multihazardous wastes — require unique management approaches for pollution prevention and waste minimization.

CONCEPT AND SCOPE

This chapter provides a comprehensive overview of all the types of waste that are generated in medical laboratories and of the various approaches that can be used to achieve the twin goals of waste minimization and pollution prevention. These approaches include:

- Reduction in the quantities of the different wastes generated through adoption of certain management practices and use of new methods and technology
- Treatment of the wastes on-site to reduce quantities of wastes requiring special off-site treatment and/or disposal
- Reuse of wastes
- Recycling of wastes

The specific approaches that one can adopt depend on the materials used, the types of testing or experiments done in the medical laboratory, the equipment used, the types of wastes generated, administration's concept of pollution prevention and waste minimization, and other factors that may be unique to each generator and each locality.

The various techniques for waste minimization and pollution prevention that are applicable to medical laboratories are discussed in detail in this chapter. Specific examples are given of techniques that can be used in each type of approach.

The chapter concludes with a section on implementing a waste reduction program. Suggestions are provided for strategies and considerations that are essential to implementing a program for achieving pollution prevention and waste minimization in medical laboratories.

MINIMIZATION OF WASTE GENERATION THROUGH MANAGEMENT PRACTICES

The generation of wastes in medical laboratories can be minimized through two general approaches — the adoption of appropriate management practices and the use of new laboratory methods and technology.

Effective management practices include:

- Formalizing a strict definition of waste types
- Implementing source separation of the different waste types
- Instituting a product substitution policy

STRICT DEFINITION OF WASTE TYPES

Clinical laboratory personnel have been aware of the potential danger and transmission of infectious agents from waste generated in their laboratories for years. Many laboratory-associated infections have been acquired from infectious aerosols and needlestick injuries. These risks to laboratory workers are evident by conversion on the tuberculin skin test for tuberculosis and on serological tests for hepatitis B virus. Of recent concern in the handling of infectious waste is the increased incidence of multiply resistant *Mycobacterium tuberculosis* and of HIV infections.[2]

Defining waste types as infectious is difficult because there is no precise way to assess the microbial content. Federal, state, and local regulatory definitions of infectious waste vary. Categories of infectious waste designated by the Centers for Disease Control (CDC), the U.S. Environmental Protection Agency (USEPA), and the Occupational Safety and Health Administration (OSHA) are listed in Table 2. The CDC defines four categories of *infective waste*.[3,4] They are microbiological waste (including cultures and stocks), blood and blood products, pathological waste, and contaminated sharps. In addition to these categories, the USEPA designated as *regulated medical waste* (for the purpose of medical waste tracking) also waste from patients isolated with a communicable disease, animal waste, and

Table 2 Categories of Infectious Waste

Category	CDC[3]	EPA[4]	OSHA[5]
Microbiological waste	Yes	Yes[a]	Maybe[b]
Pathology waste	Yes	Yes[c]	Maybe[b]
Blood and blood products	Yes	Yes[d]	Yes
Contaminated sharps	Yes	Yes[e]	Yes
Unused sharps	No	Yes[e]	No
Isolation waste	No	Yes	No
Cultures and stocks and associated biologicals	Yes	Yes[f]	Maybe[b]
Contaminated animal carcasses, body parts, and bedding	No	Yes	Maybe[g]
Wastes from surgery and autopsy	No	Maybe[h]	Maybe[b]
Contaminated laboratory wastes	No	Maybe[i]	Maybe[b]
Dialysis unit wastes	No	Maybe[j]	Maybe[b]

[a] Includes cultures and stocks and laboratory waste or equipment contaminated with infectious agents.

[b] If this waste contains blood or other potentially infectious materials.

[c] Includes body tissues, organs, body parts, and body fluids removed during surgery, autopsy, and biopsy.

[d] Includes all human blood, serum, plasma, blood products, items saturated or soaked with blood (e.g., bandages), blood containers, and intravenous bags.

[e] Includes hypodermic needles, syringes, scalpel blades, disposable pipettes, capillary tubes, microscope slides, cover slips, and broken glass.

[f] Includes cultures and the devices used to transfer, inoculate, and mix. Also includes waste from the production of pharmaceuticals.

[g] Blood, organs or other tissues from experimental animals infected with human immunodeficiency virus (HIV) or hepatitis B virus (HBV).

[h] Includes soiled dressing, sponges, drapes, and gloves. May include other categories (see footnotes c and d).

[i] Includes specimen containers, gloves, and laboratory coats. May include other categories (see footnotes a, c, d, e, and f).

[j] Includes tubing, filters, sheets and gloves. May include other categories (see footnote d).

uncontaminated sharps.[5] OSHA defines *regulated waste* as liquid or semiliquid blood or other potentially infectious materials.[6] Other potentially infectious body fluids include semen, vaginal secretions, cerebrospinal fluid, synovial fluid, pleural fluid, pericardial fluid, peritoneal fluid, amniotic fluid, saliva, and any body fluid visibly contaminated with blood. Most hospitals follow the CDC designations for infective waste, but use overly inclusive definitions.[7]

On the basis of scientific evidence and epidemiologic considerations, only microbiological cultures and stocks and contaminated sharp instruments should be considered potential infectious waste.[8] To comply with OSHA regulations that have as their goal the prevention of bloodborne infections, blood, blood products, and body fluids as well as material heavily contaminated with these substances must also be included.[6] There is no epidemiologic evidence, however, that hospital waste has caused disease in the community.[8]

Within the constraints of local, state, and federal regulations, minimizing waste designated as infectious will reduce the cost of disposing of waste from medical laboratories. A strict definition of waste types is essential for achieving waste minimization. At the same time, a combination of basic biosafety, good microbiologic technique, barrier protection, and an awareness of the risk of infection should protect the laboratory worker from the risk of acquiring an infectious disease from medical laboratory waste.

SOURCE SEPARATION OF THE DIFFERENT WASTESTREAMS

The best method for separating the different types of waste is by source separation, that is, at the point of discard where materials are discarded and thereby become waste. With this approach, each item is classified at the time of discard as a specific category of waste, and it is then discarded directly into the specific container that is designated for that particular type of waste or wastestream. The different wastestreams are then kept segregated to ensure proper management in accordance with waste minimization policies and regulatory requirements for disposal.

Implementation of Source Separation

The first step in implementing a source separation program is to define the different wastestreams and the particular types of waste that belong in each wastestream. In general, there are six wastestreams in medical laboratories that are relevant to this discussion of pollution prevention and waste minimization:

- Infectious wastes
- Chemical wastes
- Radioactive wastes
- Multihazardous wastes
- Wastewater
- Ordinary waste that is amenable to reuse or recycling

Infectious Wastes

All waste classified as infectious by your particular institution — whether because of regulatory requirements or policy — should be directed into the infectious wastestream. Classification of infectious wastes is discussed in detail in the section Strict Definition of Waste Types. The infectious wastestream may be further divided to accommodate those components that are managed in different ways or that require special containers.

Designated containers should be used for the different types of infectious waste. For example, red plastic bags are suitable for general infectious waste whereas discarded sharps should be placed directly into special, rigid-sharps containers. Sharps containers or other special containers should be used for certain glass items (such as slides, cover slips, and micropipettes) as well as broken glass.

Containers for infectious wastes should be available in proximity to the point of use and discard.[9,19] Infectious waste containers should be distinctive. Options include the use of red plastic bags or other red containers, the universal biohazard symbol, and fluorescent orange or orange-red labels.[10,20]

Chemical Wastes

Chemical wastes should be separated at the time of disposal and kept separate according to the type of chemical and the appropriate management method. Institutional policies should be in place for:

- Chemicals that are suitable for disposal to the sanitary sewer system
- Chemicals that can be redistilled for reuse
- Solvents that can be incinerated for heat recovery
- Hazardous chemicals that can be treated to neutralize or otherwise eliminate the hazard
- Hazardous chemical wastes that are regulated under RCRA or TSCA and that cannot be treated or disposed of on-site

The first type (i.e., chemicals that can be poured down the drain to the sanitary sewer system) includes those chemicals that are not hazardous, not regulated, and not detrimental to the operation of the wastewater treatment plant. These chemicals should be disposed of by prompt pouring into the drain. Such immediate disposal prevents the possibility of their becoming mixed with other chemicals that require treatment or special handling (which would unnecessarily increase the quantities of such wastes).

For the other categories of chemical wastes listed above, on-site treatment, reuse, and recycling are discussed later in this chapter.

Source separation is an important management practice. It keeps each waste type separate to minimize the quantities generated, to prevent mixing, and to permit optimum recovery.

Radioactive Wastes

Source separation of radioactive wastes is an important method for minimizing the quantity of radioactive waste generated in medical laboratories. Strict adherence to source separation policies prevents the volume of this wastestream from being increased by the introduction of nonradioactive wastes. At the same time it ensures that all wastes that should be managed as radioactive waste do enter this wastestream.

Multihazardous Wastes

Special management is needed for multihazardous wastes, and it is therefore essential that such wastes be kept separate from all other wastes. This is best achieved by source separation. Source separation also helps to minimize the quantity of this type of waste by preventing the introduction into this wastestream of those wastes that do not need such special management.

Wastewater

Most wastewater can be safely discharged into the sanitary sewer system. Certain types, however, including acidic and basic wastewaters must be treated before disposal. It is common practice to separate such wastewaters by directing them into a holding tank where they can be neutralized before discharge to the sanitary sewer system. This combination of source separation and treatment of the wastewater within the facility minimizes the quantity of waste that would otherwise have to be sent off-site for treatment and disposal.

In addition, many types of liquid wastes can be legally and safely poured down the drain for disposal to the sanitary sewer system. These include certain untreated infectious wastes, the products of infectious waste treatment (such as autoclaved wastes, chemically treated wastes, and the residue from chemical treatment), and specified quantities of certain radioisotopes. Source separation of these wastes for disposal with wastewater minimizes the quantities of infectious and of radioactive waste that would otherwise require special handling and management.

Ordinary Waste Amenable to Reuse or Recycling

This category consists of those types of waste generated in medical laboratories that do not require any special management. Nevertheless, items such as paper can be recycled, thereby minimizing the quantity of such waste needing disposal. This is best achieved by a source separation program that directs recyclable paper into a separate wastestream while keeping out of this stream other materials that would make the paper less marketable.

Aluminum cans and glass bottles are other types of waste that can be recycled. Although biosafety considerations preclude eating and drinking in medical laboratories, these wastes are generated in other parts of the facility. Source separation of these wastes maximizes the quantities available for recycling while minimizing the quantities that are sent unnecessarily to the landfill.

Many packaging materials and containers are also suitable for recycling or reuse. Source separation of these materials is important in achieving this goal.

Benefits of Source Separation

Source separation is the best approach for separating the different types of waste into the different wastestreams. There are various benefits that derive from source separation. These include:

- Safety and minimization of potential for exposure
- Minimization of waste quantities
- Efficiency
- Regulatory compliance
- Reduction in liability
- Cost reduction

Safety — Separation at the source is the safest method of separating the different wastestreams because it eliminates the need for waste sorting and the accompanying potential for exposure to hazards during waste-sorting operations.

Waste minimization — Source separation constitutes the surest method for securing separation of the wastestreams because the user of the material is the person who is most knowledgeable about the nature of the waste and its potential hazards. Strict definition of waste types combined with source separation constitute the best system for ensuring that only the designated wastes are included in each wastestream that will receive special handling and management.

Efficiency — Source separation is most efficient because the waste materials are handled only once at the time of discard. This system thereby eliminates the need for subsequent sorting of the different wastes as well as the costs associated with such extra handling.

Regulatory compliance — There are regulations applicable to many medical wastes (especially the infectious wastes, hazardous chemical wastes, radioactive wastes, and multihazardous wastes), and these wastestreams must have special handling and management because of the regulatory requirements. Source separation greatly increases the probability that all wastes will be placed into the proper wastestreams, each of which can then be managed appropriately. The policy and practice of source separation thereby increases regulatory compliance.

Reduction in liability — Source separation minimizes the potential for exposure of waste handlers because the waste is discarded directly into designated containers that are appropriate for each type of waste. This reduces the potential for exposure and the associated liability that is inherent in waste sorting and in handling waste that is improperly packaged. Source separation also reduces the liability associated with noncompliance with regulatory requirements.

Cost reduction — Source separation minimizes the quantities of ordinary waste that are unnecessarily included in certain wastestreams, thereby avoiding the incremental costs of the special handling given to these wastestreams. The increased safety, greater efficiency, greater regulatory compliance, and reduced liability discussed above all contribute to the reduction in costs that results from source separation.

PRODUCT SUBSTITUTION

Another management approach to waste minimization in the medical laboratory is product substitution. This involves reviewing all materials that are used in the laboratory to ascertain if the substitution of one product for another will result in a reduction in the quantities of wastes that are generated. Of particular concern are those wastestreams that require special handling or treatment because of their inherent hazards, pollution potential, and/or the relevant regulatory requirements. Waste management practices should include evaluation of the possibilities for reductions in waste generation that might be achieved via product substitution.

In the area of infectious wastes, the principal approach to achieving this goal is the substitution of reusable for disposable items. Substituting the use of nonhazardous materials to minimize the generation of multihazardous wastes and radioactive wastes is discussed in a section later in this chapter because such substitution usually results from use of new instrumentation or new or alternative diagnostic methods.

Substituting reusable for disposable (single-use) items is an obvious way to reduce waste generation — instead of being discarded, the items in question are cleaned (often sterilized) and then reused, thereby greatly reducing the volume of infectious waste. The range of disposable products used in the medical laboratory is great, including such diverse items as test tubes, culture plates, microscope slides, pipettes, needles and syringes, aprons and gowns, and wipes.

In theory, it is feasible to substitute reusables for all disposable items (after all, at one time only reusables were used).[21] In practice, however, this alternative must be evaluated with careful consideration of all relevant issues including practicality, availability of personnel for reprocessing activities,[22] quality assurance, quality control, biosafety, and the relative costs of disposable items and reusable items (taking into account also the costs of disposal and reprocessing).

Disposable, single-use items gained popularity and came into wide use because of certain features, especially the benefits they offer of quality assurance and quality control (QA/QC) and elimination of the need for reprocessing used items. QA/QC considerations are especially important with sterile items and culture media. Reprocessing has accompanying direct and indirect costs (e.g., personnel, cleaning materials, equipment, and space requirements). There is also the issue of predicting future labor costs in relation to demographics and the probable availability of needed additional personnel. There has been some movement toward a return to the use of reusable items, but there is not yet a clear pattern or even consensus about the relative costs.[11]

Another important factor is biosafety. It is essential to evaluate the potential for worker exposure to infectious agents during reprocessing procedures and to consider the associated occupational risks before deciding to switch back to reusable items.

The use of single-use items has many important advantages including QA/QC, elimination of labor-intensive reprocessing, and decreased costs as the disposables market share increased.[23] These advantages have not changed since this type of product was first introduced to the market. The principal disadvantage of using disposable items is that such use greatly increases the quantity of infectious waste that is generated in the medical laboratory.

Before any decision is made to substitute reusable items for all or some of the disposable items used in the medical laboratory, all of these factors should be carefully evaluated. The decision should be based on the advantages and disadvantages of product substitution for each disposable item. The option of substituting reusable for disposable items must be carefully evaluated as it would apply to each individual medical laboratory. The relevant factors are unique for each laboratory, and the results of such an evaluation will not necessarily be identical for each laboratory and facility.

MINIMIZATION OF WASTE GENERATION THROUGH NEW TECHNOLOGY

Minimization of waste generation can also be achieved through the use of new technology. The primary driving forces for technological changes in the medical laboratory have been the needs to increase productivity and to reduce costs. Manufacturers have also instituted such changes to reduce quantities of raw materials used as a cost-savings measure. Side benefits have been reduction in the quantities of wastes (infectious, chemical, and radioactive) generated as well as reduction in occupational hazards (deriving from the handling of smaller quantities of infectious, hazardous, and radioactive materials and wastes).

Another aspect of pollution prevention and waste minimization in the medical laboratory has been progress toward use of low-toxicity reagents. This has resulted mainly from governmental regulatory pressure. The RCRA regulations governing the management and disposal of hazardous wastes[12] introduce compliance, liability, and cost considerations. Safety regulations such as the OSHA standard on occupational exposure to hazardous chemicals in laboratories[13] have encouraged the elimination of toxic chemicals whenever possible.

The developments in the laboratory result from the introduction of new technology and new methods for laboratory testing. These include miniaturization, product substitution (such as using nonhazardous reagents or nonradioactive methods), and other types of process modification.

For ease of applicability, this discussion is based primarily on type of laboratory testing rather than on the type of technological modification. There are sections on microbiology, immunology/serology, and chemistry/hematology. These sections are followed by overviews of product substitution, the use of nonradiometric methods, and the paperless laboratory.

MICROBIOLOGY

A number of new systems for infectious disease testing have evolved that range from manual to fully automatic walkaway systems. As a result, more tests can be performed with smaller sample sizes. An additional benefit of automated systems is that worker safety is increased by minimizing exposure to infectious agents and hazardous chemicals.

Classical identification and antimicrobial susceptibility test methods for microorganisms have been miniaturized over the years. Micromethods have evolved from reagent-impregnated paper discs and microtube methods to completely miniaturized identification systems. Examples of such systems include API system (bioMerieux Vitek, Inc.), Biolog system (Biolog, Inc.), BBL Crystal system (Becton Dickinson Microbiology Systems), and Rapid ID systems (Innovative Diagnostic Systems, Inc.). Combinations of identification and antimicrobial test panels were developed which made laboratory testing more efficient as well as reduced the quantities of waste generated. Examples of automated walkaway identification and susceptibility instruments are the MicroScan system (Baxter Diagnostics, Inc.) and the Vitek system (bioMerieux Vitek, Inc.). Although both systems are analytically reliable, the MicroScan system generates considerably more waste. The Vitek system utilizes test cards rather than standard microdilution panels; this requires smaller test volumes and fewer inoculum preparation devices.

Rapid screening tests are frequently used in the clinical laboratory to rapidly eliminate specimens that need not be cultured. They provide accurate results in a cost-effective manner, increase efficiency in the laboratory, and often generate less waste than conventional methods. For example, rapid urine screen methods include enzyme dipstick, filtration, bioluminescence, chemiluminescence, and photometry. Rapid screen tests for bacterial identification include butyrate esterase disc test for *Moraxella catarrhalis,* PYR disc test for *Streptococcus pyogenes* and Group D enterococci, and rapid urease test for *Helicobacter pylori.*

The current trend in antimicrobial susceptibility testing is to screen for resistant microorganisms rather than to generate susceptibility profiles. Examples of screening tests include the β-lactamase test for *Haemophilus influenzae* and the oxacillin agar screen test for *Staphylococcus aureus* and *Streptococcus pneumoniae.*

The Microbial Identification System, MIS (Microbial ID, Inc.) is a fully automated gas-liquid chromatography system for the rapid identification of microorganisms. This system is based on identifying microorganisms by their unique fatty acid profiles. Only small samples of bacterial growth are needed for analysis of chromatographic patterns; however, the solvents that are used for analysis generate hazardous wastestreams.

Recently, DNA probe-based tests have been introduced into the laboratory. One DNA test offers a practical alternative to confirmatory tests which may require many serologic and physiologic tests. The Accuprobe culture confirmation test for mycobacteria (Gen-Probe, Inc.) not only eliminates conventional biochemical testing, but it also reduces considerably the time required for identification. With the development of DNA amplification techniques, direct specimen testing will further reduce infectious wastestreams by eliminating traditional culture methods.

IMMUNOLOGY/SEROLOGY

Many diagnostic serology and immunology laboratories perform a variety of manual and semimanual tests which generate varied wastestreams. Consolidation of these methods not only increases the efficiency of testing, but it also minimizes the quantities of waste generated. The MicroTrak XL system (Syva Co.) and ACCESS Immunoassay system (Sanofi Diagnostics Pasteur, Inc.) are examples of two fully automated random access systems for infectious disease testing. These analyzers conduct several immunoassays simultaneously, performing the entire range of tasks from sample pipetting to calculation of results. These analyzers require small microwell strips, and the liquid waste is captured in a single reservoir.

The Vitas system (bioMerieux Vitek, Inc.) is the most comprehensive immunoanalyzer. Its wide range of assays includes serology, immunochemistry, antigen detection, and drug testing. In this automated system, all the wastestreams (infectious and chemical) are contained within the reagent strip.

CHEMISTRY/HEMATOLOGY

Automated chemistry and hematology analyzers are now available that require only microsamples of specimen and that provide high volume test capacity. Use of these analyzers greatly reduces the volume of infectious waste generated. The Technicon Immunoassay system (Miles, Inc.) is an example of a random access chemistry analyzer that requires microliter sample volumes. Hematology analyzers include the Technicon H1E Multi-Species system (Miles, Inc.) and Cobas Argos system (Roche Diagnostic Systems, Inc.).

Manufacturers of chemistry analyzers have improved instrument design to reduce or eliminate the quantities of hazardous materials used and hazardous wastes generated; this reduces handling of these materials and wastes, disposal needs, and disposal costs. Chemical reagents are often prepackaged to minimize quality control testing and wastage.

Examples of wet chemistry analyzers are the Hitachi (Boehringer Mannheim Corp.) and Dimension AR (Dupont) systems. In the Hitachi system, the waste is separated into three streams. The first reagent stream is segregated for hazardous disposal. The remaining wastestreams are discharged into the sewer. The Dimension AR system encapsulates its waste into self-sealing cuvettes, thus eliminating potential occupational exposures and minimizing waste handling and disposal costs.

Dry chemistry analyzers include the Paramax (Baxter Diagnostics, Inc.) and Ektachem (Eastman Kodak Co.) systems. In the Paramax system, dry reagents and specimens are placed in preformed cuvettes that are self-sealing at the end of the analysis. The Kodak dry slide technology eliminates the generation of hazardous waste.

Emerging technologies currently under development are rapid, less invasive, reagent-free, portable chemistry analyzers for use at the point of care. These small analyzers range from biosensors to optimal waveguides capable of measuring analytes such as sodium, potassium, chloride, blood urea nitrogen (BUN), glucose, and hematocrit. The i-STAT (i-STAT Corp.) analyzer is already in use for bedside testing. Two drops of whole blood are placed in a sample cartridge, which is then inserted into the analyzer. In the cartridge the sample flows across a series of biosensors comprised of ion-selective electrodes attached to silicon chips. The biosensors emit electrical impulses that make contact with the analyzer, which then translates these impulses into quantitative measurements.

Other devices under development include the use of near-infrared spectroscopy for the testing of glucose. Light at wavelengths just beyond visible red are beamed through a patient's finger. The method requires neither specimens nor reagents.

PRODUCT SUBSTITUTION

There has been an effort to substitute nontoxic or less-toxic reagents for the hazardous reagents traditionally used in many laboratory tests. For example:

- Reagents have been formulated to substitute for mercury-containing reagents. For example, ion-selective electrodes have replaced use of mercuric nitrate for chloride determination in body fluids.
- Sodium lauryl sulfate provides an alternative to cyanide-based methods for automated hemoglobin analysis.
- Cupric sulfate can be substituted for mercuric chloride as a fixative for the PVA/trichrome stain used in the parasitology laboratory.

USE OF NONRADIOACTIVE METHODS

The use of radioactive test methods has declined mainly due to waste handling and disposal problems. Although considered less sensitive, nonradioactive assays have contributed to an increase in clinically useful methods. Nonradioactive assays offer longer shelf life, shorter analytical time, and no special recordkeeping or disposal requirements.

Chemiluminescent compounds have proved to be particularly sensitive labels in enzyme immunoassays and DNA probe assays.

Recent advances in blood culture instrumentation systems have led to the development of several nonradioactive microbial detection methods. In the Bactec 9240 (Becton Dickinson Diagnostic Instrument Systems) and BacT/Alert (Organaon Teknika Corp.) systems, fluorogenic and colorimetric sensors, respectively, measure microbial growth. The ESP system (Difco Laboratories) detects both the consumption and production of all gases by microorganisms. One strategy for minimizing waste and maximizing recovery of clinically important microorganisms is to use more blood for culture (10 to 20 ml per bottle) and to reduce the number of bottle sets used.

THE PAPERLESS LABORATORY

The paperless laboratory is an alternative to manually recording information. For example, in the Sunquest (Sunquest Information Systems) paperless microbiology system, all culture observations and workup activities are entered on-line instead of onto a hard-copy workcard. The system provides more efficient workload statistics and retrieval of data on a particular process.

A comprehensive audit trail (e.g., culture review report of all results) can be printed or downloaded to a personal computer prior to scheduled system downtimes and at prescheduled times during the day. Retrieval of printed information, however, may require the same or even greater quantities of paper as processing without a computer.

TREATMENT OF WASTES ON-SITE TO MINIMIZE SPECIAL DISPOSAL REQUIREMENTS

Treatment of wastes on-site, i.e., at the generating facility, provides an option for minimizing the quantities of wastes that are sent off-site for treatment and disposal. This approach to managing wastes that require special handling, especially those that are subject to regulatory requirements, is often cost effective and beneficial.

The treated waste can sometimes be disposed of as ordinary waste. This depends on the type of waste and the treatment process as well as the specific federal and state regulations that apply to each type of waste (see Table 3). In addition, some federal agencies have issued guidelines that affect the management and disposal of wastes from medical laboratories (Table 4).

Wastes can often be treated on-site more cheaply than the charges for commercial off-site treatment. On-site treatment often exempts the waste from further regulation, thereby saving the cost of regulatory compliance. In addition, on-site treatment allows the generating facility to maintain control over the waste-treatment process, an approach often considered advisable to reduce liability.

On-site treatment can be used as a method of minimizing the quantities of wastes from medical laboratories that must be handled and managed in special ways. This approach must be tailored to the specific types of waste that are generated in the medical laboratory. The discussion in this section focuses on infectious wastes, chemical wastes, radioactive wastes, multihazardous wastes, and wastewater.

INFECTIOUS WASTES

Infectious wastes are especially amenable to treatment on-site. The wastes can be treated in the laboratory itself or elsewhere in the facility. In many facilities,

Table 3 Federal Regulations Affecting the Management and Disposal of Wastes from Medical Laboratories

Waste type	Federal regulation	Agency[a]	Citation
Infectious	Medical waste tracking	EPA	40 *CFR* 259
	Bloodborne pathogens standard	OSHA	29 *CFR* 1910.1030
Chemical	Hazardous waste management	EPA	40 *CFR* 260–268
Radioactive	Radioactive waste disposal	NRC	10 *CFR* 20.301–311
Wastewater	Effluent guidelines and standards	EPA	40 *CFR* 116, 122, 401, 403

[a] EPA: Environmental Protection Agency; NRC: Nuclear Regulatory Commission; OSHA: Occupational Safety and Health Administration.

Table 4 Federal Guidelines on the Management and Disposal of Infectious Wastes

Topic	Agency[a]
Infectious waste management	EPA[b]
	EPA[c]
Infective waste management	CDC[d]
	CDC[e]
	CDC[f]
Waste disposal in microbiological and biomedical laboratories	CDC and NIH[g]

[a] CDC: Centers for Disease Control and Prevention; EPA: Environmental Protection Agency; and NIH: National Institutes of Health.

[b] U.S. Environmental Protection Agency, *Draft Manual for Infectious Waste Management*, USEPA, Washington, D.C., 1982, #SW-957.

[c] U.S. Environmental Protection Agency, *EPA Guide for Infectious Waste Management*, USEPA, Washington, D.C., 1986, #EPA/530-SW-86-014.

[d] Centers for Disease Control, Recommendations for prevention of HIV transmission in health-care settings, *MMWR*, 36(2S), 1987.

[e] Centers for Disease Control, Guidelines for prevention of transmission of human immunodeficiency virus and hepatitis B virus to health-care and public-safety workers, *MMWR*, 38(S-6), 1989.

[f] Garner, J.S. and Favero, M.S., *CDC Guidelines for Prevention and Control of Nosocomial Infections. Guideline for Handwashing and Environmental Control*, U.S. Department of Health and Human Services, Centers for Disease Control, Atlanta, 1985, #DHHS-99-1117.

[g] Centers for Disease Control and National Institutes of Health, *Biosafety in Microbiological and Biomedical Laboratories*, 3rd ed., U.S. Department of Health and Human Services, Washington, D.C., 1993, #DHHS-(CDC)93-8395.

infectious wastes from medical laboratories (and from other sources within the facility) are routinely treated on-site before the waste is sent from the facility for disposal.

Readers are referred to the *Manual of Clinical Microbiology* for more information on waste treatment.[14] See Table 5 for a comparison of the different treatment technologies that are applicable for on-site treatment of infectious wastes.

In most states, infectious waste that is treated[24] is no longer subject to regulation, that is, it can be managed and disposed of as ordinary waste. Therefore, on-site treatment minimizes the quantity of infectious waste that must be sent off-site for treatment and disposal.

However, some states have adopted medical waste tracking regulations,[25] and in these jurisdictions the waste must be both treated and destroyed (i.e., rendered nonrecognizable) if it is to be exempt from regulation. Therefore, the advantages of on-site treatment can be realized in these states only if the selected method of treatment both treats and destroys the waste.

Treatment within the Laboratory

Most medical laboratories have steam sterilizers (autoclaves) that are used to sterilize certain infectious wastes. This type of treatment is usually used for microbiology cultures and certain associated wastes to sterilize the wastes before they leave the laboratory. The treated (sterilized) waste can then be disposed of as ordinary waste.

It should be noted that steam sterilization treats the infectious waste but does not destroy it. Therefore, sterilized waste would still be subject to the tracking requirements in those jurisdictions that have such regulations.

A recent market development is the introduction of small-scale treatment equipment designed specifically for use within the laboratory to treat the relatively small quantities of infectious wastes that are generated in a medical laboratory. Some are bench-top models, others are floor models, and some devices are available in both sizes.

One such device is a small grinder that provides simultaneous grinding and chemical treatment of the infectious waste, thereby satisfying both the treatment and destruction requirements of certain jurisdictions (see above). Another small device treats the infectious waste with vapor-phase hydrogen peroxide. Small, bench-top models of autoclaves are also available.

Treatment within the Facility

Medical laboratories are not always free-standing, but rather are frequently part of a larger facility — such as a hospital, university, or pharmaceutical company — that generates infectious wastes at other sites within the facility. It is then often preferable to realize economy of scale by treating all the infectious waste that is generated at the facility at one place within the facility (which is usually not within the medical laboratory).

As was mentioned above, on-site treatment minimizes the quantity of infectious waste that must be sent off-site for treatment and disposal (with the accompanying extra costs for special management, treatment, and disposal). On-site treatment options are diverse, including incineration, steam sterilization, chemical treatment, and microwave treatment. Various factors should be considered in selecting a method and the equipment model to be used for treatment of infectious waste on-site — including regulatory requirements, treatment effectiveness, safety of operation, difficulty of operation, ease of maintenance, space requirements, and costs (capital as well as operation and maintenance, see Table 5).

Table 5 Comparison of On-site Regulated Medical Waste Treatment Technologies

	Traditional methods		Alternative mechanical-chemical methods of treatment[a]		
Factor	Incineration	Steam/ autoclave	Mechanical/sodium hypochlorite[b]	Mechanical/chlorine dioxide[c]	Mechanical/ peracetic acid[d]
Applicability[e]	All, mixed	Most	Most	Most[f]	Most
Equipment operation	Complex	Easy	Moderate	Easy	Easy
Operator requirement	Highly skilled	Trained	Trained	Trained	Trained
Load standardization	Needed[g]	Needed[g]	Needed	Needed	Needed[g]
Capacity	1200 lb/h	200 lb/h	1000 lb/h	600 lb/h	20 lb/h
Effect of treatment	Waste burned	Waste unchanged	Shredded/ground	Shredded	Pulverized
Volume reduction	85–95%	30%	Up to 85–95%	Up to 85–95%	Up to 85%
Potential side benefits	Energy recovery	None	Yes[h]	Yes[h]	Yes[h,i]
Disposal of residue					
Liquids	None[j]	Sanitary sewer[k]	Sanitary sewer[k,l]	Sanitary sewer[k,l]	Sanitary sewer[k]
Solids	Hazardous ash[m]	Sanitary landfill[n]	Sanitary landfill	Sanitary landfill	Sanitary landfill

	Alternative mechanical-physical methods of treatment				
Factor	Mechanical/ steam[o]	Mechanical/ dry heat[p]	Mechanical/ microwave[q]	Compaction/ steam[r]	Compression/ infrared[s]
Applicability	Most	Most	Most[f,t]	Most[f]	Most[f,t]
Equipment operation	Easy	Moderate	Easy	Easy	Easy
Operator requirement	Trained	Trained	Trained	Trained	Trained
Load standardization	Needed[g]	Needed[g]	Needed[g,u]	Needed[g]	Needed[g,u]
Capacity	300 lb/h	2,000 lb/h	550 lb/h	400 lb/h	50 lb/h
Effect of treatment	Shredded	Shredded/ground	Shredded	Compacted	Compressed
Volume reduction	80%	Up to 85%	Up to 85%	50%	80–90%
Potential side benefits	Yes[h,v]	Yes[h,v]	Yes[h]	Yes[h]	Yes[h]
Disposal of residue					
Liquids	None[j]	Sanitary sewer[k]	None[j]	Sanitary sewer[k]	None[j]
Solids	Sanitary landfill	Sanitary landfill	Sanitary landfill	Sanitary landfill	Sanitary landfill

[a] Information provided by manufacturers.
[b] Medical SafeTEC, Inc., Indianapolis, model Z1200.
[c] Winfield Environmental, Escondido, CA, Winfield Condor™ model HR600.
[d] STERIS/ECOMED, Mentor, OH, and Indianapolis, IN.
[e] To infectious waste types. Waste segregation is needed to eliminate pathological, chemotherapeutic, and nontreatable waste from the wastestream.
[f] No bulk fluids greater than 10% by weight.
[g] To eliminate bulk fluids such as dialysis fluid.
[h] No harmful by-products; lower volume to commercial wastestreams.
[i] Point of generation treatment.
[j] Moisture retained in solids or held within the unit.
[k] Low volume or intermittent drain to sanitary sewer.
[l] Potential formation of carcinogenic compounds in chlorine-based systems.
[m] RCRA-permitted landfill.
[n] Potential problem with recognizable red bag waste.
[o] S.A.S. Systems, Inc., Houston, G.T.H. Roland ZDA-M3.
[p] BioMed Waste Systems, Inc., Boston.
[q] ABB Environmental Services, Inc., ABB Sanitec, model HGA-250-S.
[r] San-I-Pak pacific, Inc., Tracy, CA, model MARK VII.
[s] Medifor-X Corporation, Redding, CT, DISPOZ-ALL 2000.
[t] Metallic content must be less than 1% by weight.
[u] To separate sharps from waste.
[v] Plastic recovery for recycling.

Adapted from Marsik, F. J. and Denys, G. A., Sterilization, decontamination, and disinfection procedures for the microbiology laboratory, *Manual of Clinical Microbiology*, 6th ed., Murray, T. R., Baron, E. J., Pfaller, M. A., Tenover, F. C., and Yolken, R. H., Ed., American Society for Microbiology, Washington, D.C., 1994.

The mechanisms, advantages, and disadvantages of the different treatment options have been discussed in detail in other publications.[14-16] The discussion in this section is limited to representative examples of the different technologies that are being used for on-site treatment of infectious wastes.

Available methods for on-site treatment of infectious waste include incineration, steam sterilization, thermal inactivation, gas/vapor sterilization, irradiation sterilization, and chemical disinfection. The waste is sometimes shredded first to ensure proper action by the treating agent. The two most widely used treatment methods are incineration and steam sterilization.

Incineration is a process that burns waste under controlled conditions into a noncombustible ash. It would appear to be the ideal way to destroy the hazardous components of waste as well as to reduce both the weight and the volume of waste (by 85 to 95%). Regulations on air emissions and ash disposal, however, have increased the complexity and cost of operations. Because of stringent permitting and licensing requirements, especially in populous areas, incineration is now prohibited or in limited use in many areas of the country.

Many hospitals and microbiology laboratories treat waste using a steam sterilizer (autoclave). Waste is exposed to steam and pressure at a required temperature for a specified period of time. Steam sterilization does not physically destroy the waste, and the volume is reduced by only about 30%.

Several alternative treatment methods have evolved to improve decontamination and reduce waste volume. These technologies combine mechanical destruction of the waste with heat or chemical treatment.

Mechanical/chemical treatment is a process that uses a two-stage shredding and grinding operation and mixes the waste with a disinfectant solution. In the Medical SafeTEC, Inc., system, infectious waste first passes through a preshredder and then is reduced into smaller pieces in a high-speed hammermill. The waste is sprayed with a sodium hypochlorite solution during the shredding process. The treated solid material is captured in a rotary separator, transferred up an auger conveyer, and then deposited in a waste cart. The liquids are either discharged into the sanitary sewer or recirculated through the system. The entire system is maintained under negative pressure, and the air is discharged through HEPA filters. This treatment process renders the solid material unrecognizable and reduces the volume up to 85%. The treated material is classified and handled as general refuse.

The IWP-1000 infectious waste processor (Mediclean Technology, Inc.) is another example of a mechanical/chemical treatment system. Medical laboratory waste is fed into a cutting chamber which granulates the waste to a 3/8-in. particle size. Aqueous chlorine dioxide is simultaneously injected into the cutting chamber to decontaminate the waste.

Microwave/shredding treatment is a process that shreds waste then exposes it to microwave-generated heat. In the ABB Sanitec, Inc., system, infectious waste is fed into a hopper for shredding and sprayed with steam. The waste moves up a screw conveyor past microwave generators, which heat the waste at 203 to 212°F for 30 min with multiple microwaves. The decontaminated waste is ejected into a waste container for final disposal. Air is evacuated from the hopper through a prefilter and HEPA filters. The physically destroyed waste is reduced in volume up to 80%.

Steam/compaction treatment uses an autoclave combined with a compactor. In the San-I-Pak, Inc., system, infectious waste is steam treated in a sterilizer chamber and then compressed in a compaction chamber before final disposal. Although compaction reduces waste volume approximately 50%, the waste is not physically destroyed.

Waste such as antineoplastic (chemotherapy) agents, toxic chemicals (formaldehyde-containing specimens), and radioisotopes cannot be rendered nonhazardous by the chemical disinfection, microwave, or steam sterilization treatment systems. Therefore, these wastestreams must be properly separated and minimized if alternatives to incineration are used for waste treatment.

Other approaches to the treatment and disposal of infectious waste are currently in use, or under development. These treatment technologies include chemical, irradiation, mechanical, and thermal processes.

CHEMICAL WASTES

For many chemical wastes that are generated in medical laboratories, on-site treatment provides the best, most economical management approach. In addition, although the USEPA has adopted a cradle-to-grave approach for the management of hazardous wastes (i.e., those subject to RCRA regulation), some RCRA-regulated chemical wastes can be treated on-site to exempt them from further regulation.

Treatment options that minimize the disposal of chemical wastes include recycling for reuse (such as redistillation of used solvents, and incineration of solvents with heat recovery), neutralization of acids and bases, and chemical treatment to destroy or precipitate out the hazardous constituents. In addition, some chemicals can be safely discarded directly to the sanitary sewer system.

All these options are effective in minimizing the quantities of chemical wastes needing disposal. Without prior treatment, these wastes would be subject to regulatory requirements and the added costs of regulatory compliance.

For more details on the treatment methods applicable to specific chemical wastes, please refer to other chapters in this book. For solvent distillation, see Chapters 13 and 20; for chemical treatment, Chapter 16; and for disposal to the sanitary sewer, Chapter 9.

RADIOACTIVE WASTES

Radioisotopes are often used in medical laboratories for various diagnostic procedures, experiments, and other uses. The radioactive wastes that result from such use present special problems in handling and management because of their radioactivity. All radioactive wastes must be transferred only to an authorized recipient, and disposal costs have increased as disposal capacity in the country has become more limited. Therefore, management techniques that minimize the quantities of radioactive wastes shipped from the facility for disposal are certainly cost effective and advantageous.

The waste-disposal regulations of the Nuclear Regulatory Commission do provide less-stringent options for certain radioisotopes, and a good management plan will take full advantage of these options (see Table 3). For example, there are provisions that allow sewer disposal and incineration of certain radioisotopes in specified limited quantities and concentrations. There are also special allowances for the disposal of hydrogen-3 and carbon-14. These provisions do not, however, cover all the types or quantities of radioactive wastes that are generated in medical laboratories, and a waste minimization program is essential to minimize pollution and to maximize savings in disposal costs.

Most radioisotopes that are used in medical laboratories have relatively short half-lives. On-site storage to reduce the level of radioactivity is usually the simplest, easiest, most cost-effective approach to managing these wastes by minimizing the quantities of radioactive wastes that require special disposal. Under this management approach, each radioisotope is stored for seven half-lives. During this period, the level of radioactivity decays to background level, and the waste can then be disposed of as ordinary waste without any special management or disposal requirements.

There are a number of regulatory requirements that pertain to the storage of radioactive wastes. These include use of containers that provide proper shielding, secure storage areas with access limited to authorized personnel, specified labeling of containers and posting of the storage area, and stringent recordkeeping.

MULTIHAZARDOUS WASTES

Multihazardous wastes present the greatest challenge for waste management in general and waste minimization in particular. Because there is more than one hazard in the waste, there cannot be a uniform method for managing all such wastes. Nevertheless, most multihazardous wastes generated in medical laboratories are amenable to treatment on-site. This management approach is preferable because it avoids the dilemma of disposal of waste that is subject to diverse and sometimes conflicting regulatory requirements. In addition, this approach can minimize the quantities of waste that are shipped from the generating facility for off-site treatment and/or disposal.

Each multihazardous waste must be evaluated on an individual basis. The first step is ascertaining the hazards posed by the waste. These hazards are then ranked, with priority being assigned to that which constitutes the greatest hazard. An appropriate management scheme can then be developed.[15]

When the multihazardous waste has a radioactive component, it is the radioactivity that is almost always the hazard of greatest concern. The best management approach is usually to store the waste for a period that is seven times the half-life of the constituent radioisotope. At the end of such a storage period, radioactivity is no longer a concern, and the other hazard(s) in the waste can then be addressed. For example, if the waste is also infectious, it can then be treated to eliminate the infectiousness so that the waste can be disposed of as ordinary waste.

WASTEWATER

The laboratory sink that connects to the sanitary sewer system provides a convenient site for disposing of many of the liquid and semiliquid wastes that are generated in the medical laboratory. This option may be valid for certain infectious wastes as well as some chemical and radioactive wastes. However, before this method of disposal is used, two factors must be considered. These are the regulatory requirements and occupational safety.

There are various regulatory constraints that are relevant to wastewater. These constraints apply to the constituents that are introduced with the discharged wastewater into the sewer system, and they may limit the quantities and concentrations of wastes that can be disposed of in this way. The constraints include federal and state regulations as well as the agreement that the facility has with the local wastewater treatment authority.[26] The option of disposing of certain wastes from the medical laboratory together with the wastewater effluent from the facility is viable only if such disposal is fully compatible with all relevant regulations and other constraints.

Another important consideration is occupational safety. This method of waste disposal must be safe, and the act of pouring the waste down the drain must not present any hazards to the person who is doing this. Splashing and aerosolization during disposal can expose workers to pathogens, hazardous chemicals, and radioactivity in the wastes. Use of special devices and personal protective equipment can reduce the risk of exposure, but this practice must be used with great caution.

REUSE OF WASTES

An ideal method of achieving waste reduction is to reuse the waste. If the waste can be used again — either as is or after some reprocessing — it becomes a useful material rather than a waste in need of disposal. Examples of reuse in the medical laboratory include reuse of disposable items, solvents, treatment media, and packaging materials. Each of these is discussed in this section.

REUSE OF DISPOSABLE ITEMS

There has recently been a trend to reevaluate the use of disposable, single-use products. This evaluation has been twofold: the possibility of substituting reusable for single-use items and the possibility of processing single-use products for reuse.

Many disposable single-use products cannot be reprocessed. Because of their intended one-time use, many are made of more flimsy materials that are not compatible with the rigors of cleaning and processing for reuse. Others, however, can be cleaned and reused in the laboratory. Examples include glass pipettes, tubes, flasks, and syringes.

Evaluation of this approach must include consideration of practicality, availability of personnel for reprocessing, QA/QC concerns, biosafety, and the relative costs and risks of disposal and reprocessing. See Product Substitution above for a more-detailed discussion of the relevant factors.

SOLVENTS

Various solvents are used routinely in the medical laboratory, and, therefore, waste solvents are frequently generated in fairly large quantities. Management of waste solvents is further complicated by the fact that many are hazardous wastes and therefore subject to special regulations as well as additional disposal costs. Any program that minimizes the quantities of solvents that must be sent off-site for disposal is beneficial.

There are two principal alternatives for minimizing the quantities of waste solvents generated in the medical laboratory that must be disposed of. These are redistillation of waste solvents for reuse and incineration of certain waste solvents for heat recovery.

Redistillation and Reuse

Many solvents that are used in the medical laboratory can be redistilled and reused in similar or other procedures. (Waste alcohol, for example, is especially suitable for this management approach.) Manufacturers have responded to this potential by developing solvent stills that are designed specifically for use in laboratories. See Chapters 15 and 20 for more detailed discussions of solvent distillation.

QA/QC considerations are, of course, important. It is essential that the redistilled product have the requisite characteristics and degree of purity for the particular use to which it is put, either in the same laboratory or elsewhere in the facility.

Incineration with Heat Recovery

Certain waste solvents are RCRA hazardous only because of their flammability. The best management approach for these waste solvents is frequently incineration with heat recovery — an activity that is permitted under the RCRA regulations. This practice eliminates disposal costs while accruing the benefit of recovering the heat content of the solvents for use in the production of hot water or steam for the facility.

ZERO DISCHARGE THROUGH RECIRCULATION OF TREATMENT EFFLUENT

Some alternative waste treatment technologies have been developed that minimize liquid effluent discharged to the sanitary sewer. In two mechanical/chemical treatment units — the Medical SafeTEC, Inc. model Z1,200 (Indianapolis) and the Mediclean model IWP-1000™ (Mediclean Technology, Inc., West Warwick, RI) — process wastewater is recirculated and mixed with fresh chemical disinfectant. Some mechanical/thermal systems rely on recirculated steam for treatment (see Table 5).

PACKAGING MATERIALS

Medical laboratories may be unique in having available new types of packaging materials. Manufacturers are now offering products that are packaged in such a way that the packaging materials/containers are usable afterward in the laboratory. Examples include insulated boxes and styrofoam test tube racks.

RECYCLING OF WASTES

Waste recycling is another option for waste minimization in the medical laboratory. Because of the biohazards that are inherent in infectious waste, these wastes should not be recycled before they have been treated to eliminate or minimize the biohazard.

Other wastes that are generated in the medical laboratory are not unique to this particular type of laboratory. Among these wastes are some that can be easily recycled, including paper, solvents, and packaging materials.

PAPER

Paper recycling is not a new phenomenon; rather, it has become commonplace in many home, business, and industrial settings. This type of recycling is easily implemented in medical laboratories, and it is a worthwhile endeavor because large quantities of paper are usually generated.

The principal necessities for paper recycling are recycling containers and the space to keep them. Source separation is an essential element for successful paper recycling because the value and market depend on the quality and uniformity of the recycled paper.

SOLVENT REDISTILLATION

Various solvents that are used in laboratories are amenable to recycling, that is, they can be reused after they are redistilled because redistillation restores purity by removing contaminants. This has been discussed previously (see also Chapters 15 and 20 for more information on solvent redistillation).

PACKAGING MATERIALS

Packaging materials are frequently separated for recycling. These include corrugated cartons, plastic wrappers, styrofoam peanuts, and some types of plastic containers. Although source separation is the most efficient way of separating these different types of materials, space limitations in the laboratory may allow the use of only one container for discarded packaging materials and thereby necessitate subsequent sorting of these materials.

IMPLEMENTING A WASTE REDUCTION PROGRAM

MANAGEMENT PROGRAM

Medical laboratories should establish a waste management program that addresses minimization of all types of waste generated. The College of American Pathologists Commission on Laboratory Accreditation specifically recommends that minimization of both infectious medical waste and hazardous chemical waste should be an integral part of continuous quality improvement.[17]

Strategies to minimize infectious and hazardous waste include:

- Segregate infectious waste from the noninfectious general trash
- Collect blood and other clinical specimens in smaller containers
- Aliquot blood samples instead of obtaining separate sample tubes for each test
- Substitute micromethods for conventional procedures
- Increase the use of screening tests
- Consolidate test methodologies
- Substitute nonradioactive (e.g., chemiluminescence) for radioactive tests
- Store low-level radioactive waste (short half-life) on-site for decay
- Track all purchased hazardous materials
- Eliminate use of mercury-based fixatives and reagents
- Consider use of alternative clearing agents and less-toxic reagents
- Distill formaldehyde for reuse
- Distill xylene for reuse or direct to a reclamation facility
- Consider alternative waste treatment and disposal options
- Develop an education program to promote recycling, product reuse, and waste reduction

A waste minimization program should be based on regulatory requirements and institutional policy. Employees should be encouraged to participate in the program and be recognized for their suggestions and for achieving laboratory goals.

The first component of a waste minimization program is to identify and quantify all laboratory waste. Hazardous and infectious waste should be segregated and discarded directly into appropriate containers to minimize waste handling. Hazardous chemicals should be tracked by reviewing purchase orders and waste disposal contracts; a materials balance study can then be used to determine missing quantities of chemicals and to identify unknown hazardous waste. Multihazardous waste must be segregated and directed to the appropriate treatment procedure based on the relative severity of the hazards.

Other components of the waste minimization program include procedures for transport and storage of infectious, radioactive, and chemical wastes. There should be a contingency plan for treatment of waste when equipment fails or personnel problems arise. Emergency planning should establish procedures to be used in the event of an accidental spill or loss of containment. Training and educational programs should be developed and implemented for all personnel handling waste. Recordkeeping is essential to document regulatory compliance and to evaluate the success of the program. Review of accident and incident reports can also identify potential problems and deficiencies in the waste minimization program.

A successful waste minimization program is a continuous process with emphasis on source reduction, reuse, and recycling. The program and its goals should be reviewed annually. Selection of new technology and procedures for laboratory testing should be made with consideration of the types and quantities of wastes generated.

PURCHASING STRATEGIES

Medical laboratories should work with suppliers and vendors to initiate programs and procedures to increase recycling, minimize waste, and reduce quantities of product packaging materials used. An example of customers working with a supplier is the Baxter Healthcare Corporation ACCESS™ program, through which corporate customers receive help from Waste Management, Inc. to manage their waste. Baxter is also working with customers to establish environmental goals for product packaging. These goals include:

- Elimination of inner liners and boxes
- Elimination of toxic metals in ink, dyes, and adhesives
- Elimination of chlorine-bleached paper and paperboard
- Elimination of foam packaging made with chlorofluorocarbons
- Maximized use of recycled fiber in corrugated shipping containers
- Reduction in primary, secondary, and tertiary packaging where feasible
- Promotion of bulk shipping and reusable containers
- Adoption of guidelines for the evaluation of new packaging designs

In some institutions, environmental concerns may impact on purchasing decisions. See Chapter 17 for a more general overview of this subject.

EMPLOYEE AWARENESS PROGRAM

Implementation of any waste management program is easier and more effective if the program has the support of the employees.[18] Indeed, the impetus for waste minimization and recycling has often come from the employees. Therefore, an employee awareness program can be very useful in obtaining the support and cooperation of the employees, a step that is essential to the success of the waste minimization program.

The awareness program should provide information about the purpose, scope, and benefits of the waste minimization program. Benefits include those accruing to the environment, the employer, and the employee. Specific information about the wastes that can be reused or recycled will also be of interest, and it should be included in the awareness program because it will serve to enlist employee support.

Other relevant information — the procedures for handling each of the wastes and the importance of following these procedures to obtain maximum benefit from the waste minimization program — is best presented as part of the training program for all employees who generate, collect, or handle wastes that are affected by the waste minimization program.

EMPLOYEE TRAINING

Employee training is an essential element in the implementation of a waste reduction program. It is logically included in the general training for waste generators and handlers. The waste minimization portion of the training can be the principal topic when the waste minimization program is introduced or as a topic of special interest in routine training sessions.

Employees must be trained if the waste minimization program is to be successful. All employees who have any contact with the wastes that are subject to waste minimization efforts, that is, those who generate such wastes as well as those who collect or handle the wastes, should be provided this training.

In the medical laboratory, the single most important aspect of a waste minimization program is source separation of the different wastestreams to allow subsequent special handling of those wastes that are amenable for reuse or recycling. Employees must understand:

- Which wastes can be reused or recycled
- The need for source separation to avoid sorting and to maximize the returns from waste minimization efforts
- How to differentiate and to segregate the different wastestreams
- The designated procedures and why it is important to follow them
- How to maximize the benefits of the program
- The rationale and importance of the waste minimization program

Some training (or at least an informal orientation session) should also be provided to employees in the purchasing department so that they will understand the new policies and procedures that are adopted as part of the waste minimization program (see Purchasing Strategies).

For the waste minimization program to be successful, all employees must understand the rationale for and the benefits of the program. These benefits include savings in waste disposal costs, reduction in occupational hazards (because smaller quantities of infectious and hazardous wastes are being produced and handled), and the environmental benefits derived from reduction in use of resources as well as reduction in the quantities of waste generated that need disposal.

COST SAVINGS: USING THE BOTTOM LINE TO OBTAIN THE SUPPORT OF THE ADMINISTRATION

The typical reaction of an administrator to the suggestion of a new program or a request for support in implementing a new program is, "How much will it cost me?" You should expect this type of reaction when you propose a waste minimization program. Therefore, it is best to prepare the economic side of your proposal because that will be the most convincing argument that you can present.

Waste minimization in the medical laboratory has practical benefits in terms of cost savings. These include reduced costs for:

- The purchase of supplies when discarded items are recycled and reused instead of new items
- The purchase of solvents when redistilled solvents are used instead of new chemicals
- The treatment, management, and disposal of smaller quantities of infectious waste
- The treatment, management, storage, and disposal of smaller quantities of hazardous and radioactive wastes
- Regulatory compliance for the management and disposal of smaller quantities of infectious, hazardous, and radioactive wastes

There may also be income from the sale of recyclable materials such as waste paper, corrugated cartons, and certain waste solvents.

When the quantities of wastes generated in medical laboratories are minimized, additional benefits can accrue. For example, when smaller quantities of wastes are handled, the risks of occupational exposures and accidents are reduced and there may be a corresponding reduction in workers compensation claims and costs.

There are also the varied benefits that derive from the adoption of new technology. In addition to the waste minimization aspects, there are also the benefits of efficiency and cost savings from using rapid screening, miniaturized test systems, and alternative test methods.

Additional benefits of a waste minimization program that should be cited in enlisting the support of the administration include the environmental benefits that derive from waste minimization — primarily smaller quantities of waste that require disposal and reduced use of new materials and resources. Furthermore, the commitment to waste minimization can be a positive factor in an institution's relations with its community and its shareholders.

ENDNOTES

1. Centers for Disease Control, Regulations for implementing the clinical laboratory improvement amendments of 1988: a summary, *MMWR*, 41(RR-2), 9, 1992.
2. Pearson, M. L., Jereb, J. A., Frieden, T. R., Crawford, J. T., Davis, B. J., Dooley, S. W., and Jarvis, W. R., Nosocomial transmission of multidrug-resistant *Mycobacterium tuberculosis*. A risk to patients and health care workers, *Ann. Intern. Med.*, 117, 191, 1992.
3. Centers for Disease Control, Recommendations for prevention of HIV transmission in health-care settings, *MMWR*, 36(2S), 1987.
4. Centers for Disease Control, Guidelines for prevention of transmission of human immunodeficiency virus and hepatitis B virus to health-care and public-safety workers, *MMWR*, 38(S-6), 1989.
5. U.S. Environmental Protection Agency, Standards for the tracking and management of medical waste; interim final rule and request for comments, *Fed. Reg.*, 45, 12326, 1989.

6. Occupational Safety and Health Administration, Occupational exposure to bloodborne pathogens: final rule, *Fed. Reg.*, 56, 64175, 1991; *Code of Federal Regulations*, Title 29, Part 1910.1030, Bloodborne Pathogens.
7. Rutala, W. A., Odette, R. L., and Samsa, G. P., Management of infectious waste by US hospitals, *JAMA*, 262, 1635, 1989.
8. ATSDR, The Public Health Implications of Medical Waste: a Report to Congress. U.S. Department of Health and Human Services, Public Health Service, Agency for Toxic Substances and Disease Registry, Atlanta, 1990, #PB91-100271.
9. *Code of Federal Regulations,* Title 29, Part 1910.1030, Bloodborne Pathogens, Section (d) (4) (iii) (A).
10. *Code of Federal Regulations,* Title 29, Part 1910.1030, Bloodborne Pathogens, Sections (g) (1) (i) (B), (C), and (E).
11. Hospitals are returning to reusable surgical supplies, *Wall Street J.*, April 2, 1993.
12. *Code of Federal Regulations,* Title 40, Parts 260–272.
13. Occupational Safety and Health Administration, Occupational exposure to hazardous chemicals in laboratories, final rule, *Fed. Reg.*, 55, 3327, 1990.
14. Marsik, F. J. and Denys, G. A., Sterilization, decontamination and disinfection procedures for the microbiology laboratory, in *Manual of Clinical Microbiology*, 6th ed., Murray, T. R., Baron, E. J., Pfaller, M. A., Tenover, F. C., and Yolken, R. H., Eds., American Society for Microbiology, Washington, D.C., 1994.
15. Reinhardt, P. A. and Gordon, J. G., *Infectious and Medical Waste Management*, Lewis Publishers, Chelsea, MI, 1991.
16. Denys, G. A., Infectious waste management, in *Encyclopedia of Microbiology*, Vol. 2, Academic Press, New York, 1992, 493.
17. Hoeltge, G. A., How to reduce hazardous laboratory waste, *CAP Today*, 6(6), 46, 1992.
18. Bartley, J., Employee-driven waste management, *J. Healthcare Mater. Manage.*, 9(3), 28, 1991.
19. OSHA regulations for control of occupational exposure to bloodborne pathogens requires that sharps containers be located as close as feasible to the immediate area where sharps are used.[9]
20. These are the labeling specifications in the OSHA bloodborne pathogen regulations.[10] For consistency and ease of waste identificaton, it is advisable to use them for all containers of infectious wastes, including those infectious wastes that are not contaminated with bloodborne pathogens.
21. The OSHA bloodborne pathogen standard prohibits the manipulation of hypodermic needles.[6] Reintroduction of reusable needles should *not* be considered because of the risks that would be encountered during their handling and reprocessing.
22. The availability of personnel is an increasingly important factor because of changing demographics and the trend to downsizing of the workforce.
23. At present, there is divergence of opinion about whether or not there is a cost differential between disposables and reusables and, if so, which type of item (disposable or reusable) benefits from such a differential.[11]
24. Regulatory requirements for treatment of infectious wastes vary with the state. Some states merely define treatment as a process that renders the waste no longer infectious. In other states, the regulations specify the treatment methods that must be used. In some of these latter states, specific approval must be obtained for use of treatment methods other than those specifically listed in the regulations.

25. At this time, medical waste tracking regulations are in effect in Connecticut, New Jersey, New York, Rhode Island, and Puerto Rico. These are the same jurisdictions that were subject to the USEPA's two-year demonstration program for medical waste tracking from June 1989 to June 1991.
26. See Chapter 9 for a detailed discussion of regulations and other considerations relevant to wastewater effluents from laboratories.

CHAPTER **12**

Minimization of Low-Level Radioactive Wastes from Laboratories

Peter C. Ashbrook, John Brandon, and Hector Mandel

CONTENTS

Introduction 195
Brief Overview of UIUC Radiation Safety Program 196
Brief Overview of Low-Level Radioactive Waste Management
at UIUC 197
Statistics for Low-Level Radioactive Waste Disposal at UIUC 199
Future Plans for Waste Minimization 203
Conclusions and Recommendations 203

INTRODUCTION

Radioactive materials are valuable tools used by laboratories in a wide variety of research applications. Although low-level radioactive wastes from laboratories rarely contain enough radioactivity to present public health hazards, the public has demanded that such wastes be stringently controlled. Off-site disposal options are few. Those sites that are open have drastically increased disposal costs over the last decade. Although groups of states in some parts of the U.S. have made progress in developing new disposal sites for radioactive wastes, most observers expect that any such new sites will not open for years. Given these public demands and the limited, expensive disposal options, minimization of radioactive wastes has received great attention by those institutions that use radioactive materials.

As with chemical waste minimization in laboratories, there are no panaceas for minimizing the generation of radioactive wastes. However, by employing a variety of pollution prevention and waste minimization strategies, an institution can effect substantial decreases in the amount of low-level radioactive wastes generated. In spite of having 200 principal investigators permitted to user radioactive materials, the University of Illinois at Urbana-Champaign (UIUC) has developed a radioactive waste management program under which an average of only two drums of radioactive waste requiring off-site disposal have been generated annually over the past decade. We will present strategies used by the UIUC as well as some that have proven successful at other institutions.

The standard strategies used to minimize wastes of all types:

- Waste reduction
- Source separation
- Recycling
- Reclamation
- On-site treatment

are also applicable to low-level radioactive wastes. The major difference between management of waste chemicals and management of low-level radioactive waste is that treatment methods to destroy radioisotopes are not economically feasible. Therefore, when low-level radioactive waste is managed for disposal, waste minimization strategies have focused on either concentrating the waste to minimize volume requiring storage or diluting the radioisotopes to levels that do not present hazards to the public.

BRIEF OVERVIEW OF UIUC RADIATION SAFETY PROGRAM

The structure of the UIUC Radiation Safety program provides an excellent framework for pursuit of waste minimization strategies. Overall policy is established by the Radiations Hazard Committee, which is made up of faculty members from various campus departments. The University has a Radiation Safety Office that provides campus-wide management of radioactive materials. Each principal investigator is responsible for usage of radioactive materials in her/his laboratory and must apply to the Radiation Safety Office for a permit to use radioactive materials. Among other things, the investigator must specifically identify the radioisotopes to be used and the types of wastes that might be produced. When all the information required by the Radiation Safety Office has been provided, the investigator is issued a permit specifying what isotopes and how much of each are permitted in the laboratory. In the unusual situation where an investigator has a dispute with the Radiation Safety Office about policy issues, the dispute is referred to the Radiation Hazards Committee for resolution.

The Radiation Safety Office controls the use of radioisotopes by controlling purchases. All purchases of radioisotopes must be approved by the Radiation Safety Office. All deliveries from vendors are made to the Radiation Safety Office, which in turn verifies that the researcher is permitted to receive each delivery and then delivers the radioisotopes directly to the researcher's laboratory. All waste disposal is closely monitored by the Radiation Safety Office so that each radioisotope is tracked from the time of receipt on campus to the time and place of final disposal.

This internal control framework has two very important functions with respect to management of low-level radioactive wastes. First, the permit system allows the University to determine how wastes will be disposed before they are generated. For example, the Radiation Safety Office will ask tough questions if it appears that mixed chemical and radioactive wastes might be generated, or if particularly large quantities of wastes might be generated. If wastes are likely to be generated for which there are no available disposal routes, the Radiation Safety Office may (and has) refuse to issue a permit to the investigator for the proposed research.

The second function of interest about the internal control framework is that very good data are available about the types, kinds, and sources of low-level radioactive waste generated on campus. Such data are fundamental to determining where to devote resources.

BRIEF OVERVIEW OF LOW-LEVEL RADIOACTIVE WASTE MANAGEMENT AT UIUC

Low-level radioactive wastes can be placed into a small number of categories:

- Aqueous wastes
- Solvent wastes
- Animal carcasses
- Dry wastes

Different strategies have been applied to each type of waste.

Aqueous wastes — The University does not collect aqueous wastes. Most aqueous low-level radioactive wastes are disposed via the sanitary sewer provided that the levels of radioactivity are below the standards set by the U.S. Nuclear Regulatory Commission (NRC) and the Illinois Department of Nuclear Safety (IDNS). Occasionally a researcher will have aqueous wastes that cannot be disposed down the drain. In the past, the researcher would have been instructed to solidify the waste with plaster of paris, and then handle the waste as dry waste (see below). More recently, the University has been looking into ways to solidify or dewater the waste without generating as much waste as is produced by the use of plaster of paris. In any event, the University has had a policy for over 10 years that no aqueous radioactive wastes are collected from researchers.

Solvent wastes — Liquid scintillation cocktail is the largest single category of solvent wastes. Traditionally, xylene or toluene have been used as the main component in the cocktail. Most liquid scintillation cocktail contains so little radioactivity that it is not regulated by the NRC. However, these cocktails are regulated as hazardous chemical waste because they meet the regulatory characteristic of ignitability. Even though the radioactive content is low enough to be exempt from regulation, liquid scintillation cocktail is frequently handled as a mixed waste because most chemical waste disposal facilities will not accept it for disposal. There has been at least one commercial facility willing to and permitted to burn liquid scintillation cocktail. However, disposal of liquid scintillation cocktail at a commercial facility is expensive and involves considerable paperwork. In response to the difficulty in disposing of liquid scintillation cocktail, the chemical manufacturers have developed environmentally friendly cocktails that can be disposed down the drain. Many institutions have required researchers to switch to one of these environmentally friendly cocktails so that they will not be left with such wastes to dispose at commercial facilities.

The UIUC has not joined its counterparts in requiring its researchers to switch to biodegradeable cocktails. In the late 1970s, the University began burning its waste cocktail as a fuel supplement at its own power plant. Burning of cocktail is specifically approved by the NRC and the IDNS, provided that the levels of radioisotopes at the point of release at the top of the stack are within specified levels. This practice is also specifically approved in the air quality permit issued to the University by the Illinois Environmental Protection Agency. Because of the small amounts burned, the practice is exempt from RCRA Part B Permit requirements under the regulations applicable to the use of utility boilers to incinerate hazardous waste.

A waste issue associated with disposal of liquid scintillation cocktail is what happens to the vials. For those not familiar with liquid scintillation cocktail, the waste is typically generated 5 to 10 ml at a time in vials, which are usually made of glass. Most institutions collect the cocktail in vials. The emptying of vials has been a significant logistical issue at many institutions. At the UIUC, all researchers who generate waste liquid scintillation cocktail are required to empty the vials into poly jerricans. The jerricans are collected by the waste management staff, emptied into a 55-gal drum or holding tank, and returned to the researcher. The empty vials are then decontaminated by the researchers and either reused or disposed of as ordinary trash.

Another potential problem with radioactive solvents occurs when the solvents contain chlorinated compounds. Even if one is able to ship liquid scintillation cocktail for off-site disposal or burn it on-site, disposal of radioactive solvents containing chlorine is another matter. At the UIUC, generation of waste radioactive solvents containing chlorine is strongly discouraged, and hence is rarely encountered.

Animal carcasses — Radioisotopes are used as tracers in various studies involving animals. After the animals are sacrificed, the carcasses are burned in a pathological incinerator on campus. This practice is specifically approved in permits from the NRC and the IDNS, provided that radioisotope releases at the top of the stack are within specified limits. The incinerator also has an air quality permit from the Illinois EPA.

Dry wastes — As a result of the above management practices, the University does not have any liquid low-level radioactive wastes that require off-site disposal. Dry wastes are separated by isotope in the laboratories where wastes are generated. This practice allows the University's waste management staff to segregate the waste in storage into separate containers for short-lived isotopes and long-lived isotopes.

Dry wastes are collected in plastic bags, and labeled with the amount of isotope present. The waste management staff collects the wastes directly from the laboratories where they are generated and stores them in 55-gal steel drums pending disposition.

Wastes containing short-lived isotopes (P-32, I-125, Cr-51) are allowed to decay to background levels. Background levels are defined as less than 0.0001 mCi of the isotope. Once at background levels, the University has the option to go through the waste, deface all the radioactive labels, and dispose of the waste as regular trash.

Wastes containing long-lived isotopes (H-3 and C-14) are evaluated for the amount of radioactive content. In virtually every case, the radioactive levels have been low enough that the waste can be burned at the pathological incinerator used for animal carcasses and be in compliance with the NRC and IDNS limits for release of these isotopes at the top of the incinerator stack. Instead of disposing of decayed waste in the trash, the University has incinerated such waste in a pathological incinerator. Wastes containing S-35 are generally allowed to decay prior to incineration; however, because of the longer half-life (87 days), such wastes are not usually decayed to background levels prior to incineration.

The incineration program has allowed the University to become virtually self-sufficient in the disposal of its radioactive wastes. Those wastes that cannot be incinerated usually contain radium, uranium, or other long-lived isotopes that do not go up the incinerator stack. Because such radioisotopes would remain in the ash, there is no point in incinerating such wastes.

Following each burn of radioactive material, the ash is monitored for radioactivity as it is removed from the incinerator. In our experience, the only ash that has been found to be radioactive was ash from the incineration of animal carcasses containing microspheres. The radioisotopes used in microspheres are relatively short lived. Ash containing microspheres is held for storage, and has eventually decayed to background levels in our experience, allowing for disposal as special waste. Analytical tests on the ash have found the levels of heavy metals to be below the limits pertaining to hazardous wastes. All ash from incinerators is regulated as "special waste" in Illinois.

STATISTICS FOR LOW-LEVEL RADIOACTIVE WASTE DISPOSAL AT UIUC

The success of the waste minimization efforts at the UIUC is illustrated by historical data. As shown in Figure 1, the number of deliveries of radioisotopes has increased steadily over the 40 years such materials have been used on campus.

Low-level radioactive waste generation at UIUC increased along with deliveries. In fiscal year 1976, the UIUC generated 197 55-gal drums of dry waste and

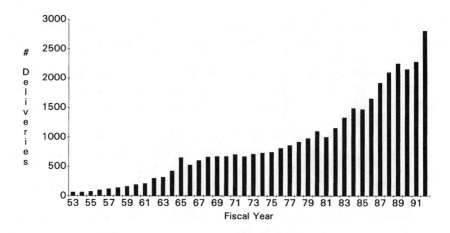

Figure 1 Number of radioisotope deliveries.

31 30-gal drums (3720 l) of liquid scintillation cocktail, all of which were disposed in a radioactive waste landfill. As long as disposal costs were low, the volume was of little concern. By the late 1970s, the UIUC Radiation Safety Office became concerned that land disposal of cocktail was environmentally unsound and could see that increasing restrictions would be placed on landfill disposal of dry wastes. As a result, UIUC adopted several strategies to reduce waste generation and disposal costs.

One of the first initiatives was obtaining permission to burn liquid scintillation cocktail at the power plant as a fuel supplement. This strategy was designed to stop the landfilling of the cocktail as well as to save money. At first the cocktail was introduced into the feed line of oil-fired boilers. Several years later, the system was changed to inject the cocktail directly into a gas-fired boiler. Production of waste liquid scintillation cocktail reached over 5600 l in fiscal year 1981.

Unlike many institutions, the problem of handling all the little glass vials was only a minor problem for the Radiation Safety Office. Researchers at UIUC have been required for many years to empty the vials into poly jerricans, which are then collected for disposal. The jerricans are emptied into a suitable storage container (drum or tank) and returned to the user. Sometime around 1980, the Radiation Safety Office implemented a policy to reduce the presence of glass vials in the dry waste. Specifically, researchers were required to decontaminate the vials and either reuse them or discard them in the normal trash.

While many institutions implemented a requirement that researchers use biodegradeable cocktails, the UIUC has not done so. Nevertheless, generation of liquid scintillation cocktail has dropped dramatically since 1980 (Figure 2).

A second strategy was the purchase of a compactor in the late 1970s to minimize storage requirements and disposal costs. Although the compactor did not minimize the amount of waste disposed, it did minimize the number of drums required for waste disposal.

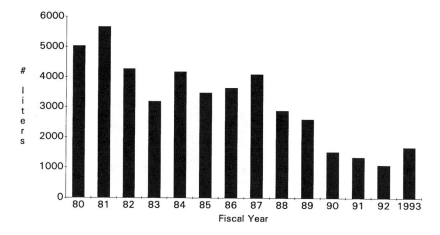

Figure 2 Amount of cocktail produced at UIUC.

It was soon apparent that the use of a compactor was only a band-aid solution. In 1980, the UIUC began a program to segregate low-level radioactive wastes containing short-lived isotopes (P-32, I-125, and Cr-51) from those with long-lived isotopes. Once decayed (defined as less than 0.0001 mCi), the radioactive labels on the wastes were removed or destroyed and the wastes were disposed with the normal trash. In the 1980s, the number of drums containing short-lived radioisotopes averaged about 60 per year.

In connection with requiring researchers to segregate waste by isotopes, the Radiation Safety staff implemented a number of procedures to reduce the amount of radioactive waste generated. One of the most significant policies was the decision to require researchers to decontaminate scintillation cocktail vials instead of disposing of them as radioactive waste. In addition, the staff made efforts to educate the research community about the cost of radioactive waste disposal and to encourage them to take steps to minimize the amount of waste generated.

In the early 1980s, the University conducted a series of tests to determine if it would be feasible to incinerate wastes containing long-lived radioisotopes in the pathological incinerator. Tests using wastes spiked with known amounts of H-3 and C-14 showed that all of these two radioisotopes went up the stack, presumably as water and carbon dioxide. As a result of these tests, incineration of radioactive waste is an important component of the UIUC waste management program (see Figure 3).

As noted above, the ash from incineration is monitored for radioactivity. If the ash is found to be radioactive, it is held for decay. (The usual cause of such radioactivity is the presence of microspheres containing relatively short-lived isotopes.) Ash that is not radioactive is disposed of as special waste, which applies to all incinerator ash disposed of in Illinois, by landfilling in a special waste landfill.

Over the last few years, the number of drums collected and disposed have been roughly equal (see Figure 4), while the number of drums in storage pending either

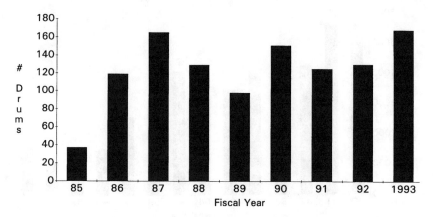

Figure 3 Number of drums incinerated.

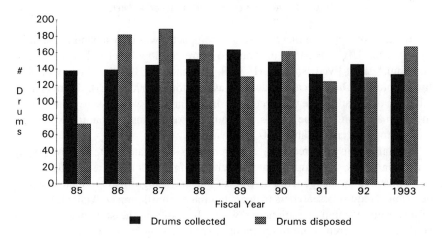

Figure 4 Number of drums collected and disposed.

on-site or off-site disposal has dropped to a steady level (Figure 5). Though not shown in a graph, the number of drums that cannot be disposed on campus has averaged about two drums per year since 1981, the last year the UIUC made a routine shipment of radioactive waste to an off-site facility.

Among those wastes that cannot be disposed on campus are naturally occurring radioactive materials (NORM). Some such wastes are excess laboratory chemicals. If suitable, excess laboratory chemicals containing uranium and thorium are placed on the list of chemicals available for redistribution, a program that is part of the UIUC chemical waste minimization program.

Recycling of radioactive wastes has been explored, but, unlike the situation with waste chemicals, options are limited. One area of success has been with one of the University's few instances of mixed wastes: radium-contaminated lead bricks. These bricks were wrapped in plastic and are now in use for shielding purposes in a campus laboratory. Although these lead bricks may eventually

MINIMIZATION OF LOW-LEVEL RADIOACTIVE WASTES

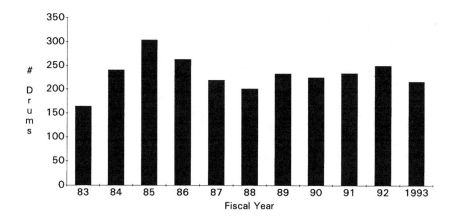

Figure 5 Drums in storage.

require disposal as mixed waste, at the present time they should not be considered waste because they are being used for a useful purpose.

FUTURE PLANS FOR WASTE MINIMIZATION

The Radiation Safety staff continues to monitor all research projects to make sure that wastes are not generated for which disposal alternatives are not available. Initiatives planned for the near future are (1) to phase out the use of plaster of paris and (2) to eliminate the disposal of large glass or metal objects, which could be decontaminated, from the trash.

CONCLUSIONS AND RECOMMENDATIONS

Through its structure for control of radioisotopes and application of a variety of initiatives over the past 15 years, the UIUC has managed to reduce the generation of low-level radioactive waste requiring off-site disposal to a trivial amount — two drums per year. While some of the options available to the UIUC may not be available or workable at all institutions, the UIUC example of exploring a wide variety of strategies can be applied universally.

We make the following recommendations:

1. Use of radioisotopes must be strictly controlled by the institution. Control of purchasing is probably the most effective method of requiring compliance with institutional policies.
2. Research protocols should be reviewed prior to experimentation. This review will determine (1) if use of radioisotopes is necessary, (2) if short-lived isotopes can be used, (3) that excessive amounts of radioactive materials will not be purchased, and (4) to make sure that wastes can be disposed.

3. Consideration should be given to requiring the use of biodegradeable (drain disposable) cocktails unless your institution has an alternate method of disposal.
4. Wastes containing short-lived isotopes should be separated from those containing long-lived isotopes. After decay, such wastes may be disposed as normal trash. (Be sure to deface all radioactive labels.)
5. Consider the possibility of burning wastes containing tritium and C-14 in a pathological incinerator, if one is readily available.
6. Educate researchers about the costs and technical difficulties of radioactive waste disposal. Encourage them to practice waste minimization techniques in all phases of their laboratory work.

**Part 4:
What Individual Laboratories Can Do**

CHAPTER 13

The Microscale Chemistry Laboratory

Ronald M. Pike, Zvi Szafran, and Mono M. Singh

CONTENTS

Introduction 207
History of the Modern Microscale Chemistry Laboratory 209
Microscale Equipment and Techniques 211
    Reaction Vessels 211
    Glassware Assembly 211
    Stirring 212
    Heating 212
    Measurement of Chemicals 212
    Transferring and Filtering of Liquids 213
    Crystallization 213
    Separation Techniques 214
    Other Techniques 214
    Changed Reaction Conditions 215
The Effects of Microscale Chemistry in the Instructional Laboratory 215
Applications of Microscale Chemistry to Industry 217
    Conversion Costs to a Microscale Program 218
Conclusion 219
Endnotes 219

INTRODUCTION

The generation of chemical wastes, the disposal of these materials, and their impact on the environment are subjects of increasing concern in the U.S. and around the world. Over the past 2 decades, there has been a growing awareness of the various problems associated with chemical usage. Chemical manufacturers generate thousands of tons of chemical waste each year. Traditionally, these wastes were handled by incineration (causing air pollution), disposal through sewage systems (creating water pollution), or disposal to landfills (polluting ground water and soil).

Various regulations have been passed by both local and federal government agencies to deal with these large volumes of wastes. Since the publication of *Silent Spring* by Rachel Carson,[1] billions of dollars have been allocated and spent in an effort to clean up chemical wastes. The Superfund program is just one of the latest manifestations of this effort. In many cases, wastes have been illegally dumped. In other cases, even when materials were disposed of *following every regulatory law* and incorporating *the best technology* in existence at the time, problems still developed. "Permanently" sealed landfills leaked. "Permanently" sealed canisters of nuclear waste corroded, spilling toxic materials into the oceans. "Perfectly safe" materials turned out not to be so safe after all.

As the population has become more aware of chemical waste disposal problems, the traditional ways of dealing with wastes are no longer feasible. Siting of new incinerators or waste disposal dumps are increasingly problematic. The chemical industry has begun to recognize the importance of full-cost accounting as a means of attacking this very serious problem.[2] Is it any wonder that the emphasis of the regulatory agencies has shifted from regulation of the disposal site to *elimination of the waste at the generation site?* Allowing chemical fumes to pass into the atmosphere though laboratory ventilation systems is no longer a viable option. Many of the chemicals that are in common use are associated with fire and explosions hazards. Others react violently under mild conditions. Still others are exposure hazards, being carcinogenic, mutagenic, irritants etc. Increasingly long listings of TLVs, LD50s, LDLos, and so on can be found in the chemical and governmental regulatory literature.

In addition to the problems of chemical waste disposal, there has been an increasing awareness of the need to reduce the exposure of workers (both laboratory and manufacturing) to chemicals. The legally acceptable exposure levels have consistently declined over the past decades, while the number of chemicals that must be monitored has sharply increased. Maintenance of a healthful laboratory and manufacturing environment has become a top priority. This is also very expensive, requiring large scrubbing, air handling and heating capacities, not to mention liability costs.

The recent trend in favor of reduction of chemical wastes at the *source* is most directly accomplished by sharply reducing the usage of chemicals. In the laboratory setting, the reduction in scale (to approximately 50 to 150 mg of solids and 50 to 2000 µl of liquids) is known as microscale chemistry. In this way, microscale chemistry amounts to a total quality management (TQM) approach to chemical waste handling. The ideas of chemical use reduction, air quality improvement, exposure limitation, recycling, and waste reduction need to be introduced in the chemical laboratory curriculum at the most elementary levels, ranging from the general science offerings of elementary schools, to introductory chemistry in high schools and to the college- and university-level courses. This major effort is needed to impact the training of future generations of scientists (especially chemists and engineers), so that they will become familiar with the techniques and equipment necessary to work with microquantities of chemicals. Microscale chemistry will accomplish a *cultural change* in the way scientists view the usage of chemicals. Persons trained in this way will have a major impact on the chemical industry of the future, and as informed consumers and voters in the wider society.

HISTORY OF THE MODERN MICROSCALE CHEMISTRY LABORATORY

The feasability of carrying out chemistry experiments at microscale levels was established in the mid 1800s, principally in central Europe with the work of Emish and Pregl. Pregl received the Nobel Prize for his microscale work in 1923. In the U.S., microscale chemistry gained a foothold shortly after World War II. Cheronis and Ma at Brooklyn College, Benedetti-Pichler and Schneider at Queens College, and Stock at the University of Connecticut taught microscale techniques and published several papers and texts.[3]

These programs, however, were not widely adopted. This was due to several factors:

- The concerns we have for the the environment today were not in vogue before 1970
- There was little concern for the laboratory air quality, as risks associated with chemical exposure were not known or seriously underestimated
- Costs of chemicals were low and cleanup consisted of all waste being dumped down the drain
- Only small numbers of students were involved in the upper-level courses where the microscale techniques were used, who could handle the complicated manipulations involved
- The lack of accurate analytical balances and measuring devices restricted the utility of microscale techniques; the equipment that was available did not lend itself to rapid manipulation of chemicals

Thus, microanalysis was highly delicate, tedious, time consuming, and specialized. The technique was confined to a very limited and restricted area of reseach in graduate schools and industrial laboratories. For new techniques to become widespread, there must be practical reasons for their adoption, and it must be technologically possible to implement them. It was not until the 1980s, when environmental concerns had risen to the forefront and the electronic milligram balance became available, that using microscale experiments at the introductroy level of instruction became a reality.

Microscale chemisty was developed for the introductory laboratory by Dana W. Mayo and Samuel S. Butcher (from Bowdoin College), and Ronald M. Pike (from Merrimack College, who was on Sabbatical leave at Bowdoin). Bowdoin College was faced with a difficult problem because of the poor air quality in its organic chemistry laboratory. Students were experiencing headaches and nausea, especially during the Grignard experiments (which involved the use of large quantities of diethyl ether). After the complaints reached the president of the college, W. F. Enteman, the chemistry department was directed to find out how much it would cost to retrofit the laboratory to improve the air quality. The amount was about $300,000, leading to the inevitable question, "Isn't there a better way?" Hence, the birth of microscale chemistry in its modern form.

Until this point, reactions had been carried out at the traditional laboratory scale of 5 to 20 g of solid and 50 to 500 ml of solvent. After several weeks of deliberation, the discussion focused on how much the scale of chemical usage could be reduced in the introductory laboratory. What was the smallest amount of chemicals that an introductory student could efficiently manipulate and still "see," "feel," and learn the chemistry involved? What changes in laboratory techniques would be necessary at this small scale? What new glassware and analytical equipment would become necessary?

The necessary techniques and materials were then developed for the organic chemistry laboratory during 1982–1984. The first teaching tests occurred at Bowdoin and Merrimack Colleges in 1983. The tests proved to be highly successful, with the sophomore level students rapidly adapting to the new techniques. It was found that students could readily accommodate microscale manipulations at this level. Preliminary results were reported at the national meeting of the American Chemical Society in 1984,[4] and were published in the *Journal of Chemical Education*.[5] The first microscale chemistry textbook, *Microscale Organic Laboratory*, appeared in 1986.[6] Many others have followed since then.[7] This work directly led to the growth of a mini-industry in the area of microscale chemistry, with at least seven glass companies manufacturing microscale kits and glassware, as well as several chemical companies now marketing microscale quantities of chemicals.

The success of the organic laboratory program led Zvi Szafran, Ronald M. Pike, and Mono M. Singh (at Merrimack College) to investigate the implementation of microscale techniques to inorganic chemistry. There had been an increasing tendency toward elimination of laboratories in inorganic chemistry, due to excessive cost of chemicals and equipment, being unable to perform interesting laboratories in a safe and efficient manner, and the problems of waste disposal. By converting the inorganic chemistry laboratory to the microscale level, it became possible to widely expand the range of experimental coverage to include such important areas as organometallic chemistry of the heavy metals, catalysis, and bioinorganic chemistry. The results of the work in inorganic chemistry were first presented at the Biennial Conference on Chemical Education (Purdue University, Lafayette, IN) in 1988, and were published in the *Journal of Chemical Education* in 1989.[8] The textbook *Microscale Inorganic Chemistry: A Comprehensive Laboratory Experience* appeared in 1991, and is the only such laboratory textbook currently in print in the area of inorganic chemistry.[9]

With microscale chemistry courses increasing to a current number of about 1000 in the U.S. alone, microscale chemistry was clearly an idea whose time had come. Microscale chemistry was currently affecting more than 100,000 students each year in the U.S. and has begun to be implemented around the world. Microscale programs are now in operation in Canada, Mexico, France, England, Czechoslovakia, Finland, the Peoples Republic of China, and South Africa. Most recently, the National Microscale Center has been established at Merrimack College, to further the worldwide dissemination of microscale chemistry technology, and to encourage the incorporation of these techniques into the elementary and high school curricula.

MICROSCALE EQUIPMENT AND TECHNIQUES

Since performing an experiment at the microscale level involves using a sharply reduced (by a factor of about 100) quantity of solid reagent and solvent, it is tempting to think that all one needs to do is reduce the quantity of chemicals and shrink the glassware. In general, this is not true. Experimental techniques at the microscale level are considerably different from their conventional scale equivalents. In many cases, the microscale glassware will have significant modifications relative to their larger counterparts. Reaction conditions are also significantly altered.

In designing microscale equipment, there are a number of considerations that must be borne in mind relative to the convential scale. With a 100-fold reduction in the quantity of solvent, the problems of surface wetting and reagent handling require serious consideration.

REACTION VESSELS

Consider as "simple" a piece of equipment as a reaction vessel. Traditionally, reaction vessels are round bottomed, and connected to condensers, distillation heads, etc. via greased ground glass joints. If any separation of liquid phases is to be carried out within the vessel, it will be difficult to impossible to see the phase interface with this quantity of material (50 to 500 µl) in a round-bottom flask. Reactions are therefore typically carried out in conical vials, shown in Figure 1. The flat bottom of the vial gives it more stability (the vial will not roll over) than a round bottom would, crucial when the equipment is that small. The tapered cone within the vial gives a greater height to the same quantity of liquid relative to a nontapered vial. Even when the quantity of one phase is relatively small, the taper allows a sharp interface to be seen, improving the ease of the extraction process.

GLASSWARE ASSEMBLY

The conical vial is threaded on the top of the outside, and has a female ground glass joint (TS 14/10) on the inside. Condensers and other equipment that might be attached to the conical vial are connected in two ways. First, there is a vacuum-tight TS 14/10 ground glass male joint, which fits within the female joint without

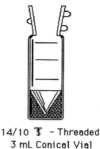

14/10 ℑ - Threaded
3 mL Conical Vial

Figure 1

needing any grease as a lubricant. Second, a threaded plastic cap supported by an O-ring sits on the shoulder of the male joint. This cap screws onto the threads of the conical vial. Collectively, this is called an O-Ring Cap Seal Connector™. A great advantage of this type of connector is that the entire glassware assemblage can be held by a single three-prong clamp, as shown in Figure 2. Since the glassware is small, there is not much space available for additional clamps, etc., even if one wished to use them, necessitating this type of connection.

STIRRING

If stirring is required, specialized techniques must be employed. It would be difficult to use a conventional mechanical stirrer with this size of equipment. Instead, liquids are stirred magnetically, using spin vanes (see Figure 2). Since the reaction vial is conical, the vanes must be designed to fit into the cone. To spin the vane, a magnetic-stirring hot plate is employed (which can also act as a heat source for reactions). If a round-bottom flask is used, micro spin bars are employed for this purpose.

HEATING

As mentioned above, magnetic-stirring hot plates can serve as a heat source for microscale reactions (Figure 2). Since the surface area in contact with the hot plate is small, it is desirable to have the reaction vessel positioned in a device that increases the contact area. There are two common devices used: the sand bath (held in a small crystallizing dish or other such container as shown in Figure 2) and the metal block (which is made from any good conductor, such as Al or Cu). Alternatively, gas microburners are available, and can be conveniently employed using a fireproof wire gauze mat provided the solvent being used is not flammable. For safety reasons in introductory laboratories, the use of oil baths should be avoided.

MEASUREMENT OF CHEMICALS

The digital electronic balance has revolutionized the use of microquantities of solids. Masses as low as 0.1 mg can be efficiently and rapidly weighed using

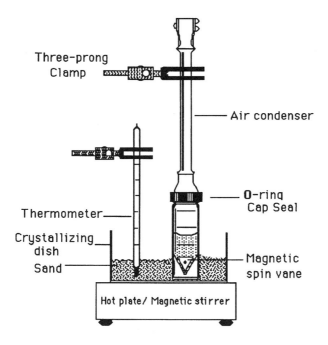

Figure 2 Glassware assemblage illustrating stirring and heating.

commercially available automatic-taring balances. Small quantities of liquids (up to 1 ml) are conveniently dispensed using automatic delivery pipets. A wide variety of these devices are commercially available, making measurement of quantities as low as 10 µl routine. Larger quantities of liquids are handled in the traditional manner, i.e., using graduated pipets and pipet pumps (or bulbs).

TRANSFERRING AND FILTERING OF LIQUIDS

Dust, dirt, and other filterable impurities can cause a contamination disaster at the micro level, and must be removed. The operations of transferring liquids and liquid filtration are combined by use of a Pasteur filter pipet. These are prepared by inserting a small wisp of cotton (balled up into a small plug) into the long tip of the pipet using a copper wire for tamping.[6,7a] The plug acts as a filter, and also guards against back pressure for volatile liquids. The long, narrow tip of the pipet allows the last trace of liquid to be captured. Thus, transfers can be made efficiently without undue loss of materials.

CRYSTALLIZATION

Microscale crystallizations (for 10 to 100 mg of solid) can be conveniently carried out using a Craig tube. The Craig tube consists of two parts, a test-tube-like glass receptacle and a Teflon™ barrel. The glass receptacle is imperfectly ground on the inside to allow a loose fit with the barrel, allowing passage of the mother liquor. In this process, the solid to be recrystallized is dissolved in a minimum amount of hot solvent within the glass receptacle (using a sand bath/ magnetic stirring hot plate as the heat source). The mixture is rapidly triturated using a microspatula to avoid bumping, and to help effect dissolution. The receptacle is then placed in a small beaker, the Teflon™ barrel is inserted (to keep out dust), and allowed to cool slowly to room temperature (or below, if an ice bath is used). A round-bottom test tube is placed over the assemblage, which is then inverted. Centrifugation removes the mother liquor from the recrystallized product.[6,7a]

SEPARATION TECHNIQUES

Liquid solutions can be separated using distillation techniques at the microscale level. Several methods are available for quantities of liquids from 500 µl to 2 ml. If the liquids boil at widely different temperatures, a Hickman still (commercially available with and without side arms) can be used. The still has a collar that can hold (on average) 500 µl of distillate, which is removed using a long-tip Pasteur pipet. For more difficult separations, a spinning-band distillation column is effective. The spinning band is made of Teflon™, and has a magnet embedded in it at the tip, which in turn is cut to fit in the cone of a conical vial. The band is spun magnetically using a magnetic-stirring hot plate, which also serves as the heat source for the distillation. The spinning band helps partition the liquids, and separations of components boiling as closely together as 15°C are routinely accomplishable.[6,7a,10]

For even smaller quantities (down to 5 to 10 µl), purification can be effected by the use of preparative gas chromatography. The material to be separated is injected into the GC. Special stainless steel heat sink fittings are available to attach to the output of the chromatograph. At one end, the fitting is threaded on the inside to screw onto the outlet port of the GC. At the other end, the fitting emulates a female ground glass joint on the inside, and is threaded on the outside to accept an O-Ring Cap Seal Connector™ of a glass collector. The collector is a 10-cm long glass tube, with two blown-out bulbs to accommodate the eluent. Once the separation has been accomplished, the glass collector is removed from the fitting, inserted into a small conical vial (using the same Cap Seal), inserted into a test tube, and centrifuged. The product liquid is thereby transferred into the conical vial for further work.[6,7a,11]

OTHER TECHNIQUES

Any technique that can be accomplished conventionally can also be carried out at the microscale level, using a little imagination and readily available equipment. Examples include extraction (using a Pasteur filter pipet in place of a separatory funnel), titration (using a 1- or 2-ml calibrated glass pipet fitted with a disposable tip),[12a] sublimation (using a small sublimator, made from a filter flask and a test tube used as a cold finger[6,7a,7b]), and determination of densities of liquids using a micropycnometer.

CHANGED REACTION CONDITIONS

When the quantity of solid reagent employed in a chemical reaction is reduced, the surface area to bulk ratio increases sharply. This results in a proportionally larger reaction interface, which in turn results in faster reaction times. Coupled with time savings generated by easier material measurements, faster glassware assembly, and shortened workups, a time savings of approximately 50% is readily achieved.[13]

This improvement in reaction rate has several potential drawbacks, however. Undesirable reactions (such as oxidation and hydrolysis) also occur more readily. In cases where this is a problem, reaction assemblages should be protected under inert atmosphere or using $CaCl_2$ drying tubes.[6,7a,9] A pedagogical "problem" is caused by the lack of spot heating. Efficient stirring readily eliminates problems of spot heating at the microscale level. At larger industrial levels, this cannot be solved as easily. Students trained at the microscale level must be informed of the problems of mass and heat transfer, which are generally not observed at the microlevel.

THE EFFECTS OF MICROSCALE CHEMISTRY IN THE INSTRUCTIONAL LABORATORY

Conversion to microscale chemistry has a profound impact for the better in the instructional laboratory.

One of the major advantages of microscale chemistry is the greatly enhanced safety aspect within the laboratory. It is obvious that air quality is markedly upgraded as the quantity of solvents and other volatile substance is reduced by a factor of 100 to 1000 from the conventional laboratory work scale. This is especially beneficial to those laboratories that do not possess high-efficiency ventilation systems and have limited funds for upgrading their present facilities.

Another important advantage is the reduced risk of fire or explosion afforded by the reduced quantities of chemicals used. Over the past 8 years of operation of microscale organic and inorganic laboratories, there have been no reports of accidents of this type. Currently, without microscale, 30,000 educational institutions generate more than 4000 metric tons of hazardous wastes per year in the U.S. alone.[14] The cost of disposing of this waste is astronomical. At Texas A&M University, for example, more than $10,000 is spent each month in disposing of laboratory wastes.[15]

Conversion to microscale chemistry markedly reduces the amount and costs of chemical waste disposal. In the academic sector alone, this reduction in scale (by a conservative factor of 100) if fully implemented, would amount to an elimination of 3960 metric tons of toxic waste annually, and a cost savings of hundreds of millions of dollars per year. Robin Woods, a spokesperson for the federal Environmental Protection Agency stated "Waste minimization is a big priority at the EPA these days. We spent the last 15 to 18 years controlling pollution that is already generated. *Now we are trying to prevent it in the first place*." The microscale chemistry approach is a major way of doing exactly that.

Table 1 Product Generation Relative to Product Need

Product	Approximate amount generated	Amount needed for analysis	% disposed
NH_4BF_4	5.0 g	0.1 g for ^{19}F-NMR	98
SnI_4	2.7 g	5 mg for melting point	99+
$(NH_4)_2PbCl_6$	5.0 g	2 g for stability tests[a]	60
$CpFe(CO)_2CH_3$	2.0 g	5 mg for melting point, 0.1 g for ^{1}H-NMR 10 mg for IR	94
$(C_2H_5)_4Sn$	10 g +	0.1 g for ^{1}H-NMR[b]	99
$Cu(NH_3)_4SO_4$	15 g	50 mg for solubility test	99+
BF_3	40 g	12 g to make adduct[a]	70

[a] Amount used in published experiment. Much less could have been used to do the same analysis or reaction at the microscale.

[b] Some product used in second reaction, also at large scale.

An interesting question is, "Is anything lost by producing that much less product?" In the educational laboratory, in experiments carried out in the organic or inorganic area, the product produced must be used in various characterization tests or have utility in a subsequent reaction. Given current analytical techniques, the amount of product required is quite small, as shown in Table 1, which summarizes the amouts of product produced relative to product need following conventional laboratory procedures.[17]

It is clear that most of the product generated in the experiments was never employed for any useful purpose. The obvious lesson to be learned is "Never make more product than is required for subsequent work or characterization." The data in Table 1 also dramatically illustrate the wisdom of using small quantities of all reagents, and for *recycling to the maximum degree possibile for toxic materials*. This recycling aspect is especially critical when designing new experiments for the chemical laboratory.

In addition to cost savings associated with decreased chemical use and decreased waste disposal, there are substantial cost savings associated with improved laboratory and worker safety and decreased liability costs. The healthfulness of the laboratory environment is improved for potentially millions of students and teachers.

In addition to cost savings, microscale chemistry also offers the promise of improved pedagogy. Studies done by the late Miles Pickering showed a time savings factor of approximately 2 in time trial studies of the Grignard reaction carried out at Princeton University.[13] This time savings is of key importance in incorporating microscale chemistry in the elementary and high school levels, where time limitations for laboratories are significant. The chemical cost savings allow a wider variety of reagents to be used in the laboratory. The areas of organometallic chemistry, catalysis, and heavy metal chemistry now become affordable.[8,17] Thus, an increased variety of experiments now become possible using a greater variety of reagents and materials, due to lowered cost and time requirements.

The microscale concept is equally adaptable to cover both the qualitative and quantitative aspects of the general chemistry laboratory. Experiments illustrating

Table 2 1985 Survey of Hazardous Waste Generated

SIC Code	Description	Solid waste (tons)	Wastewater (tons)
281	Industrial inorganic chemicals	243,208	24,456,702
282	Plastics and synthetic resins	226,051	23,092,523
283	Drugs	269,263	8,293,092
284	Soaps, detergents, etc.	6,374	6,380
285	Paints, varnishes, etc.	90,833	34,888
286	Industrial organic chemicals	1,782,271	112,000,970
287	Agricultural chemicals	152,129	7,970,867
289	Miscellaneous	122,613	2,156,785
2800	Chemical manufacturing, general	38,054	32,237,299
	Total	2,930,796	210,249,506

the areas of stoichiometry, equilibria, thermodynamics, electrochemistry, etc. have been published.[18] An inorganic qualitative analysis scheme without the use of the toxic hydrogen sulfide or thioacetamide reagents is available.[18a]

Use of smaller quantities of chemicals makes on-site recycling a more viable option, reducing waste production even further. Storage facility needs are reduced. Reduced chemical usage translates into less chemical being purchased, less needing to be stored, and less decomposing due to short shelf life. Should an accident occur, there is also less chemical present to cause any problem to the surroundings, as well as firefighting personnel, police, and the public at large.

In conclusion, laboratory operational costs are reduced in many ways. There is less expenditure for starting materials, for storing bulk chemicals, for waste disposal, for liability insurance and coverage, and workmen's compensation costs due to chemically related illness. Conversely, the quality of the laboratory is improved, due to lessened exposure to toxic materials, improved air quality, and a wider variety of laboratory experiments being accomplishable in a shorter period of time.

APPLICATIONS OF MICROSCALE CHEMISTRY TO INDUSTRY

Chemical industry consumes and disposes of much larger quantities of chemicals and solvents than in the academic sector, as summarized in Table 2[19] for companies belonging to the Chemical Manufacturers Association (CMA). Since CMA companies make up about 58% of the chemical industry, this would indicate approximately 5 million tons of solid waste, and 360 million tons of wastewater.

If microscale techniques were adopted in industry, a similar reduction (by a factor of 100) in chemical usage and disposal would be possible in most research and development laboratories. In the manufacturing plants themselves, material substitution and waste minimization promise to reduce wastes substantially.

The 3M Company stated the following key points relative to waste minimization:

> It takes resources to remove pollution. Pollution removal generates residue. It takes more resources to dispose of this residue and in the process create more pollution.

The combined total of 1,900 3M projects has resulted in eliminating annually the discharge of 110,000 tons of air pollutants, 13,000 tons of water pollutants and 260,000 tons of sludge of which 18,000 tons are hazardous, along with the prevention of approximately 1.6 billion gallons of wastewater. Cost savings to 3M have been more than $292 million. These cost savings are for pollution control facilities that did not have to be built, for less pollution control operating costs, for reduced manufacturing costs, and for retained sales of products that might have been taken off the market as environmentally unacceptable.

This did not include the savings due to enhanced safety and reduced exposure, which would improve work-environment healthfulness for 1.07 million workers in the chemical industry.[20] Chemists trained in microscale techniques become more cognizant of the chemical wastes that they produce, and ways to minimize them. If even a 5% reduction in industrial hazardous waste was accomplished via microscale chemistry technique application, this would amount to the elimination of 250,000 tons of solid waste, as well as 18 million tons of waste water. This would amount to a cost savings of $820 million (1985 dollars, on the same basis as 3M). Again, this does not include the savings due to enhanced safety and reduced exposure.

CONVERSION COSTS TO A MICROSCALE PROGRAM

In both the academic and the industrial sectors, conversion to a microscale program requires an initial outlay of funds. New glassware must be purchased (costing about $200 per student or worker), as well as more sophisticated analytical equipment (transfer pipets, balances, spectrophotometers, etc.). How quickly can these costs be amortized?

An important factor in funding the conversion of chemistry programs to the microscale level is the large savings in operational costs for laboratories, due to decreased chemical, glassware, insurance, storage, energy, and waste disposal costs, discussed above. Direct cash savings can typically pay back the conversion costs in a period of 6 months to 2 years, depending on the size and scope of the program. The cost savings continue to accrue from this point. Energy savings and cost savings for not needing to build laboratories with better air-handling capacities are difficult to estimate. The benefits in terms of improved healthfulness in the laboratory cannot be measured in financial terms. A simplified funding analysis for a college or community college academic program is shown in Table 3, on the basis of 20 students per laboratory section.

It has been our experience that on average there is a direct cost reduction of $2000 in chemical costs and $3000 in waste disposal cost per 20-student section per year. On this basis, a small academic program of 1 section per year will have a payback time of 2.62 years, a medium-sized program (5 sections per year) will have a payback time of 0.91 years, and a larger program (20 sections per year) will have a payback time of 0.69 years. Even this rapid payback can be improved if the students share the microscale glassware — the assumption above being that each student will be supplied with his own individual kit. In addition, the labora-

Table 3 Microscale Laboratory Costs (New Costs for Going Micro, $)

	No. of Students		
Item	20 Single section	100 Five sections	200 Ten sections
Glassware (1/student)	2,400	12,000	24,000
Micro balances (4)	4,000	4,000	4,000
Automatic delivery pipets (4)	500	500	500
Melting point appararatus (2)	2,200	2,200	2,200
Magnetic stirring hot plates (20)	4,000	4,000	4,000
Total	13,100	22,700	34,700

tory equipment needed for organic chemistry is essentially the same as is used in inorganic chemistry. The two laboratories can share space, cutting costs even further. In high-school programs, implementation costs are lower due to lessened glassware requirements.

CONCLUSIONS

The advent of the microscale concept has revolutionized the training methods used in the introductory instructional labortory and also in the research and/or product development areas. The impact of this method on the environmental aspects of air pollution and waste generation and disposal will be seen in the years to come.

ENDNOTES

1. Carson, R., *Silent Spring,* Houghton Mifflin, Boston, 1962.
2. For example see, *Chem. Eng. News,* January 11, 1993, p. 8.
3. See Stock, J. T., *J. Chem. Educ.,* 67, 898, 1990 and references therein.
4. (a) Pike, R. M., Mayo, D. W., Butcher, S. S., Hotham, J. R., Foote, C. M., and Page, D. S., *An Introductory Microscale Organic Laboratory Program,* 187th ACS National Meeting, St. Louis, April, 1984; (b) Butcher, S. S., Mayo, D. W., Hebert, S. M., Hotham, J. R., Foote, C. M., Page D. S., and Pike, R. M., *Microscale Organic Laboratory Program: Air Quality Aspects,* 187th ACS National Meeting, St. Louis, MO, April, 1984.
5. (a) Butcher, S. S., Mayo, D. W., Pike, R. M., Foote, C. M., Hotham, J. R., Page D. S., *J. Chem. Educ.,* 62, 147, 1985; (b) Mayo, D. W., Butcher, S. S., Pike, R. M., Foote, C. M., Hotham, J. R., and Page, D. S., *J. Chem. Educ.,* 62, 149, 1985.
6. Mayo, D. W., Pike, R. M., and Butcher, S. S., *Microscale Organic Laboratory,* 1st ed., John Wiley & Sons, New York, 1986.
7. (a) Mayo, D. W., Pike, R. M., Butcher, S. S., and Trumper, P. K., *Microscale Techniques for the Organic Laboratory,* John Wiley & Sons, New York, 1991; (b) Williamson, K. L., *Macroscale and Microscale Organic Experiments,* Lexington, MA, 1989; (c) Pavia, D. L., Lampman, G. M., Kriz, G. S., and Engel, R. G., *Introduction to Organic Laboratory Techniques, A Microscale Approach,* W.B. Saunders, Philadelphia, 1990; (d) Rodig, O. R., Bell, C. E., Jr., Clark, A. K.,

Organic Chemistry Laboratory: Standard and Microscale Experiments, W.B. Saunders, Philadelphia, 1990; (e) Nimitz, J. S., *Experiments in Organic Chemistry, From Microscale to Macroscale,* Prentice-Hall, Englewood Cliffs, NJ, 1991; (f) Landgrebe, J. A., *Theory and Practice in the Organic Laboratory with Microscale and Standard Scale Experiments,* 4th ed., Brooks/Cole, Pacific Grove, CA, 1993.

8. Szafran, Z., Pike, R. M., and Singh, M. M., *10th Biennial Conference on Chemical Education,* Purdue University, August 1–4, 1988; paper 362, *J. Chem. Educ.,* 66, A263, 1989.
9. Szafran, Z., Pike, R. M., and Singh, M. M., *Microscale Inorganic Chemistry: A Comprehensive Laboratory Experience,* John Wiley & Sons, New York, 1991.
10. Mayo, D. W. and Pike, R. M., Spinning Band Fractional Distillation Column, U.S. Patent 4,770,746, September 13, 1988.
11. Mayo, D. W. and Pike, R. M., Gas Chromatography Collection Device and Process, U.S. Patent 4,730,480, March 15, 1988.
12. (a) Singh, M. M., Szafran, Z., and Pike, R. M., *J. Chem. Educ.,* 68, A125, 1991; (b) Singh, M. M., Szafran, Z., and Pike, R. M., *J. Chem. Educ.,* 70, A36, 1993.
13. Pickering, M., and LePrade, J. E., *J. Chem. Educ.,* 63, 535, 1986.
14. *Chem. Eng. News,* May 22, 1989.
15. Private communication: Dr. Donald Clark, Health and Safety Officer, Texas A. & M. University, American Chemical Society National Meeting, Dallas, 1989.
16. Mandt, D. K., *J. Chem. Educ.,* 70, 59, 1993.
17. Singh, M. M., Szafran, Z., and Pike, R. M., *J. Chem. Educ.,* 68, A125, 1991; (b) Singh, M. M., Szafran, Z., and Pike, R. M., *J. Chem. Educ.,* 67, A261, 1990.
18. (a) Szafran, Z., Pike, R. M., Foster, J. C., *Microscale General Chemistry Laboratory with Selected Macroscale Experiments,* John Wiley & Sons, New York, 1993; (b) Russo, T., *Microscale Chemistry for High School General Chemistry,* Kemtec Educational, Kensington, MD, 1986; (c) Mills, J. L. and Hampton, M. D., *Microscale Laboratory Manual for General Chemistry,* Random House, New York, 1988; (d) Ehrenkranz, D. and Mauch, J. J., *Chemistry in Microscale: A Set of Microscale Laboratory Experiments with Teacher Guides,* Kendall/Hunt Publishing, Dubuque, IA, 1990; (e) *Microscale Chemistry,* Curriculum module by The Woodrow Wilson National Fellowship Foundation Chemistry Institute, 1987.
19. Wentz, C. A., *Hazardous Waste Management,* McGraw-Hill, New York, 1989, 3.
20. Wentz, C. A., *Hazardous Waste Management,* McGraw-Hill, New York, 1989, 120.

CHAPTER **14**

At the Lab Bench: Finding the Right Balance in Source Separation

Peter A. Reinhardt

CONTENTS

Introduction 222
Waste Separation Definitions 222
The Benefits of Source Separation 223
    Choose a Suitable Disposal Method for Each Waste 223
    Manage Waste Safely 224
    Facilitate Compliance and Minimize Regulatory Complexity 224
    Keep Disposal Costs in Step 225
    Recycle and Minimize Waste 225
    Maintain and Maximize Waste Management Options 226
The Costs of Source Separation 226
    Assess and Analyze Waste 226
    Collect and Contain Waste 227
    Train Laboratory Personnel and Waste Handlers 228
    Separation Errors and Public Relations 229
Finding the Balance in Source Separation 230
    One Modern Extreme: Minimal Source Separation 230
    Define Separate Hazardous Wastestreams 230
    Criteria for Separating and Combining Wastestreams 231
Implementing Source Separation Laboratories 233
    Train Laboratory Personnel 233
    Use the Sanitary Sewer and Normal Trash Properly 234
    Supply Easy-to-Use Waste Collection Containers 234
    Consider Waste Collection Stations 235
    Identify Waste Collection Containers Distinctively 235
    Provide User-Friendly Waste Collection Services 236

Monitor and Minimize Errors 236
Beyond the Lab Bench 236
    Waste Segregation: Keeping Wastes Separate 236
    When to Mix Waste 237
Conclusion 237
Endnotes 237

INTRODUCTION

If you separate wastes by type you can reduce disposal costs and make waste management safer and easier. Optimal disposal methods can differ by waste type, so separation allows you to use the optimal method for each waste type.

Keeping wastes separate can be difficult and costly in terms of training, errors, and maintaining separate systems for waste collection and management.

These viewpoints, which seem contradictory, illustrate the dilemma of waste separation. While some people worry that multiple receptacles for separate wastes will fill laboratories, others feel it isn't worth the trouble. A better perspective is to understand the benefits and costs of keeping wastes separate and seek a balance in differentiating waste types to create a sensible collection system. This chapter should help find that balance for your laboratory.

WASTE SEPARATION DEFINITIONS

This chapter uses several terms, defined below, that sometimes have different meanings. The glossary of this book has additional and more detailed definitions.

Source separation is the practice of separating different types of wastes at their point of generation and discard (i.e., separation at the source of generation).

Wastestream is a type or class of waste with similar characteristics, usually generated on a regular basis. A wastestream may be delimited by the special way in which it is managed. Hospitals generate many distinct wastestreams, such as discarded newspapers and hypodermic needles. On a larger scale, all hospital waste could be considered a single hospital wastestream, which differs in character from an industrial wastestream. Identification and division of distinct wastestreams is the subject of this chapter.

Waste segregation is used here to describe the practice of keeping different wastestreams segregated from the time they are placed in the collection container through transport, storage, treatment, and disposal. Separation and segregation are not used in this chapter to mean the sorting of wastes after they have been combined, which is usually difficult or hazardous. To maintain waste management options, it is important that distinct wastestreams be kept segregated when they are handled and transported.

Mixed or *mixing* is used in this chapter to mean the combining of two or more waste containers or wastestreams, which is their plain meaning. The term "mixed waste," also has a legal meaning (which is not used in this chapter) of waste that is both regulated as chemically hazardous by the U.S. Environmental Protection Agency (EPA) and waste that is regulated as radioactive by the U.S. Nuclear Regulatory Commission (NRC). Commingling and bulking are other terms used for mixing wastes after collection, and are discussed in Chapter 19.

Normal trash refers to nonhazardous dry garbage, refuse, and other waste that is typically disposed of in a local sanitary landfill or in an incinerator for energy recovery. These wastes are commonly referred to as *solid wastes*.

THE BENEFITS OF SOURCE SEPARATION

To illustrate the various degrees of source separation, consider one extreme. Imagine a laboratory where there is no separation of wastes. This imaginary laboratory has only one trash container where all unwanted chemicals, radioactive materials, and biological wastes are disposed of.

Such a laboratory may have been common in the 1950s or 1960s, and some laboratories may have such a disposal system today. A clinical laboratory might discard all of its wastes in a red bag to be sent to a medical waste incinerator. However, it is improper to discard a broken mercury thermometer in the red bag. Mercury is not destroyed by incineration, but instead is emitted to the air. Conversely, it is unnecessarily expensive to dispose of packaging materials, discarded reports, newspapers, pop cans, and other normal trash as medical waste.

This illustrates the benefits of keeping wastes separate. It would be very unusual for a lab to have such a homogeneous wastestream that all its wastes can be safely treated in the same manner. As the risks of the traditional disposal methods (i.e., sewer, landfill, and incinerator) are better understood, the trend is to manage our wastes more carefully and efficiently. This has been fueled by specialization in disposal methods and recycling, which relies on keeping wastes separate by their physical and chemical characteristics and hazardous properties.

CHOOSE A SUITABLE DISPOSAL METHOD FOR EACH WASTE

Consider the many routes of waste management that are available to a laboratory, such as the sanitary sewer, normal trash, and autoclaving. The ultimate method of managing waste may not be obvious to laboratory personnel, but it is likely that the laboratory's safety office or environmental services staff have access to a chemical waste hauler who can send it to a chemical incinerator or other specialized commercial disposal facilities.

There is no single disposal method that is appropriate for all wastes. Each disposal method is limited in the types of wastes that it can accept. An autoclave does not alter the hazard of radioactive waste. Flammable solvents should not be disposed of in the sanitary sewer. Some chemicals disposed of in a normal trash can contaminate the groundwater beneath the solid waste landfill to which it is sent.

Even incinerators differ in their capabilities. Chemical waste incinerators are designed to be extremely efficient; some can destroy 99.9999% of the organic chemicals fed into them. Medical waste incinerators easily destroy pathogens but do not attain that efficiency; instead they are operated to combust contaminated paper, plastics, and putrescible wastes.

These differences in disposal capabilities are often the impetus for waste separation. To manage waste properly, match each waste type with an appropriate means of disposal.

MANAGE WASTE SAFELY

Safety is another important reason for source separation. When incompatible chemicals are collected or stored in the same container there is risk of fire, explosion, or toxic gas generation. Highly exothermic reactions are possible when incompatible laboratory chemical waste is mixed. Chapter 19 discusses safe waste commingling in detail. Further, containers and caps need to be compatible with their contents.

The U.S. Occupational Health and Safety Administration (OSHA) standard for bloodborne pathogens requires that contaminated sharps be collected separately from other regulated waste.[1] If not collected in a puncture-resistant container, hypodermic needles and other sharps can injure waste handlers, and in the process may expose them to chemicals or disease.

FACILITATE COMPLIANCE AND MINIMIZE REGULATORY COMPLEXITY

Environmental laws are the motivation for separation of many types of wastes. The NRC has long required licensees to keep certain low-level radioactive waste separate from other wastes for shipment to a site dedicated for the disposal of such wastes. Laboratories began to collect flammable solvents separately when drain disposal was found to be unsafe and became much more restricted by law. In 1980 EPA began regulating hazardous chemical waste and made disposal as normal trash illegal. This prompted laboratories to set up a system to collect hazardous chemical waste separately from normal trash.

One reason to keep different wastes separate is to minimize the amount of regulated waste. Most laws do not allow dilution as a means of reducing the hazardous characteristic of a regulated waste. Mixtures of regulated and nonregulated waste must be managed in their entirety; dilution or careless mixing simply results in a voluminous regulated wastestream, and commensurately higher waste management costs and regulatory risks.

Perhaps the most difficult to manage laboratory wastes are those that have multiple hazards and are regulated by more that one set of laws. Radioactive waste that is chemically hazardous is especially difficult to manage as specified by NRC and EPA, whose rules can conflict and interfere with safe management. Source separation is the first step in avoiding these problematic wastes; do not unnecessarily mix wastes that have radioactive, biological, or hazardous chemical characteristics.

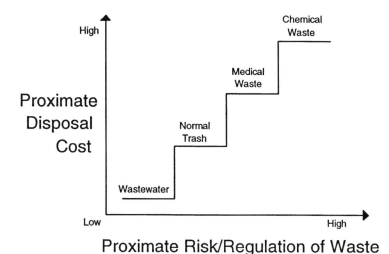

Figure 1 Risk vs. cost is a step function.

KEEP DISPOSAL COSTS IN STEP

Waste types differ in their characteristics, risks, and degree of regulation. As regulations and risks increase, so generally does the cost of waste management and disposal. Nonhazardous wastes can be disposed of in the sanitary sewer and the normal trash, which are the least expensive disposal routes. Hypodermic needles, which can cause wounds and transmit diseases, must be decontaminated and destroyed in many states. Mercury can threaten public health if not properly disposed of, so it must be managed as a hazardous chemical waste by federal law, usually through a specialized disposal firm. The most expensive waste to dispose of is usually multihazardous waste.

Due to the way waste is regulated, the relationship between risk and disposal cost is a step function (see Figure 1). A small change in risk, such as one degree of flash point, one tenth of a pH, or one part per million can change an unregulated waste to an EPA-regulated chemical waste. This is called a "notch effect." There may be little difference in the risk of certain medical wastes and normal trash, but the disposal cost difference can be large.

Beware of wastes whose characteristics place it near a regulatory step or notch. Separate wastes on the basis of risk to avoid "taking a step up" in increased waste management costs. Keep needles out of the normal trash, and keep broken mercury thermometers out of sharps receptacles. Above the notch, a small reduction in risk can result in large cost savings.

RECYCLE AND MINIMIZE WASTE

Most people are first introduced to source separation when their community asks them to keep certain household wastes and recyclables out of their general trash. This minimizes the amount of wastes being placed in solid waste landfills or being combusted as refuse derived fuel. Recycling conserves resources. Laboratories should participate in their institution's and community's recycling programs. If they can be recycled and are not contaminated to regulated or unsafe levels, keep aluminum cans, recyclable paper, magazines, glass, polystyrene, and other materials out of red bags and containers for contaminated labware.

Keeping organic solvents separate by type can facilitate recovery, distillation, and reuse. The energy value of bulk nonchlorinated organic solvents can be inexpensively recovered when used as fuel in a cement kiln or other boiler/industrial furnace that meets EPA requirements (i.e., fuel blending). Histopathology laboratories that keep their spent xylene apart from other organic solvents can distill and reuse it on-site. When organic solvents are mixed indiscriminately, azeotropes can form that make distillation impractical.

MAINTAIN AND MAXIMIZE WASTE MANAGEMENT OPTIONS

Source separation maximizes your waste management options. When wastes are kept separate by type, management methods can be tailored to each type to maximize safety and suitability of the disposal method, and to minimize cost. Separation of laboratory wastes makes recycling and recovery possible, just as it does for aluminum cans and office paper.

When different wastes are mixed, disposal options are often lost. Even a few milliliters of PCBs may make waste flammable liquids unsuitable for fuel blending.

After wastes have been mixed, most are very difficult, costly, or hazardous to separate. Careless mixing of wastes carries a high cost in limiting potential waste management methods. In the laboratory, separation of waste types is best accomplished by source separation rather than attempting to separate wastes after they have been collected mixed and commingled.

THE COSTS OF SOURCE SEPARATION

Now imagine the other extreme in source separation: a laboratory with hundreds of waste receptacles, one for every type of waste that may be generated: waste alcohols, waste alkanes, waste chlorinated solvents, etc.; tritium (^{3}H)-labeled waste, carbon-14 (^{14}C)-labeled waste; used plastic pipette tips, used glass pipettes — you get the picture. Scientists fear that this laboratory may be in the not-too-distant future.

In addition to the above benefits, there are several difficulties to overcome when implementing waste separation. These difficulties result in either a direct cost to the laboratory or require additional labor of the people who work there.

ASSESS AND ANALYZE WASTE

It is difficult, costly, and time consuming to determine a waste's type or classification (see Figure 2). OSHA's Bloodborne Pathogens Standard

FINDING THE RIGHT BALANCE IN SOURCE SEPARATION

Doonesbury   BY GARRY TRUDEAU

Figure 2 Waste assessment is difficult.

(29 CFR 1910.1034) regulates wastes that, "would release blood or other potentially infectious materials in a liquid or semi-liquid state if compressed," as well as other waste types. To completely separate regulated waste and normal trash, a medical technologist would need to carefully evaluate every item. When in a hurry or when time is valuable, it is a common practice to err on the side of caution and dispose of most clinical laboratory waste in the red bag.

Likewise, EPA's Toxic Characteristic Leachate Procedure (TCLP), which estimates a waste's potential to release certain toxic chemicals in a landfill, is an expensive analytical test. Wastes that fail the TCLP test are regulated by EPA as hazardous waste. Small wastestreams are costly to test, so if a waste is known or suspected to contain a TCLP-listed chemical it is usually most efficient to save the cost of the test by assuming the waste would fail and manage it as regulated by EPA.

Although additional assessment and analysis is a cost of waste separation, failure to separate can result in high analysis costs as well. Most treatment and disposal facilities require an analysis or other characterization to ensure that received wastes are acceptable for their processes.

COLLECT AND CONTAIN WASTE

The additional receptacles for separate wastestreams is a cost to the laboratory and institution. In turn, source separation requires staff to find, select, and use separate containers for each waste type, which translates to additional labor costs. Likewise, waste handlers must take additional efforts for each new wastestream. Perhaps a scientist, technician, or physician (whose time is valuable and may be better spent at another task) should not be troubled with selecting the proper waste container.

As described below, the most effective source separation systems employ dedicated waste containers and visual and spatial cues to make separation easier. Thus, space (and possibly clutter) is another cost of waste separation. Laboratory personnel may add benchtop collectors (see Table 1), and waste handlers may prefer temporary accumulation areas on floors and wings (see Figure 3). To keep wastes separate, additional space is needed at the building's loading dock and institution's storage facility.

Table 1 Example of Waste Collection Containers

400-ft² laboratory		600-ft² laboratory	
Waste Collection Containers for Biohazardous Waste			
Sharps container for hypodermic needles	1	Sharps container for hypodermic needles	3
Can for petri plates	1	Wastebasket for dry waste marked "biohazardous"	1
		Bag of dry waste	1
		Wastebasket for biohazardous pipettes	1
		Wastebasket for biohazardous paper	3
		Box for biohazardous glass	2
Waste Collection Containers for Chemical Waste			
5-gal jug for flammable organic solvents	1	5-gal jug for acrylamide solutions	2
5-gal jug for chlorinated organic solvents	1	5-gal jug for chlorinated organic solvents	1
Bottle of stock chemical marked "waste"	1	5-gal jug for TTE solutions	2
		Bottle of phenol waste	1
		Bottles of ethidium bromide waste	3
		Benchtop collector for liquids	1
		Used oil	1
Waste Collection Containers for Radioactive Waste			
1-gal bottle for aqueous radioactive solutions	1	5-gal jug for aqueous H-3 solutions	1
Box of liquid scintillation vials	1	5-gal jug for aqueous S-35 solutions	1
		Box of solid (dry) S-35 waste	1
Waste Collection Containers for Other Solid Waste			
Benchtop collectors (cans) for pipette tips	7	Benchtop collectors (cans) for pipette tips	2
Box for pipette tips from benchtop collectors	1	Box for bleach-treated pipettes	1
Pasteur pipettes	2		
Glass for recycling	1		
Paper for recycling	1		
Styrofoam peanuts for recycling	1		
Waste baskets for normal trash	2		
Total	22	Total	28

Note: The above are lists of waste collection containers found in two veterinary medicine research laboratories at the University of Wisconsin–Madison in February 1993.

TRAIN LABORATORY PERSONNEL AND WASTE HANDLERS

As described below, successful source separation requires the training or informing of staff who generate, handle, and manage waste. Training is a cost to the institution in time spent for policymaking, preparation of informational materials, and learning. Training costs increase as more wastes are separated and more errors are likely. For laboratory personnel, training costs are particularly high (and typically not fully valued) due to their large number and the fact that their full cooperation and understanding is necessary for source separation.

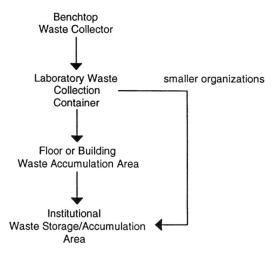

Figure 3 Physical pathway of waste collection.

SEPARATION ERRORS AND PUBLIC RELATIONS

Separation errors can be costly in terms of personal injury, environmental harm, noncompliance, and deterioration of community relations, such as when a hypodermic needle or a chemical hazardous waste is mistakenly disposed of in the normal trash. The risk of separation errors increase with the number of separate wastestreams.

Institutions that utilize the normal trash and sanitary sewer for nonhazardous waste risk noncompliance when staff make separation errors. This may be a reason that a sizable number of institutions avoid these practices. Results of the National Laboratory Waste Survey (see Chapter 2) indicate that between 20 and 30% of laboratory populations surveyed do not dispose of any chemical laboratory waste into the sanitary sewer, and about 60% dispose of no laboratory chemicals into the normal trash.[2]

The desire to maintain an institution's public image can result in separation policies that both increase and decrease waste management costs. When nonhazardous wastes are kept out of the normal trash, separation costs and political liability are decreased at the expense of higher disposal costs. For example, the Nuclear Regulatory Commission and some Agreement States allow licensees of radioactive material to propose criteria for designating certain wastes as radioactive (e.g., wastes containing short half-life radionuclides that have decayed to background levels). Yet some laboratories using radionuclides handle all wastes as "radioactive" for fear the public may misunderstand the effects of disposing of it in the local landfill. Likewise, hospitals may choose to dispose of innocuous items (e.g., noncontaminated syringe barrels) as infectious waste to avoid its presence in the normal trash and the perception of improper disposal practices.

At least one large corporate research facility has a policy to incinerate almost all of its laboratory wastes, and has an on-site incinerator licensed for a broad array of chemical, infectious, and radioactive wastes. Laboratory wastes, including normal trash and other solid wastes, are combined in "burn boxes" for incineration. The corporate policy to not separate wastes is based on the facility's incineration capabilities, their desire to avoid liabilities associated with off-site disposal, and their wish to conceal proprietary information that may be gained from knowledge of its wastes. The company also benefits from the low training and labor costs of such a simple system and by eliminating the possibility of separation errors. The high capital and administrative costs of this option preclude its use by most laboratories, even at large institutions.

FINDING THE BALANCE IN SOURCE SEPARATION

For laboratories, the most important source separation decisions are the identification of individual waste types and their management. Separation decisions determine the degree to which laboratories separate wastes and the number of distinct wastestreams it will manage. Table 1 illustrates the separate wastestreams of two laboratories at the University of Wisconsin–Madison. Laboratories separate wastes in response to institutional requirements, experimental protocol, department policies, container availability, recycling programs, and lack of knowledge of disposal routes and the risk of mixing wastes. As a result, it is probably not unusual to find one collection container per 20 square feet of laboratory area, as was found in these labs.

The value of separating each potential wastestream should be evaluated individually, considering the costs and benefits described above (also see Table 2). Finally, the proper balance of waste separation will depend on the realistic ability to manage multiple wastestreams, the overall system complexity, and commensurate costs of separating and handling the multiple wastestreams.

ONE MODERN EXTREME: MINIMAL SOURCE SEPARATION

Table 3 lists waste types that, if generated, usually should be collected separately. While this table is not complete, it illustrates the minimal possible number of laboratory wastestreams. Although few labs will generate all of these wastes, it is relatively easy to generate several of them, and thus need a well-developed source separation system. When an institution manages the wastes of many research, teaching, and medical laboratories, source separation can involve dozens of separate wastestreams.

DEFINE SEPARATE HAZARDOUS WASTESTREAMS

As more is understood about the environmental effects of waste treatment and disposal, lawmakers require that hazardous waste be kept out of the sewer and normal trash. The occupational risks of certain wastes has also been a factor in their regulation. Thus, wastes that exhibit a hazardous property are the first

Table 2 Overview of Costs and Benefits of Source Separation

	Source separation of wastes	
	Minimal	**Maximum**
Waste collection system characteristics	Few wastestreams Few waste collection containers	Many wastestreams Many waste collection containers
Waste collection and management system requirements	Minimal requirements for collection containers	Training of laboratory personnel Specialized design for collection containers and collection system
Other costs	Risk of generating multihazard or mixed waste that is costly to dispose of Occupational, environmental, and compliance risk from waste that is improperly handled or disposed of High volume and disposal cost of hazardous waste from the addition of nonhazardous waste	Risk of separation errors Political risk of separation errors
Other benefits	Few separation errors Minimal training and labor costs	Ability to match wastestream with optimal waste management method Greatest potential for minimizing off-site disposal costs Greatest potential for recycling and waste minimization

candidates for source separation. To further limit waste and conserve resources, lawmakers require separation of recyclable wastes as well.

Still, many laboratory wastes are unique and not regulated, and are allowed in the sanitary sewer and normal trash. Staff should actively identify and assess the hazard of unregulated wastes that may have characteristics that deem special management, such as incineration. Waste assessments are also important to minimize the laboratory's liability for waste disposal.

CRITERIA FOR SEPARATING AND COMBINING WASTESTREAMS

The decision to identify and separate a new, distinct wastestream depends upon the factors described above. Consider wastestreams as candidates for separation whenever there are significant changes in laboratory procedures and activities. Beyond legal considerations, new wastestreams should generally be identified and kept separate when:

- Recycling and waste minimization is available and cost effective
- New waste management methods are available that are more appropriate, more efficient, or have lower costs for certain wastes

Table 3 Critical Wastes Requiring Separation

Waste type	Reason for source separation	Examples of problems from failure to source separate
Biohazardous, infectious	Requires specialized treatment (e.g., disinfection), ideally close to the point of generation	Untreated wastes may pose a risk of infection Most states require treatment
Flammable liquids, toxic metal salts	A separate collection system is necessary to provide an alternative to disposal in the sanitary sewer	Explosions in sewer lines Violation of sewage treatment works limits
Hazardous waste (RCRA-regulated chemical waste)	Federal law prohibits disposal as normal trash; if mixed with trash or garbage, the mixture must be disposed of as a hazardous waste	Improper disposal can result in fines. Failure to separate from normal trash results in unnecessary disposal costs
Mercury salts	Few waste treatment and disposal facilities accept mercury salts or above a very low level of mercury contamination	Waste that is contaminated with mercury from failure to separate is subject to: High disposal costs Regulatory scrutiny
Needles and other sharps (e.g., hypodermic)	Unless tightly contained, needles can easily injure waste handlers	Needlesticks For needles that may be contaminated with blood or body fluids, risk of transmission of bloodborne diseases
Normal trash	Unlike other laboratory wastes, normal trash needs no special or expensive management precautions	If not separated from chemical, radioactive, or biological wastes: Unnecessary increase in waste volume Unnecessary increase in disposal costs
Polychlorinated biphenyls (PCBs)	PCBs are regulated separately, under the Toxic Substance Control Act (TSCA); all but very low levels must be incinerated in a TSCA-approved incinerator	PCB-contaminated organic solvents (above 10 PPM) are not allowed for fuel blending or recycling
Radioactive waste	Requires specialized management, treatment, and disposal	Improper handling may pose an exposure risk The Nuclear Regulatory Commission and some states require separation
Wastewater	Unlike other laboratory liquid wastes, most wastewater needs no special or expensive treatment	If not separated from chemical, radioactive, or biological liquid wastes (e.g., waste organic solvents): Unnecessary increase in waste volume Unnecessary increase in disposal costs

- Management costs or disposal difficulties increase, especially for off-site disposal

Occasionally, it may be appropriate to combine wastestreams. Consider this when:

- Wastes are predominately of one type, such as medical waste generated in a hospital microbiology laboratory
- The consequence of separation error is very high (e.g., risk of noncompliance or exposure); separation errors are reduced with less separation; as precautionary practice, a laboratory may initially consider all laboratory chemical waste to be hazardous
- Separation errors are intractable or costly to reduce to tolerable levels
- The combined wastes can be managed easily, or the mixed constituent is so small as to be tolerable; for example, organic solvents containing a small amounts of water are usually acceptable for fuel blending
- On-site waste separation, handling, and management costs are high

Note that source separation often shifts off-site or disposal costs to on-site separation and handling costs. For example, the high cost of disposing of waste that is both chemically hazardous and radioactive may be reduced by careful, laborious, and costly separation. Evaluate source separation decisions carefully. Separation may result in lower off-site disposal bills at the expense of higher costs for separation errors and on-site management.

IMPLEMENTING SOURCE SEPARATION IN LABORATORIES

Once laboratory wastestreams have been identified, source separation requires a system to gain the cooperation of laboratory personnel and minimize separation errors.

TRAIN LABORATORY PERSONNEL

Laboratory personnel need to be trained to:

- Identify waste types and understand their differences
- Determine and assess a waste's hazardous properties and regulatory status and know when further analysis may be warranted
- Use the appropriate, designated receptacles and label properly
- Use collection containers properly — OSHA prohibits the overfilling of containers of regulated waste (i.e., waste potentially contaminated with bloodborne pathogens); EPA requires that hazardous chemical waste containers be kept closed; to prevent exposure, staff who mistakenly discard the wrong waste into a hazardous waste receptacle should not retrieve it
- Understand the importance of source separation and consequence of error, such as risks to waste handlers (e.g., getting stuck by a needle), environmental risks (emitting mercury into the air), or compliance risks (being fined for disposing of hazardous waste in the normal trash)

Be flexible when providing training and information — people learn differently. A variety of formats, distributed at different times, are more likely to be noticed. Waste separation can be discussed during employee orientation, group meetings, or experimental review. A periodic refresher is helpful, especially to correct errors and whenever the source separation system is changed (e.g., containers are changed, separated wastes are added or subtracted). Waste handling procedures can be distributed via safety manuals, brochures, occasional memos, or computer networks.

Contextual information is retained well. Tags or labels on waste containers inform and remind at the time of discard (see below), as do instructions on disposal forms. Incorporate waste management instructions when discussing related issues during chemical hygiene and exposure control training.

USE THE SANITARY SEWER AND NORMAL TRASH PROPERLY

Most laboratories utilize the sanitary sewer or dispose of waste in the normal trash, which is then combined with the institution's solid waste. Because of the high consequence of separation errors, it is critical to train personnel on the proper use of these disposal routes.

Some laboratories have conservative policies on the use of the sewer and normal trash, which contrasts with the many types of chemical, radioactive, and biological waste that can be safely disposed of by these methods. Like industry, households, and other institutions, laboratories can take advantage of these inexpensive disposal routes, so long as the waste is determined to be legally and environmentally acceptable. Work with your sewage treatment works and solid waste disposal firm to establish safe laboratory disposal procedures. A written agreement is best. Consider the perspective and safety of custodians, plumbers, waste haulers, and recycling centers in the these decisions.

SUPPLY EASY-TO-USE WASTE COLLECTION CONTAINERS

In most laboratories, waste that is not managed or disposed of in the laboratory is removed and managed by institutional support staff. For collection and containment, laboratories use either receptacles supplied by the institution, "found" containers, or containers in which the source materials are received. Unused, surplus source materials are typically disposed of in their original containers. For off-site disposal, haulers overpack or combine laboratory waste collection containers into larger containers that are suitable for shipment (i.e., meet state and federal transportation requirements).

Waste containers provided by the institution have several advantages. The institution can provide standard waste receptacles that:

- Are compatible with the specified waste
- Are of sufficient strength and integrity to minimize leaks and breakage (avoid glass if possible; the container should not tip easily)

- For flammable liquids, are fire safe
- Have tight, compatible closures to prevent leaks and releases
- Are standardized with respect to shape, color, and markings to facilitate waste separation (see below), which is especially helpful for those people who work in several locations

Depending on needs and waste types, other desirable features of provided containers may include:

- Variations in size to correspond to generation rate
- Large mouths
- Spring-loaded caps to prevent leaks, evaporation, and other releases
- A small footprint to minimize space needs
- Clear containers to allow visual inspection (e.g., to detect liquids or hypodermic needles)
- An outer tray or pan to hold drips, leaks, and spills (i.e., secondary containment)

Convenient placement and proximity are other factors that encourage source separation. The veterinary medicine laboratories studied in Table 1 used benchtop collectors to make disposal of commonly generated wastes easier. To minimize the chance of injury, OSHA requires that sharps containers be "easily accessible to personnel and located as close as is feasible to the immediate area where sharps are used or can be reasonably anticipated to be found." Convenient separation requires careful placement of waste receptacles and, in many cases, multiple collection containers in each lab.

CONSIDER WASTE COLLECTION STATIONS

As containers multiply, reducing space needs becomes important. One solution is to provide a small-footprint waste collection station that holds several containers, either stacked or in another space-efficient configuration.[3] Stations make finding waste receptacles easier. Stations also make removal by institutional waste handlers easier. For example, a wheeled collection station can simply be exchanged for an empty one. The station can provide secondary containment for liquid wastes. A station may provide more opportunities for standard point-of-discard instructions.

IDENTIFY WASTE COLLECTION CONTAINERS DISTINCTIVELY

Source separation is facilitated by visual and spatial cues. Waste receptacles should be marked, labeled, or tagged with not only the waste's name and warnings, but also with separation instructions and waste management reminders. The NRC requires that waste be labeled with a "radioactive" label. Point-of-discard information should be very brief and in a large type size. Containers that are coded by a standard shape, color, or material also facilitate subconscious compliance with separation rules. OSHA allows either labeling or color coding for regulated waste (i.e., waste potentially contaminated with bloodborne pathogens).

PROVIDE USER-FRIENDLY WASTE COLLECTION SERVICES

To prevent overfilling, waste receptacles should be removed routinely and replaced with empty containers. The responsibility for closing, sealing, and removing containers should lie with either lab personnel or waste management staff, but it should be clear. When containers are broken or torn they should be promptly overpacked or replaced. Establish a procedure for spill response. Keep in mind that separation principles are applicable to spill cleanup wastes and may influence the selection of spill absorbents or treatment agents.

MONITOR AND MINIMIZE ERRORS

Source separation errors are inevitable. It is too much to expect every person to match each waste with the correct waste receptacles at all times, especially people who are busy and whose primary responsibilities are research or testing.

Given this inevitability, a source separation system (i.e., policies, containers, training, etc.) should be designed to minimize errors, and a tolerable level of error should be defined. A logbook or other record should be kept of separation errors to discern a pattern, frequency, circumstance, or origin. An analysis of errors may suggest the need to improve training, clarify labeling, or add engineering controls.

Separation errors are a serious matter with hazardous waste. If separation errors become intolerable:

- Combine wastestreams to reduce the need for source separation
- Add engineering controls, such as a requirement to autoclave all wastes before they leave the lab; this is standard procedure for Biohazard Level III (P3) laboratories
- Monitor waste containers more closely; some institutions analyze all waste organic solvents for PCBs
- Make training mandatory and enforceable for scientists who generate high risk waste

BEYOND THE LAB BENCH

After removal from the lab, waste is typically brought to an accumulation area or institutional waste storage facility to await shipment to an off-site treatment or disposal facility (see Figure 2). Waste handlers are responsible for keeping waste types segregated and contained. In general, to prevent releases and personal exposure, containers should not be opened unless the facility is equipped with engineering controls to do so.

WASTE SEGREGATION: KEEPING WASTES SEPARATE

It is important that the institution provide sufficient and adequate space for waste accumulation and storage. Multiple wastestreams put a high space demand on loading dock and waste holding areas, which will increase if more solid wastes are separated for recycling. Attention to design and vertical storage can alleviate space pressures.

Accumulation and storage areas should be posted to encourage segregation of waste types by area. Compatibility and fire hazard should be considered; keep oxidizers away from flammable liquids. Leaking liquids can be carefully transferred to a new receptacle, but it may be best to overpack leaking containers by simply placing them in a larger package (i.e., an "overpack"). If you decide to combine wastes, prevent personal exposure by placing the unopened containers together into a larger overpack.

WHEN TO MIX WASTE

Chapter 19 describes when small containers of organic solvents and other liquids can be commingled into a larger shipping container. Even then, failure to separate an unlisted constituent (e.g., PCBs) may make the entire shipping container unacceptable for commercial disposal. Some waste managers combine powders of like chemicals to reduce volumes. Take extensive precautions to prevent personal exposure and mixing incompatibilities for all waste mixing, commingling, and bulking. For other wastes, it is rare to be able to safely mix them. The risk of mixing usually outweighs the benefits in volume reduction and easier handling.

CONCLUSION

Source separation can be a burden, but it is often a necessary burden to minimize and recycle wastes, and to manage hazardous wastes properly and safely. Cooperation of laboratory personnel is critical to successful source separation. Laboratories that are most successful are those that stress the value of source separation, involve their staff in the design of a waste management system, and find a way for all to share in the resulting benefits of conserving resources and protecting the environment.

ENDNOTES

1. U.S. Occupational Safety and Health Administration, Occupational Exposure to Bloodborne Pathogens, Title 29 Code of Federal Regulations Section 1910.1030.
2. Reinhardt, P. A. and Leonard, K. L., Choosing How to Manage Laboratory Chemical Wastes, paper presented at 11th Annual College and University Hazardous Waste Conference, Stanford University, August 9, 1993.
3. Hintzke, M. A., Evaluation and Design of the Biohazardous Waste Disposal System at the Veterinary Medicine Building, unpublished, May 5, 1993.

CHAPTER 15

Solvent Recycling by Spinning Band Distillation: Theory, Equipment, and Limitations

John A. Mangravite and R. Roger Roark, Jr.

CONTENTS

Introduction 240
Theoretical Background 240
A Comparison of Distillation Equipment 248
Terminology and Distillation Characteristics 255
    Efficiency 256
    Holdup 257
    Throughput 257
    Reflux Ratio 258
    Pressure Drop 259
    Equilibration Time 259
    Distillation Cuts 259
Characteristics of Spinning Band Stills 260
    Description of Spinning Band Stills 260
    Determination of the Number of Plates 261
    Determination of Holdup 262
    Determination of Pressure Drop 263
Safety Considerations 263
Conclusion 264
Endnotes 274

INTRODUCTION

One of the more important areas of waste minimization involves the recovery of solvents from technical processes. There are thousands of procedures used in industrial, research, university, and hospital laboratories in which solvents are used for extractions, analysis, and sample preparation. At the end of most of these procedures, the solvent usually remains. Organic solvents are the core components of these techniques and must either be recovered or disposed of in an environmentally safe manner. The range of solvent type is great. Alcohols and hydrocarbons, such as xylene and xylene substitutes are routinely used in tissue analysis. In addition, hospital labs routinely use formaldehyde in the form of formalin solutions. High-performance liquid chromatographic (HPLC) analyses require a wide variety of solvents ranging from the nonpolar hydrocarbons, such as heptane, to the polar material such as tetrahydrofuran and acetonitrile. In many cases, other solvents are combined with these principle materials which usually complicates their recovery. Even the more exotic and expensive hexafluoroisopropanol (HFIP) is in common use in HPLC procedures. EPA methods for trace analysis of pesticides and a myriad of other analytes use large quantities of dichloromethane, freons, and hydrocarbons.

Since 1975, we have been developing methods and procedures in which a majority of these waste mixtures could be recycled. In most cases, our end result was the recovery of material of sufficient purity so that it could be reused in the procedure in which it was generated. In other cases, we were able to carry out volume reduction which led to a minimization of the amount of solvent that would then be disposed of by acceptable methods.[1] In Chapter 20, we provide a variety of case studies in which solvent recycling has become a routine part of the analytical procedure in which the solvent was regenerated. It is the purpose of this chapter to discuss the theoretical background to using spinning band distillation as the method of choice for laboratory-scale recycling of solvents and to describe in detail the available equipment for accomplishing this task. We also discuss the areas in which solvent recycling is not particularly feasible.

THEORETICAL BACKGROUND

As an introduction to a discussion of the theory of distillation, we should first consider the question of why liquids boil. If a pure liquid is placed in a closed container and allowed to come in equilibrium with its vapor (after first evacuating the system), the measured pressure is the vapor pressure of the pure liquid at the temperature of the experiment. By the conditions of the experiment, if this vapor pressure is plotted as a function of temperature, the graphs shown in Figure 1 would result. These graphs show the vapor pressure behavior for four typical liquids. The point of particular interest on these graphs is the intersection of the graph when the vapor pressure equals 760 torr (1 atm pressure). The temperature at which this occurs is the normal boiling point of the pure liquid. Every individual

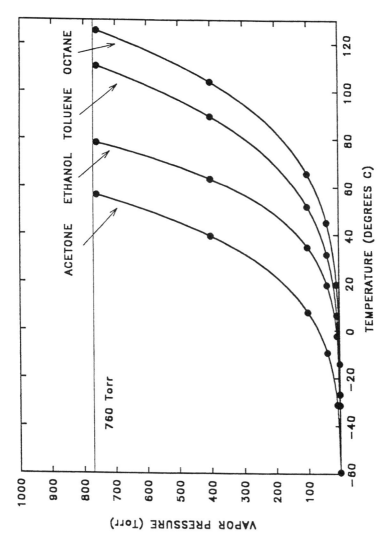

Figure 1 Vapor pressure vs. temperature for various liquids.

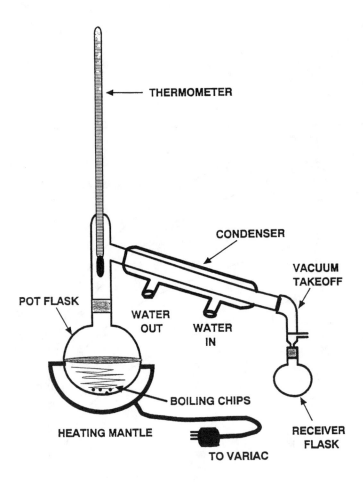

Figure 2 Simple distillation apparatus (student setup).

substance has its own characteristic boiling point and it is this property that must be considered when carrying out a distillation procedure.

The purpose of distillation is the separation of mixtures of volatile liquids or solutions. One type of procedure, *simple distillation* is useful for separating volatile liquids from nonvolatile substances. For example, if one was to place an aqueous solution of sucrose in the pot flask of the apparatus shown in Figure 2 and heated the mixture to over 100°C, water would vaporize and then be condensed in the condenser and collected as a pure material in the receiver flask. Theoretically, a 100% separation of water from the nonvolatile sucrose can be achieved in this simple distillation apparatus, although for practical reasons an amount less than this is obtained. At the end of the distillation, some of the water is found coating the walls of the apparatus and cannot be recovered. This is called *holdup*. To maximize the percent recovery of a distillation, the apparatus must be designed to minimize the holdup in a given procedure.

The situation becomes much more complicated if the mixture contains two or more volatile substances. In this case, simple distillation is not effective in achieving separation of the components of the mixture. To see that this is the case, we must consider the relationship of the vapor pressure (or *more importantly* boiling temperature) and the composition of the liquids making up the mixture. Figure 3 displays a *vapor-liquid composition* curve for a hypothetical mixture of two volatile components A (boiling point = 40°C) and B (boiling point = 60°C). The essential factor that defines a simple distillation is the equilibration of liquid and vapor phases at the boiling point. In Figure 3, the bottom line represents the composition of liquid in the mixture, while the top line represents the composition of vapor. If we place a 50:50 mixture of these materials in a simple distillation apparatus and heat it to boiling, the liquid-vapor equilibrium at the top of the distillation head would occur at a temperature about 44°C and, if the liquid was withdrawn from the apparatus (the purpose of the condenser), its composition would be 75:25. While the original mixture has been partially purified, a complete separation has not been achieved. In addition, after removal of this material, the composition of the pot contents has changed and is no longer 50:50. For the sake of discussion, let us assume that the composition has changed to 45:55. It is apparent from Figure 3 that vaporization of a 45:55 mixture would lead to a condensate of less purity that the original, let us say 70:30, and the temperature at which this is collected would have risen to 45°C. This process would continue until all of the material in the pot flask would have distilled. It is obvious that even in this simple case in which there is a 20°C difference in boiling points, simple distillation will not achieve an acceptable separation. Another way of demonstrating this result is to plot the boiling temperature vs. the amount removed (i.e., the volume distilled) as the experiment is conducted. Such a plot, a *distillation curve,* is shown in Figure 4 for the distillation of 100 ml of a 50:50 mixture of A and B. An ideal separation is obtained if 50 ml of pure A, distilling at 40°C , is collected before any B distills. After all the A is collected, the temperature would rise to 60°C and pure B would be collected. This ideal separation is shown as the theoretical curve in Figure 4. On the other hand the results obtained from an actual simple distillation of this mixture would give the experimental curve, a result very far removed from that desired.

The important factor that describes a simple distillation is the equilibration of liquid and vapor phases at the boiling point. If the vapor phase from this equilibration is condensed and revaporized, then a new equilibrium would be established in which the vapor would be richer in the lower boiling component than in the initial equilibration. If this vaporization-condensation was continued without removal of condensate for enough times (we'll call this number N), a vapor consisting of 100% pure low-boiling component could be produced. Since equilibration between vapor and liquid will occur rapidly, it is possible to achieve these multiple simple distillations in an automatic fashion using an apparatus in which a fractionating column is inserted between the pot flask and the distillation head. Such an apparatus is shown in Figure 5. This type of distillation is called *fractional distillation* and the number, N, of vapor-liquid equilibrations of which the

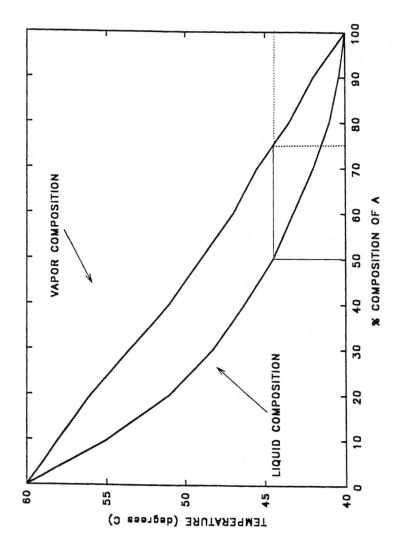

Figure 3 Vapor-liquid composition curve for a hypothetical mixture of A (boiling point — 40°C) and B (boiling point — 60°C). Simple distillation.

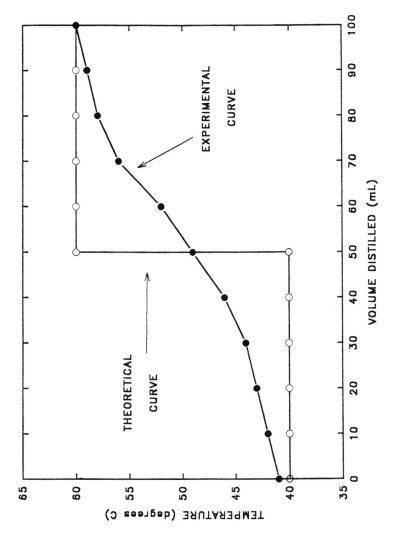

Figure 4 Results of a simple distillation of a 50:50 mixture of A and B.

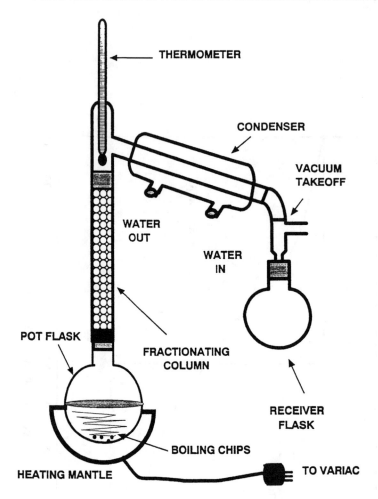

Figure 5 Fractional distillation apparatus (student setup).

apparatus is capable of achieving is described as the efficiency of the unit in terms of *theoretical plates*. One theoretical plate is defined as one liquid-vapor equilibration or one simple distillation.

The principle upon which the fractionating column is based is that a thorough mixing of rising vapor and falling liquid will occur throughout the length of a vertical tube. (The nature and characteristics of various fractionating columns will be discussed in a later section.) If we consider the original hypothetical mixture of 50:50 A and B described in Figure 3, and distill this mixture in an apparatus that has an efficiency of three theoretical plates, the result shown in Figure 6 is obtained. The original condensate (removed from the apparatus) after three equilibrations results in a material that is 100% pure A rather than the 75% obtained in the simple distillation. The temperature of this distillate would be 40°C, the normal boiling point of pure A. While the composition of the pot flask would change as before, it is obvious that five plates are sufficient to obtain pure A in

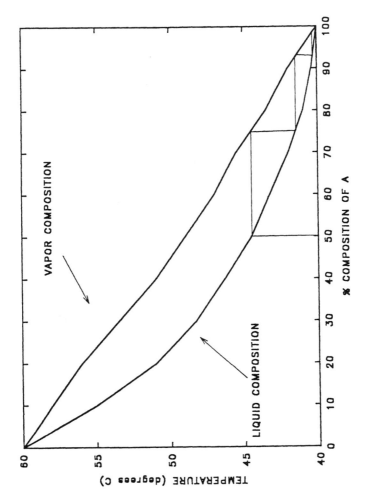

Figure 6 Vapor-liquid composition curve for a hypothetical mixture of A (boiling point — 40°C) and B (boiling point — 60°C). Fractional Distillation.

the early part of this distillation. Eventually, the composition of the pot flask will reach a point in which five plates are not enough efficiency to produce pure distillate. This would first be indicated by a rise in distillate temperature. A curve for this fractional distillation is shown in Figure 7. It is clear that a five-plate distillation of our hypothetical mixture shows behavior must closer to an ideal separation than that obtained in the simple distillation.

For real mixtures, the composition-boiling point curves do not always show the lens shape ideal behavior of our hypothetical mixture. Typical minor deviations are seen in Figures 8 and 9, with the heptane/toluene system (Figure 9) exhibiting the greatest deviation. In these cases, the overall behavior is normal and the deviations only lead to a somewhat more difficult separation in certain regions of the composition curve. Figures 10 and 11 show a much more serious deviation from ideal behavior and illustrate the phenomenon of *azeotropic behavior*. The isopropanol/water system shown in Figure 10 represents the more common situation, *a minimum boiling azeotrope*, while Figure 11 shows *a maximum boiling azeotrope*. An azeotrope is defined as a combination of miscible components such that the composition of the liquid at its boiling point is the same as the composition of the vapor in equilibrium with it, that is, the vapor and liquid lines come together at some point along the composition curve (at 68:32 mol % in the case of the isopropanol/water system). An azeotropic mixture will distill unchanged in composition regardless of the efficiency of the distillation unit. If the initial composition is to the left of the azeotropic point (e.g., 50:50 isopropanol/water), a 68:32 mol % mixture would result until all of the original isopropanol was removed. If the initial composition was to the right of the azeotropic point (e.g., 98:2 isopropanol/water), a 68:32 mixture would still result until all the water was removed. In general, maximum boiling azeotropes are rare while minimum boiling azeotropes are common, especially with mixtures containing water as one of the components. Because of the frequency of azeotropes, one can never assume that fractionation of an unknown mixture with an efficient distillation unit will yield pure compounds even if the distillation curve appears ideal. All products from distillations should be analyzed for purity using some spectroscopic or chromatographic technique.

A COMPARISON OF DISTILLATION EQUIPMENT

Distillation equipment of various types is standard equipment in chemical laboratories and the method of distillation is routinely used in industry and university settings. Ground glass equipment is routinely assigned to second-year organic chemistry students. The distillation setups utilizing this type of equipment are shown in Figures 2 and 5. For commercial and research use, however, more sophisticated equipment is necessary and it is the purpose of this section to describe this equipment and its characteristics.

The heart of a fractional distillation is the fractionating column and a wide variety of units are commercially available. For purposes of discussion, five types of columns will be compared with regard to four operational characteristics: *efficiency, holdup, throughput,* and *pressure drop.* Figures 12 through 16 depict

SOLVENT RECYCLING BY SPINNING BAND DISTILLATION 249

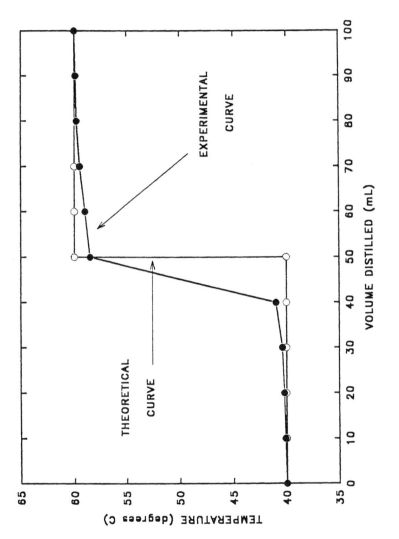

Figure 7 Results of a fractional distillation of A 50:50 mixture of A and B.

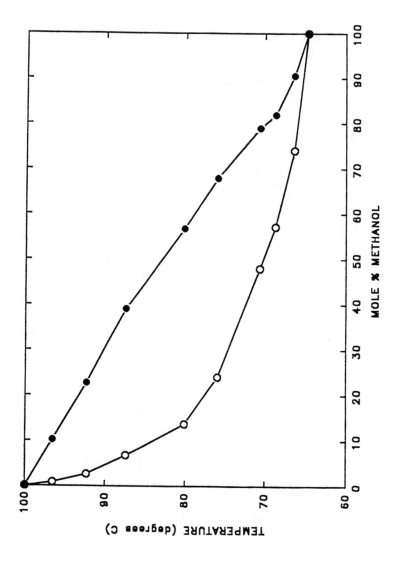

Figure 8 Vapor-liquid composition curve for methanol/water.

SOLVENT RECYCLING BY SPINNING BAND DISTILLATION 251

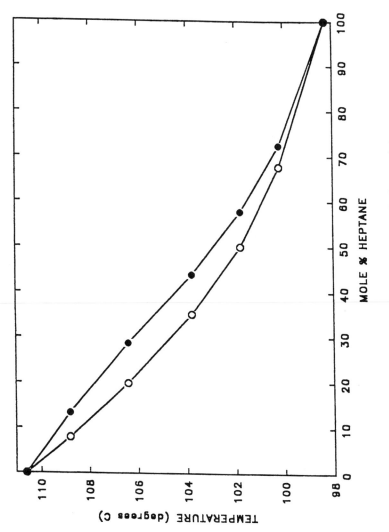

Figure 9 Vapor-liquid composition curve for heptane/toluene.

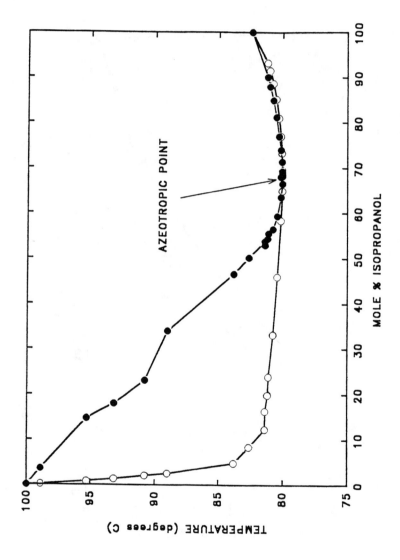

Figure 10 Vapor-liquid composition curve for isopropanol/water.

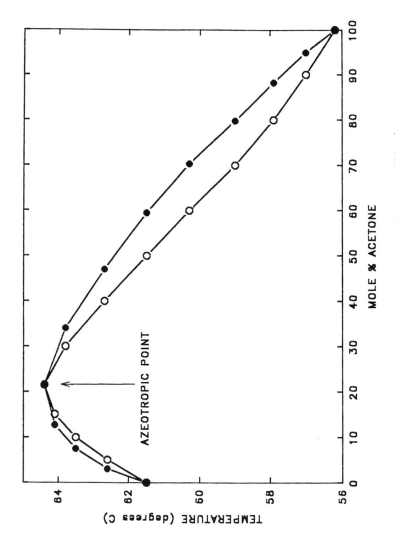

Figure 11 Vapor-liquid composition curve for acetone/chloroform.

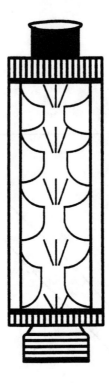

Figure 12 Vigreux column (empty tube column).

representative columns of these five types. All data comparing the characteristics of these five columns were either directly measured or extrapolated from experimental data in the literature.[2,3]

1. Empty column (Figure 12) — A typical empty column is a Vigreux column in which vapor-liquid contact is made between rising vapor and liquid falling (due to gravitational forces) down the walls and inclined indentations of the column. This type of column has had some use in semi-microdistillations and distillations under reduced pressure (vacuum distillations).
2. Packed column (Figure 13) — A packed column is a tube that is randomly filled with a packing material supported with a perforated funnel or some sort of grid. The packing material can be glass beads, rings, or spirals. Ceramic pieces, various metal turnings, and even Teflon beads have been employed. These all supply greater vapor-liquid contact than an empty tube type because of the greater surface area.
3. Plate column (Figure 14) — A typical plate column is the Bruun[4] bubble cap column. The mechanism of vapor-liquid contact in plate columns is the result of rising vapor passing through liquid that has collected on each plate. The reflux flows downward from each plate to the one below through tubes situated inside the column.
4. Stationary element column (Figure 15) — These columns contain some sort of loose, regularly arranged solid or perforated element used for vapor-liquid

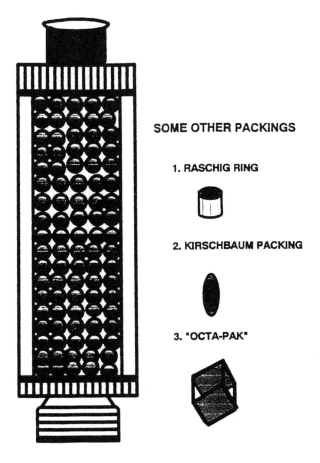

Figure 13 Packed column filled with glass beads.

contact. These elements can be made of glass or metal spirals or wire helices. The column shown in Figure 14 is a Podbielniak[5] column in which a wire helix is wound around a central core. A tight fit to the wall of the tube is essential.

5. Rotating element column (Figure 16) — The quintessential rotating element column is the *spinning band column* in which a Teflon or Monel metal band is rotated at high speeds within the column. Rising vapor makes contact with descending liquid that is physically forced down the column by the pitch of the rotating band. The forces involved in vapor-liquid contact are much greater that just gravitational forces and results in a much greater efficiency for this type of column.

TERMINOLOGY AND DISTILLATION CHARACTERISTICS

There are four major column characteristics that must be considered when planning a distillation. In addition, several operational factors must be adjusted to

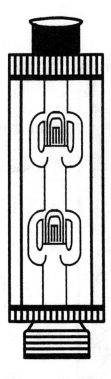

Figure 14 Bruun bubble-cap column.

achieve a desired distillation result. The experimental parameters will be discussed later in this section.

EFFICIENCY

The efficiency of a column can be described in two ways:

1. Theoretical plate (N) — The number of fully equilibrated simple distillations (vaporization-condensations) carried out in sequence. In a fractionating column, this is accomplished by allowing vapor-liquid contact to be carried out continuously in the column.
2. Height equivalent to a theoretical plate (HETP) — This represents the number of plates in a given length of column and is calculated as follows:

$$\text{HETP} = \frac{\text{Length of column (cm)}}{\text{Number of plates (N)}}$$

It is apparent from these definitions that, the larger the number of theoretical plates and the lower the HETP, the more efficient the distillation unit. For the purposes of this chapter, we will confine ourselves to talking about efficiencies in terms of theoretical plates and will not use HETP values.

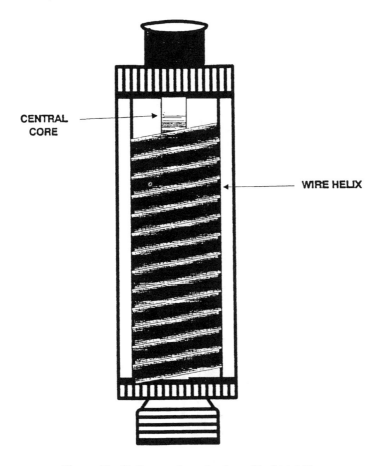

Figure 15 Stationary element column (Podbielniak).

HOLDUP

The amount of material (as liquid and vapor) that is present in the column at any time during the distillation is defined as holdup. The amount of holdup compared to the original pot charge limits the percentage that can be distilled. Optimally, holdup should be small, especially when distilling small amounts of material. The total holdup is the sum of the static holdup (material remaining when the column is at rest) and the dynamic holdup (material present during the distillation). A simple experiment is available (ASTM Method 2892)[6] to determine holdup for a given distillation unit.

THROUGHPUT

Throughput is the rate at which vapor passes through a distillation column and is usually expressed as volume per unit time. Vapor velocity and boilup rate are other terms used to describe this characteristic, although there are practical

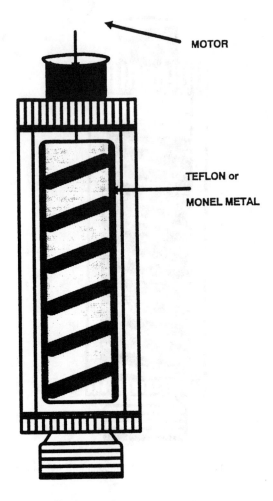

Figure 16 Spinning band column.

differences between these terms. Boilup is a measure of the amount of material that reaches the distillation head and can be measured by the amount of material returning to the column per unit time. Throughput is dependent on this boilup rate (controlled by the heat input to the pot material) and the reflux ratio.

REFLUX RATIO

The reflux ratio is a measure of the amount of material withdrawn from the column to the amount that is allowed to return to the fractionating column. In modern equipment, the reflux ratio is controlled by a valve that can be programmed to be open and closed for various time intervals. High reflux ratios will result in lower throughput regardless of the boilup rate.

PRESSURE DROP

This is the difference in pressure between the pot flask and the distillation head. At high throughputs, pressure drop increases to the point that the velocity of rising vapor prevents the fall of reflux and slugs of liquid form in the column. This phenomenon is known as *flooding* and reduces the efficiency of the column.

EQUILIBRATION TIME

To maximize the efficiency of a distillation column, an equilibrium between rising vapor and falling liquid must be established. While this equilibration requires a considerable amount of time when distillation is conducted in analytical stills (up to 200 plates), the amount of time necessary in large solvent recovery stills is minimal. For most applications, equilibration is achieved in under an hour. Experimentally, full equilibration is determined by a constancy of the head temperature. Once this has been achieved, removal of material can begin.

DISTILLATION CUTS

Since distillation is designed to separate (rectify) the various components of a mixture, various fractions or distillation cuts are obtained. These are usually determined by the boiling temperature in the head. Typically in solvent recovery applications, a *forecut* containing any low boiling contaminants is removed and discarded. A *heart cut* consisting of pure material is then obtained. The material remaining in the pot upon completion of the experiment is called the *pot residue*.

Table 1 compares the five column types described above with regard to these parameters. Since the columns vary in efficiency, the best comparison of holdup is *holdup per plate* (ml/plate). As shown in the last column of Table 1, packed and plate columns show the largest holdup due to the large surface area of packing in the former and the necessity for liquid to be present on each plate in the latter. The amount of holdup in a stationary element column is low but about tenfold higher than that in a spinning band column. The disadvantages of high holdup are twofold: it limits the amount of pot charge that can be distilled and it decreases the sharpness of separation. This latter effect is due to the concentration of a significant portion of the sample in a small area of the column which prevents the effective separation of its components.

Efficiency is measured by a low HETP. It is clear from the data in Table 1 that the stationary element and spinning band columns are the most efficient. While spinning band units of 100 and 200 theoretical plates are commercially available, for purposes of solvent recovery and reuse columns rated at 20, 30, and 50 plates are the units of choice. In the following section, the characteristics of these stills are detailed. In addition, a recently developed nonspinning band distillation unit that has rapidly become valuable in recycling efforts in hospital laboratories will be discussed in Chapter 20.

Table 1 Characteristics of Various Distillation Units

Column type	Throughput (ml/min)	HETP (cm/plate)	Holdup (ml/plate)
Vigreux column	5–10	7–12	0.5–2
Packed column	2–7	3–5	0.7–1
Plate column	1–5	1–2	0.2–0.5
Stationary element	0.5–2	0.5–3	0.01–0.03
Spinning band	3–20	0.4–3	0.01–0.03

CHARACTERISTICS OF SPINNING BAND STILLS

A large number of experiments have been carried out on spinning band units to determine their overall operational characteristics. The methods used are standard and are given in detail elsewhere.[7,8] In this section, we briefly describe the results obtained with regards to efficiency, holdup, and throughput for solvent recovery stills. It should be pointed out that the test mixtures must be of ultra high purity and the experiments must be conducted in such a manner as to minimize or eliminate the large sources of error that are possible in such studies. These sources of error are as follows:

1. All components of the test mixture must be of high purity since the presence of impurities will produce erroneous results due to their impact on the vapor-liquid composition curve. All of the components used in our studies were purified by distillation of the best commercially available sample in a 200-plate spinning band unit.
2. The test mixture chosen must be neither too easy nor too difficult to be separated on the still being tested and the vapor-liquid equilibrium curve must approach ideality.
3. The initial composition of the test mixture must give a considerable amount of the high boiling component in the head since the change in composition per plate becomes very small when the mixture is very pure in one or the other of the components. For example, if one starts with 95:5 isooctane/heptane and achieves 90% heptane in the head, the still would be rated at 171 plates. If the analytical procedure gave a 91% heptane composition in the head an efficiency of 192 plates would be calculated. A 1% difference in composition produces a 12% difference in the number of plates. On the other hand, if the starting mixture was 80:20 isooctane/heptane, a 90% heptane composition in the head would calculate out as 156 plates while a 91% composition would give 161 plates. The error here is only 3.6%. We have carefully chosen initial compositions in our tests to reduce these analytical errors.
4. An accurate analytical method is required. We use GC-FID analysis and have prepared standard calibration curves in all of our tests.

DESCRIPTION OF SPINNING BAND STILLS

Three spinning band solvent recovery units are available for laboratory solvent recycling purposes and these have been used in our work. The capacity or pot charge for these units can be as little as 1000 ml and as much as 22 l. All of these units are operated automatically under a turnkey system. The stills are of the spinning band type in which two Teflon bands (an inert material) are twisted in a helix and attached to a motor driven Teflon shaft. The ends of the band are perforated and small holes have been punched along the surface. The shaft is rotated within the column at a speed of 2200 rpm. The still is vacuum jacked and silvered to maintain adiabatic (no heat loss) conditions throughout the length of the column. The still condenser is a double-cooling type with an internal glass coil jacketed by an outside condenser. For very low boiling solvents, it is recommended that a chiller be used to maintain condenser water at a low temperature. For most situations, however, ordinary laboratory water can be used for cooling. The units are equipped with a computer controller that controls the complete operation of the experiment. Band speed, heat imput to the heating mantle, temperature monitoring (head and pot), reflux ratio, and shutdown temperature are monitored and controlled by the computer. The system undergoes constant diagnostics during a distillation run and will shut down the unit in the event of some disturbance. All of the stills are hand blown and incorporate the use of expansion rings to allow the glass flexible movement during distillation.

The principle operating feature of these units is the spinning band itself. The tight-fitting rotating band pumps returning liquid condensate down the column, depositing a thin film of liquid along the walls. This action provides the maximum exposure of the falling liquid with the heated vapor that is rising in the column. As a result, vapor is forced to contact condensate producing a very high degree of vapor-liquid equilibration. The overall process produces increased rectification demonstrated by the high number of plates and low HETP values for a given unit. In addition, holdup in the spinning band system is minimal since the descending liquid is essentially contacting only the walls of the column. High pressure drop and the attendant low throughputs are not evident since the vapor path is essentially open and not hindered by rectifying elements such as packings. These units were tested using the following experimental procedures.

DETERMINATION OF THE NUMBER OF PLATES

To ascertain the number of plates in a given distillation apparatus, it is necessary to operate the still at *total reflux*. While distillations are actually carried out at only partial reflux (material is removed from the still during distillation), it is generally assumed that the number of plates achieved at total reflux is indicative of the type of separation that can be achieved by a given distillation unit. The general procedure involves charging the pot flask with the proper test mixture, adding Teflon boiling chips, and heating the mixture until the flood point is reached. This achieves a complete wetting of the band and the column. The heating rate is then reduced until the desired boilup is obtained. In computer-controlled units the band will rotate at a preset band speed once the programmed motor-on temperature is reached. The mixture is then equilibrated for an appropriate amount of time which, in solvent recovery stills, is generally about 1 hour. The still is frequently checked during the experiment to insure that no leakage of

Table 2 Test Mixtures for Testing Spinning Band Stills

Mixture	Relative volatility
Methylcyclohexane/toluene	1.306
Benzene/1,2-dichloroethane	1.162
n-Heptane/methylcyclohexane	1.076

material from the still is occurring. This is extremely important since loss of material from the unit results in changing compositions and severely disrupts equilibrium. The unit is also checked at the completion of the experiment to ascertain the percentage of material that has been lost. At the appropriate time, a small sample (10 µl) is removed from the *head*. Simultaneously a small sample is removed from the *pot* using a syringe. The samples are then analyzed by gas chromatographic analysis using an FID detector. The number of *theoretical plates* are then calculated from the *Fenske equation*.[7]

$$N = \frac{\text{Log} \frac{(Xa \times Ya)}{(Xb \times Yb)}}{\text{Log } \alpha} - 1$$

where α = the relative volatility of low boiler to high boiler,
  Xa = the composition of high boiler in the *pot*,
  Xb = the composition of high boiler in the *head*,
  Ya = the composition of low boiler in the *head*,
  Yb = the composition of low boiler in the *pot*.

Table 2 lists the test mixtures that have been used to test the spinning band units used in this work.

DETERMINATION OF HOLDUP

Dynamic holdup is determined using the standard ASTM Method D 2892.[6] In this method, 1000 ml of a test mixture composed of 20 wt % stearic acid in ultrapure heptane is weighed and refluxed for 30 min at a boilup rate of approximately 1100 ml/h. One-milliliter samples are removed from the head and the pot and analyzed by titration with a standard 0.10 M solution of NaOH. This removal and analysis is repeated at 30-min intervals until the concentration of stearic acid in the pot stabilizes. At this point, the dynamic holdup can be calculated from the following equation:

$$\text{Holdup} = \frac{P - Po}{P} \times \frac{M}{d}$$

where P = wt % stearic acid removed from the flask,
  Po = wt % stearic acid in initial sample,

M = weight of test mixture,
d = density of n-heptane (0.688 kg/l).

DETERMINATION OF PRESSURE DROP

The pressure drop in a distillation column is measured by monitoring the pressure at the head of the column and at the pot simultaneously. The pressure differential over the column will increase with boilup rate, external pressure, and on the nature of the material being distilled. To obtain a constant comparison, we measure pressure drop at atmospheric pressure using a 50:50 methylcyclohexane/ toluene mixture at a boilup rate of 1100 ml/h. It should be noted that pressure drop increases with decreasing pressure and can lead to adverse effects such as irregular boiling and flooding. For normal solvent recycling applications, distillation under vacuum is not employed. Most solvent wastes are thermally stable at their boiling points and do not require vacuum distillation. There are, however, several examples of the recovery of thermally sensitive solvents that we will discuss in Chapter 23. For those systems, large pressure drops would lead to adverse distillation results.

SAFETY CONSIDERATIONS

The recovery of waste solvents using laboratory distillation methods has become a common procedure in recent years. There are many safety considerations that must be recognized in the area of solvent recycling including both the equipment employed and the material being recycled. The types of equipment used for this procedure vary considerably from bubble cap and packed columns constructed from ordinary laboratory glassware to large commercial units. The laboratory constructed units usually employ various controlled heating mantles or heat lamps as the heat source for the pot material and thermometers to measure temperatures. Take off is usually controlled with a manual take-off valve and cooling water is generally taken directly from the tap. Fractions are removed by manual changing of receivers. These systems are not as safe as commercial units which are generally turnkey operations. The laboratory constructed units require constant monitoring by the operator and skill in their operation. There is the possibility of water shutdown during the distillation with the resultant emission of vapor into the laboratory. Other problems associated with these units are overheating of the pot material and the possibility of heating to dryness. These units are generally rack mounted and not properly ventilated. There is also the possiblity of the contamination of fractions due to the necessity of manually changing flasks. Due to these inherent problems, these systems are not recommended for routine solvent recovery.

The commercial units are generally turnkey operations in which heating, take off, and shutdown are microprocessor controlled. In addition, some units have redundant systems that constantly check all of the safety aspects of the distillation and shut the unit down in the event of a malfunction. A well-designed commercial distillation system is totally self-contained and safe. It also requires very little operator attention. If a laboratory is planning to carry out routine solvent recycling, it is strongly recommended that a commercial unit with the following basic safety features be employed:

1. Monitoring of head and pot temperatures
2. Control of heat input to the heating mantle
3. Monitoring and control of condenser (cooling) water; for very volatile solvents, a chiller is strongly recommended
4. Monitoring and control of distillation fractions to identify and collect the heart cut and send the forecut to waste
5. Ventilation capability to eliminate any buildup of harmful vapors

The solvent being recycled may have inherent safety problems associated with its chemical nature that must be identified and made known to all the personnel involved in the distillation. In addition, proper laboratory procedures must be put in place in any solvent recovery program to monitor and control the wastes involved. Since most solvents are flammable and toxic and some can form harmful peroxides, it is important that the equipment be placed in an area of the laboratory with proper ventilation and that no ignition sources are present. There must be proper storage facilities available for both the waste material and the recovered solvent. All waste materials must be kept segregated from each other. Mixing of wastes not only leads to situations in which the components may not be easily separated, but more importantly can cause dangerous safety problems due to chemical incompatibility. The case of peroxide-forming materials is special. Presently, the major peroxide-forming solvent that is routinely being recycled is tetrahydrofuran (THF). This solvent reacts with atmospheric oxygen to produce dangerous peroxide which may explode in a distillation residue. It is recommended that, before distilling this solvent, the waste be tested for the presence of peroxide by mixing a few milliliters of the solvent with an aqueous acidic KI solution. Addition of one or two drops of starch solution will turn blue if peroxide is present. In the event of a positive test, the waste should be treated with copper (I) chloride before distilling. We will discuss the details of this recovery in Chapter 20.

CONCLUSION

In this chapter, we have presented the theoretical aspects for the use of distillation as a means of waste minimization by solvent recovery and have described the characteristics of the equipment by which this could be accomplished. We have also discussed the safety aspects that must be addressed when employing this technique on a routine basis in the laboratory. In Chapter 20, we discuss a number of applications of this technique and present analytical data describing the quality of the recovered material.

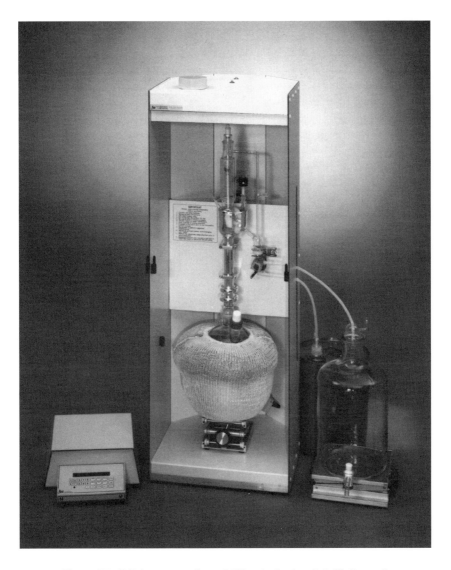

Figure 17 B/R Instrument Corp. 8300 spinning band distillation unit.

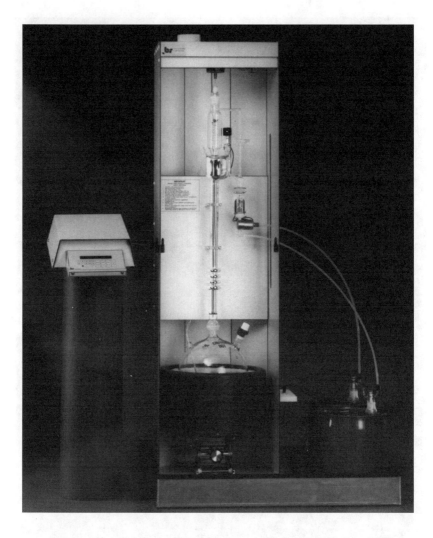

Figure 18 B/R Instrument Corp. 8400 spinning band distillation unit.

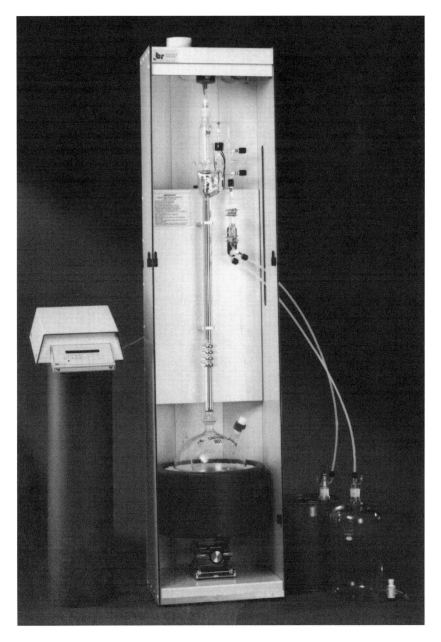

Figure 19 B/R Instrument Corp. 8600 spinning band distillation unit.

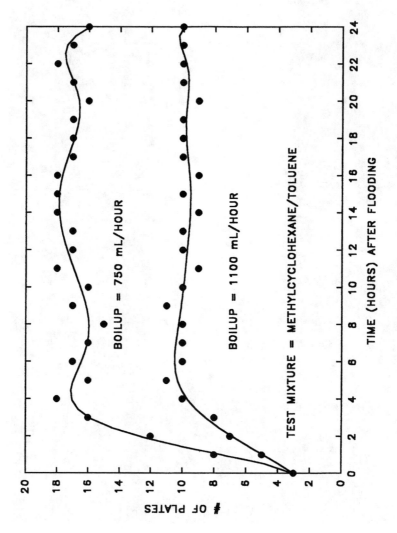

Figure 20   Efficiency vs. equilibration time for the 8300 still.

SOLVENT RECYCLING BY SPINNING BAND DISTILLATION

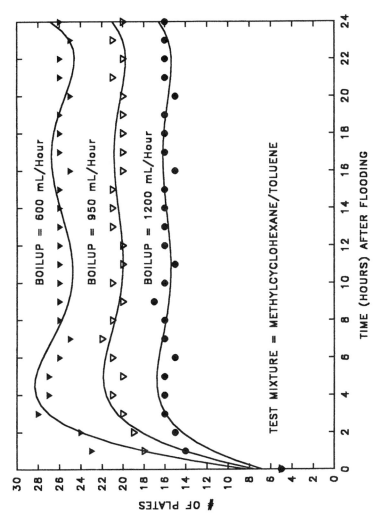

Figure 21 Efficiency vs. equilibration time for the 8400 still.

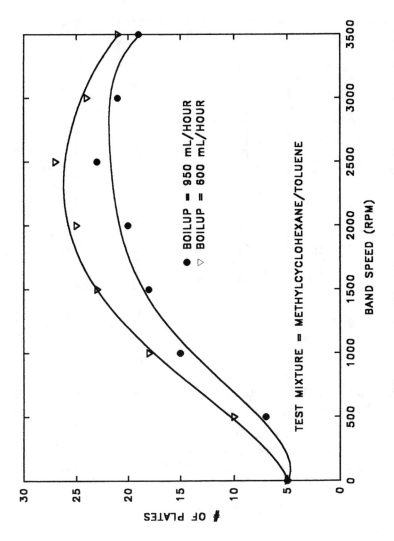

Figure 22 Efficiency vs. band speed for the 8400 still. Equilibration Time = 24 hours.

SOLVENT RECYCLING BY SPINNING BAND DISTILLATION 271

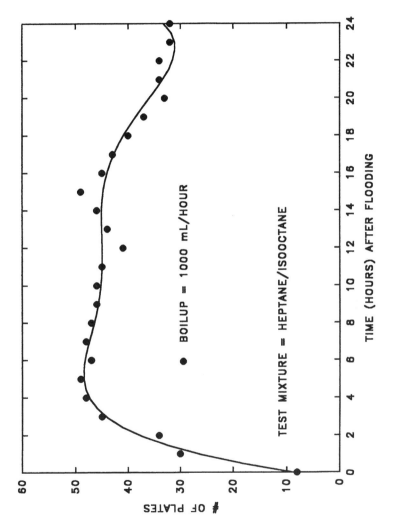

Figure 23 Efficiency vs. equilibration time for the 8600 still.

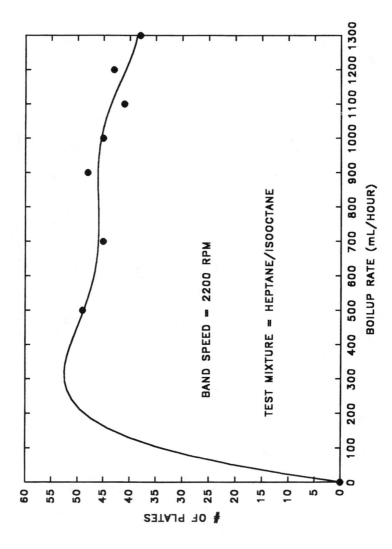

Figure 24 Efficiency vs. boilup rate for the 8600 still. Equilibration Time = 2 hours.

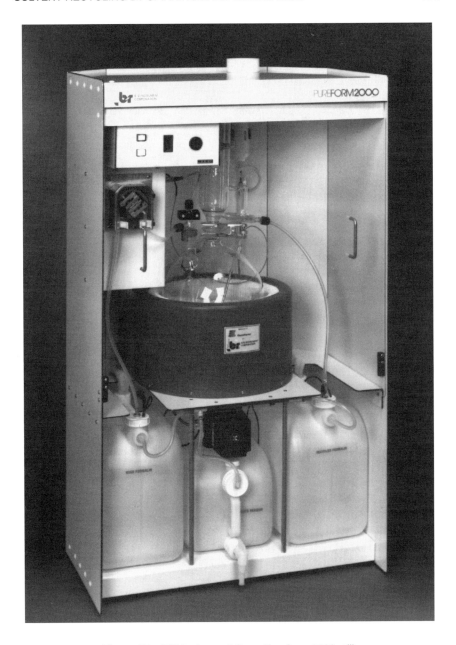

Figure 25 B/R Instrument Corp. Pureform 2000 still.

ENDNOTES

1. *Prudent Practices for Handling, Storage and Disposal of Hazardous Chemicals in Laboratories,* National Academy of Sciences, Washington, D.C., 1983. [New edition pending.]
2. Krell, E., *Handbook of Laboratory Distillation,* Elsevier Scientific, Amsterdam, 1982.
3. Weissberger, Ed., *Technique of Organic Chemistry,* Vol. 4, Interscience, New York, 1951.
4. Bruun, A., An all glass bubble cap column, *Ind. Eng. Chem., Anal. Ed.,* 9, 192, 1937.
5. Podbelniak, E., Different heli-grid columns, *Ind. Eng. Chem., Anal. Ed.,* 13, 1639, 1941.
6. *1988 Annual Book of ASTM Standards,* ASTM Method D2892, American Society for Testing and Materials, Philadelphia, 1988, 538.
7. Rose, A. and Rose, E., *Technique of Organic Chemistry,* Vol. 4, Interscience, New York, 1951, 30.
8. Baker, R. H., Barkenbus, C., and Riswell, C. A., A large spinning band fractionating column, *Ind. Eng. Chem., Anal. Ed.,* 12, 468, 1940.

CHAPTER **16**

Chemical Treatment Methods to Minimize Waste

Margaret-Ann Armour

CONTENTS

Introduction 276
General Waste Minimization Reactions 277
    Acid/Base Neutralization 277
    Volume Reduction of Dilute Solutions of Inorganic Salts 278
    Destruction of Hydroperoxides in Ethers and Alkenes 278
Specific Disposal Reactions for some Organic Chemicals 279
    Oxalic Acid, Sodium Oxalate, and Oxalyl Chloride 279
    N-Bromosuccinimide and N-Chlorosuccinimide 279
    Decomposition of Dimethyl Sulfate and Diethyl Sulfate 280
    N,N-Dimethylformamide and N,N-Dimethylacetamide 280
    Solutions of Picric Acid 281
    Nitriles 282
    Organic Azides 282
    Aromatic Amines 283
Specific Disposal Reactions for some Inorganic Chemicals 284
    Iodine 284
    Bromine 284
    Oxidizing Agents 284
    Cyanides 284
    Inorganic Azides 285
Metal-Containing Wastes 286
    Solutions Containing Heavy Metal Ions 286
    Solutions Containing Mercury Ions 287
    Solutions Containing Chromium Ions 287
    Metal Carbonyls 288

Treatment of Spills 288
Conclusion 289
Acknowledgements 290
Endnotes 290

INTRODUCTION

The generation of small quantities of a wide variety of waste and surplus chemicals in laboratories can present a difficult and expensive disposal problem. These wastes result from practical chemical laboratory classes in schools, colleges, and universities and from work in hospital dispensaries, in quality control and research laboratories, and in small industries that use chemical processes. As awareness has increased of the potential health and environmental problems that such wastes may cause, appropriate methods for handling them have been developed. The responsibility of the generator of the waste to ensure its safe disposal has been emphasized, so that random mixing of wastes in containers to be removed by others is no longer acceptable. Historically the first step taken to improve the handling of the waste in the laboratory was to require the generator to segregate and carefully label the material before it was removed for treatment. Next, the disposal of waste was recognized to be only a part of an overall management of chemicals, from their purchase through their use to their disposal.[1] Finally, emphasis was placed on minimizing the waste to the greatest extent possible.[2] This encouraged both an examination of the processes by which the waste was being generated and the identification of methods to reduce the volume of waste that had to be shipped for treatment. A variety of strategies have emerged as a result of these exercises. The move to microscale experiments has had a major impact on the amount of waste produced in teaching laboratories[3,4] and many chemical processes have been adapted to result in much smaller wastestreams. Most laboratories distill some used solvents. Especially in teaching institutions, this makes possible the reuse of these solvents, if not in applications such as high-performance liquid chromatography in research laboratories, at least in undergraduate experiments. Other specific strategies for recycling and reusing chemical wastes are described elsewhere in this book.

In laboratories nowadays, strategies to prevent or reduce the generation of waste should be considered first. However, in many cases it is not possible to completely eliminate the waste. A second approach may then be considered. A way of reducing the volume of waste that must be removed from the laboratory for treatment is by conversion of the waste ideally to nontoxic and environmentally acceptable products, but at least to materials that are less hazardous than the original, at the site of generation. In today's society, with its emphasis on environmental stewardship, it is common for bench chemists to be concerned about the wastes they produce and to be willing to take responsibility for eliminating or reducing hazardous properties before the wastes leave their control. This approach has many benefits. The chemist can accurately characterize the waste and so can also determine the most appropriate treatment procedure. It also removes the need to package and transport the waste elsewhere for disposal. The technology to allow chemical treatment of a variety of toxic and hazardous wastes has been developed over a number of years and information on "disposal methods" has been included in books and chemical catalogs.[5-8]

To encourage laboratory workers to treat small quantities of waste at the bench, it is necessary to have a practical and dependable procedure that is documented so that it can be followed both easily and reproducibly. Since 1982, we have been testing in the laboratory potential treatment methods for individual chemicals and have documented the acceptable ones in a format similar to that used by "organic syntheses."[9-13] In several instances, the methods have been checked in an independent laboratory.

The disposal methods have been tested for the rate of disappearance of the toxic material, the degree of conversion to nontoxic products, the nature and identity of the products, the practicality of the method, and the ease of reproducibility. Methods were selected that were shown to be safe to the operator, reliable, and reproducible. Greater than 99% of the starting material had to be reacted under the conditions and length of time given. As far as possible, the products of the reaction were identified: where this was not practical since a large number of fragments was produced, and especially when the chemical being treated was a suspect carcinogen, the reaction mixture was submitted for an Ames test which detected whether the material formed had excess mutagenicity over background.

In the following paragraphs, general techniques for the neutralization of acids and bases, the reduction of volumes of aqueous solutions, the destruction of hydroperoxides in peroxidizable solvents, and methods for the disposal of specific chemicals will be described. The procedures are designed to be used by a person with some training in the handling of chemicals.

While performing the procedures, appropriate protective clothing must be worn. This includes goggles, gloves, and a laboratory coat or apron. Work in a fume hood behind a protective screen where possible. Additional safety precautions are noted with specific procedures.

It should be noted that the disposal of hazardous chemicals, by whatever method, must be performed in accordance with local regulations. Many institutions have an environmental health and safety office which can provide information and advice. Otherwise, contact the local wastewater treatment authority to determine if discharges will be acceptable.

GENERAL WASTE MINIMIZATION REACTIONS

ACID/BASE NEUTRALIZATION

Waste or surplus dilute mineral acids such as hydrochloric acid, sulfuric acid, nitric acid, or acetic acid can be neutralized with waste or surplus dilute ($\approx 5\%$ or

1 M) solutions of sodium hydroxide, potassium hydroxide, or other dilute base. The acid solutions should be added to water until they are not more than 1 M in strength. The base is slowly added to the acid while stirring or in such a way that efficient mixing occurs. If necessary, ice is added to the solution to maintain the temperature below 40°C. The pH of the final mixed solutions should be about 7 or neutral to litmus. If only waste acid is available for disposal, it can be neutralized by dilution to 1 M or less followed by slow addition of either solid sodium carbonate (soda ash) or 5% aqueous sodium hydroxide with efficient mixing until the solution is neutral to litmus. Waste base is treated similarly. Thus, solutions of sodium or potassium hydroxide or other base are diluted to less than 5% with water and dilute waste or surplus acid added slowly and with stirring and cooling if necessary. If only waste base is available, it should be diluted to 5% or less with water and dilute (1 M) hydrochloric acid or acetic acid added slowly with stirring. The neutral solution is washed into the drain.

VOLUME REDUCTION OF DILUTE SOLUTIONS OF INORGANIC SALTS

When an experiment yields a large volume of dilute aqueous solution containing toxic heavy metal ions that may not be washed into the drain, the solution can be placed in a large evaporating dish (or other container providing a large surface area of the solution) in the fume hood and allowed to evaporate. The residue is packaged and labeled for disposal. Note that this technique can be used only for aqueous solutions and not for solutions containing organic solvents.

DESTRUCTION OF HYDROPEROXIDES IN ETHERS AND ALKENES

Hydroperoxides form readily in ethers and alkenes in the presence of air and light. Before attempting to either distill for reuse or dispose of waste or surplus quantities of these liquids, tests should be performed to detect whether or not peroxide is present and, if so, the peroxide should be removed. The presence of peroxides can be detected readily and dependably using the following well-accepted method.

To a sample of the ether or alkene (0.5 to 1.0 ml) is added an equal volume of a solution of potassium iodide in glacial acetic acid (about 100 mg/ml). A yellow to brown color indicates the presence of peroxides. The test is sensitive to 0.5 mg of peroxide per milliliter of ether or alkene. If shown to be present, peroxides can be removed by shaking the liquid (100 ml) in a separatory funnel for 5 min with about 20 ml of a freshly prepared 50% aqueous solution of sodium metabisulfite, $Na_2S_2O_5$. The advantage of this method is that it can be used for water-miscible ethers such as tetrahydrofuran and dioxane since two easily separated layers are formed. The organic layer, peroxide-free, can be dried to the required level, first over sodium sulfate and then, if necessary, using a more-powerful drying agent appropriate to the solvent.

SPECIFIC DISPOSAL REACTIONS FOR SOME ORGANIC CHEMICALS

OXALIC ACID, SODIUM OXALATE, AND OXALYL CHLORIDE

Oxalic acid and its derivatives are highly toxic. Sodium oxalate and the free acid are water soluble and are decomposed to carbon dioxide, carbon monoxide, and water by heating with concentrated sulfuric acid.

$$HOOC\text{-}COOH + H_2SO_4 \rightarrow H_2O + CO_2 + CO + H_2SO_4$$

This procedure should be performed in the fume hood and contact of the skin with concentrated sulfuric acid must be avoided. Oxalic acid or sodium oxalate (5 g) is added to 25 ml of concentrated sulfuric acid in a 100-ml round-bottomed flask. Using an oil bath or a heating mantle, the mixture is maintained at 80 to 100°C for 30 min. The sulfuric acid can be reused since the only nonvolatile product of the decomposition is a small quantity of water. If it is desired to dispose of the acid, allow the liquid to cool to room temperature then slowly and carefully pour it into a pail of cold water and ice, neutralize with dilute (≈5%) sodium hydroxide solution or sodium carbonate, and wash into the drain.

Oxalyl chloride (1 ml) is converted to oxalic acid by slowly adding it to cold water (3 ml) in a round-bottomed flask and allowing the mixture to stand for 1 h. To decompose the oxalic acid, concentrated sulfuric acid (10 ml for each 1 ml of starting chloride) is slowly added with stirring and the solution heated at 80 to 100°C as described previously for oxalic acid.

N-BROMOSUCCINIMIDE AND *N*-CHLOROSUCCINIMIDE

N-Bromosuccinimide and *N*-chlorosuccinimide find frequent use in the laboratory as brominating and chlorinating agents, respectively. They must be handled with care since they cause skin burns and severely irritate the respiratory system. A practical disposal method for waste quantities of these compounds is reduction with sodium sulfite solution.

N-Bromosuccinimide + Na_2SO_3 + H_2O ⟶ Succinimide + Na_2SO_4 + HBr

N-Chlorosuccinimide + Na_2SO_3 + H_2O ⟶ Succinimide + Na_2SO_4 + HCl

Thus, in the fume hood, the *N*-bromosuccinimide or *N*-chlorosuccinimide (5 g) is slowly added to a vigorously stirred 20% aqueous solution of sodium sulfite (30 ml). The reduction takes place almost immediately and the product mixture can be washed into the drain.

DECOMPOSITION OF DIMETHYL SULFATE AND DIETHYL SULFATE

Alkylating agents such as dimethyl sulfate and diethyl sulfate are highly toxic to living systems. The dimethyl ester is listed as a carcinogen by the Environmental Protection Agency, and the diethyl ester as a suspect carcinogen. Surplus or waste material should be destroyed at once and completely. This can be accomplished by hydrolyzing these materials in sodium hydroxide solution to sodium sulfate and methanol or ethanol, respectively.

$$(CH_3)_2SO_4 + 2NaOH \rightarrow 2CH_3OH + Na_2SO_4$$
Dimethyl sulfate $\qquad\qquad$ Methanol

$$(CH_3CH_2)_2SO_4 + 2NaOH \rightarrow 2CH_3CH_2OH + Na_2SO_4$$
Diethyl sulfate $\qquad\qquad$ Ethanol

To destroy 100 ml of dimethyl sulfate or diethyl sulfate, 500 ml of a 20% aqueous solution of sodium hydroxide is placed in a 1-l round-bottomed flask. In the fume hood the dialkyl sulfate is added to the flask, a reflux condenser fitted, and the mixture heated on the steam bath with stirring for 4 h. The cooled products are neutralized with dilute hydrochloric acid and washed into the drain.

N,N-DIMETHYLFORMAMIDE AND *N,N*-DIMETHYLACETAMIDE

N,N-Dimethylformamide in both the vapor and liquid forms irritates the skin and eyes, but, more seriously, prolonged inhalation of the vapor has been shown to cause liver damage in experimental animals. *N,N*-Dimethylacetamide also is harmful by inhalation and on contact with the skin and is an experimental teratogen which may cause liver damage on prolonged exposure. Both of these compounds are used as specialized solvents in the laboratory. They can be smoothly decomposed by heating with aqueous base.

$$HCON(CH_3)_2 + NaOH \rightarrow (CH_3)_2NH + HCOONa$$
Dimethylformamide $\qquad\qquad$ Dimethylamine + Sodium formate

$$CH_3CON(CH_3)_2 + NaOH \rightarrow (CH_3)_2NH + CH_3COONa$$
Dimethylacetamide $\qquad\qquad$ Dimethylamine + Sodium acetate

$$(CH_3)_2NH + HCl \rightarrow (CH_3)_2NH \cdot HCl$$

Dimethylamine   Dimethylamine hydrochloride

A 10% aqueous solution of sodium hydroxide (400 ml) is placed in a 1-l round-bottomed flask. In the fume hood, 40 ml of dimethyl formamide or dimethylacetamide is added and the flask fitted with a reflux condenser. The mixture is heated at the boiling point for 2 h. The cooled residue is neutralized with dilute hydrochloric acid and washed into the drain.

SOLUTIONS OF PICRIC ACID

When dry, solid picric acid is a powerful explosive. If a bottle of the chemical has dried out it should not be moved or handled, but the appropriate disposal authorities contacted. In dilute aqueous solution, the hazard arises from the possibility of evaporation of the solution on stoppers, caps, or lids of containers, since the explosive can be detonated by friction. Picric acid is smoothly reduced with tin and hydrochloric acid to triaminophenol, which is no longer explosive.

$$\text{Picric acid} + 9Sn + 18HCl \longrightarrow \text{Triaminophenol} + 6H_2O + 9SnCl_2$$

The reaction should be performed behind a shatterproof screen.

The picric acid (1 g) is placed into a three-neck round-bottomed flask fitted with a dropping funnel and condenser. Any traces of acid on glassware or equipment are rinsed into the flask using about 20 ml of water. To the solution is added 4 g of tin and the mixture stirred magnetically. Into the dropping funnel is placed 15 ml of concentrated hydrochloric acid and, while cooling the flask in an ice-water bath and with stirring, the hydrochloric acid is added dropwise. The first few milliliters of hydrochloric acid should be added slowly since the initial reaction is vigorous; the rate of addition may be increased as the reaction moderates. When all of the hydrochloric acid has been added, the mixture is allowed to warm to room temperature then heated under reflux for 1 h to complete the reduction. The unreacted tin is filtered and the precipitate washed with $2\,M$ hydrochloric acid (10 ml) (the residual tin can be reused). The filtrate is neutralized with 10% sodium hydroxide solution and refiltered. The tin chloride precipitate can be treated as normal garbage. The solution that contains 2,4,6-triaminophenol can be packaged, labeled and sent for incineration or it can be treated with acidic potassium permanganate solution to decompose the triaminophenol. In the latter procedure, to the solution is added slowly and cautiously 50 ml of 3 M sulfuric acid containing 12 g of potassium permanganate. After standing at room temperature for 24 h, solid sodium bisulfite is added until a clear solution is obtained. The liquid is neutralized with 10% sodium hydroxide solution and poured into the drain. This method has been used to decompose batches of up to 8.5 g of picric acid at one time.

It has been found that very dilute solutions of picric acid (including large volumes, e.g., 45-gal drums) can be reduced by acidifying to pH 2 or less with concentrated hydrochloric acid and allowing the solution to stand in the presence of granulated tin for 14 days. The color gradually darkens from yellow to brown, and the complete disappearance of the picric acid can be determined by thin-layer chromatography of the solution on silica gel plates using methanol:toluene:glacial acetic acid, 8:45:4 as eluant. Picric acid has an R_f value of about 0.3 and the bright yellow spot is easily visible. The detection limit can be increased by developing the plate in iodine vapor. Thus tin powder can be added to 45-gal drums of acidified dilute picric acid solutions and the acid is reduced to 2,4,6-triaminophenol. When it can be shown that no picric acid is present in the solution, contractors are usually willing to ship the material.

NITRILES

Organic nitriles are toxic. Waste solutions containing these compounds are treated by hydrolyzing the nitrile to a nontoxic acid. For example, benzonitrile (1 g) is converted to benzoic acid by heating under reflux with 10% potassium hydroxide in ethanol solution (30 ml) for 3 h. The cold solution is neutralized with dilute hydrochloric acid and washed into the drain.

$$\text{C}_6\text{H}_5\text{CN} \xrightarrow{\text{ethanolic KOH}} \text{C}_6\text{H}_5\text{COOH}$$

Note that, although they also contain a cyano group, this method cannot be used for inorganic nitriles such as sodium cyanide, which do not react with ethanolic potassium hydroxide solution (see Cyanides).

ORGANIC AZIDES

Azides are explosive on heating. Oxidation of inorganic azides by ceric ammonium nitrate (see Inorganic Azides) is a useful method of disposal. However, for organic azides this is a slow and unsatisfactory process. Preferable is a reduction to the corresponding amine with tin and hydrochloric acid. The procedure is detailed.

For solutions of organic azides, work in the fume hood behind a safety shield.

$$C_6H_5N_3 + Sn + 3HCl \rightarrow C_6H_5NH_3{}^+Cl^- + SnCl_2 + N_2$$

Phenyl azide

Slowly add the azide (1 g) to a stirred mixture of granular tin (6 g) in concentrated hydrochloric acid (100 ml). Continue stirring for 30 min. Cautiously decant the solution into a pail of cold water. Wash the residual tin with water and reuse. To the aqueous solution in the pail add 10 g of potassium permanganate and stir until the purple solid has dissolved. Allow to stand at room temperature overnight during which time the aniline hydrochloride is decomposed. Add solid sodium metabisulfite to reduce both the manganese dioxide formed and any excess potassium permanganate, neutralize the solution with dilute($\approx 5\%$) aqueous sodium hydroxide or with soda ash, and wash into the drain.

AROMATIC AMINES

Many aromatic amines have been shown to cause cancer in rodents and are confirmed or suspect human carcinogens. They can be rendered physiologically inactive by removal of the amino group. For example 4-aminobiphenyl **1** is converted to biphenyl **2** by treatment with sodium nitrite and then hypophosphorous acid.

Thus, to 1.0 g of 4-aminobiphenyl in a 125-ml Erlenmeyer flask, is added a mixture of 0.8 ml of water and 2.5 ml of concentrated hydrochloric acid. The mixture is stirred for 10 to 15 min until a homogeneous slurry is formed. The slurry is cooled to 0°C in an ice-salt bath and a solution of 1.0 g of sodium nitrite in 2.5 ml of water is added dropwise at such a rate that the temperature of the mixture does not rise above 5°C. After stirring for 1 h, 13 ml of ice-cold 50% hypophosphorous acid is added slowly. Some foaming may occur. When addition of the acid is complete the mixture is stirred for 18 h at room temperature. The solid is collected by filtration, the filtrate washed into the drain with a large volume of water, and the solid (biphenyl **2**) discarded with normal refuse or disposed of by burning. Other aromatic amines such as 2-aminofluorene **3** can be converted to the corresponding hydrocarbon by a similar reaction. In this case, the product is fluorene **4**.

SPECIFIC DIPOSAL REACTIONS FOR SOME INORGANIC CHEMICALS

IODINE

Iodine vapor is harmful and the solid burns the skin. It is reduced to sodium iodide by reaction with sodium thiosulfate and sodium carbonate.

$$I_2 + Na_2S_2O_3 + Na_2CO_3 \rightarrow 2NaI + Na_2SO_4 + S + CO_2$$

In the fume hood, iodine (5 g) is added to a solution of sodium thiosulfate (11 g) and sodium carbonate (1 g) in water (300 ml). The mixture is stirred at room temperature until all of the iodine has dissolved and the solution is colorless. The resulting solution is neutralized with sodium carbonate and washed into the drain.

BROMINE

Bromine is very toxic by inhalation and causes severe burns if spilled on the skin. It can be reduced to sodium bromide by reaction with sodium bisulfite solution.

$$Br_2 + 2NaHSO_3 \rightarrow 2NaBr + H_2SO_4 + SO_2$$

In the fume hood, bromine (5 ml) is added to a large excess of water (1 l). Slowly, a freshly prepared 10% solution of sodium bisulfite (about 120 ml) is added until all color disappears. The solution is neutralized with sodium carbonate and washed into the drain.

OXIDIZING AGENTS

Solutions of compounds such as potassium permanganate, sodium chlorate, sodium periodate, and sodium persulfate should be reduced before being discarded into the drain to avoid uncontrolled reactions in the sewer system. The reduction can be accomplished by treatment with a freshly prepared 10% aqueous solution of sodium bisulfite or metabisulfite. The use of sodium metabisulfite is preferred because of the greater stability of this salt. Precise quantities and conditions for these reactions are detailed in Table 1.

If the concentration of the oxidizing agent to be destroyed is greater than that listed in Table 1, the solution is diluted with water until the stated concentration is reached.

CYANIDES

Highly toxic aqueous solutions of sodium or potassium cyanide are oxidized to nontoxic cyanates by reaction with household bleach (5% sodium hypochlorite

CHEMICAL TREATMENT METHODS TO MINIMIZE WASTE

Table 1 Conditions for the Reduction of Oxidizing Agents with Sodium Metabisulfite

Oxidizing agent present in wastestream	Quantity and concentration of oxidizing agent in aqueous solution	Quantity of 10% aqueous sodium metabisulfite (L)	Comments
Potassium permanganate	1 l of 6%	1.3	Solution becomes colorless
Sodium chlorate	1 l of 10%	1.8	50% excess reducing agent added
Sodium periodate	1 l of 9.5%	1.7	Solution becomes pale yellow
Sodium persulfate	1 l of 10%	0.5	10% excess reducing agent added

solution). These solutions must not be acidified since this results in formation of the very highly toxic hydrocyanic acid.

$$CN^- + ClO^- \rightarrow CNO^- + Cl^-$$

Solutions of sodium or potassium cyanide should be diluted with water to a concentration not greater than 2%. (This dilution is advisable to avoid the possibility of a delayed rapid reaction of the cyanide with the bleach.) Work in the fume hood. To each 50 ml of solution is added 5 ml of 10% sodium hydroxide solution followed by 60 to 70 ml of household bleach. The solution can be tested for the continued presence of cyanide as follows. About 1 ml of the solution is removed and placed in a test tube. Two drops of a freshly prepared 5% aqueous ferrous sulfate solution are added and the mixture is boiled for 30 s. After cooling to room temperature, two drops of 1% ferric chloride solution are added. The mixture is acidified to litmus with 6 M hydrochloric acid. If cyanide is still present, a deep blue precipitate forms. Concentrations of cyanide greater than 1 ppm can be detected. If the test is positive, more bleach is added to the cyanide solution, and the test repeated. When the blue precipitate no longer forms, the solution can be washed into the drain.

INORGANIC AZIDES

Like organic azides, inorganic azides are explosive on heating. Acidification of inorganic azide solutions must be avoided since it results in the formation of hydrogen azide, a highly toxic and extremely explosive gas.

Inorganic azides are readily oxidized by ceric ammonium nitrate solution. Work in the fume hood and behind a safety shield. The concentration of the azide solution should be about 1 g/100 ml of water. Prepare a quantity of 5.5% ceric ammonium nitrate solution that is four times the volume of the azide solution. Slowly add the ceric ammonium nitrate solution to the azide solution and stir for 1 h. At the end of the reaction time, the solution should show the orange color of ceric ammonium nitrate. If the color has faded, add a small portion of 5.5% ceric ammonium nitrate solution (about 10 ml for each 100 ml of original azide solution) and stir the mixture for 30 min. Continue to add these small portions until the orange color persists after stirring. The solution can be washed into the drain.

METAL-CONTAINING WASTES

SOLUTIONS CONTAINING HEAVY METAL IONS

Increasingly, the disposal into landfills of aqueous solutions containing heavy metal salts is being banned. The metals can be precipitated as insoluble salts which are acceptable for disposal. The insoluble salt of choice has often been the sulfide. This requires the use of highly toxic reagents such as hydrogen sulfide, sodium sulfide, ammonium sulfide, or thioacetamide. To avoid the use of these reagents, a number of heavy-metal ions can be precipitated as silicates. These salts show similar solubility properties to the sulfides in neutral, acidic, and basic aqueous solutions. Thus, the effect of acid rain on the leaching of silicates is comparable to that of the sulfides. Also, natural ores often contain metals in the form of silicates. In some cases, for complete precipitation, the pH has to be controlled. As an example, the method is described in detail for solutions containing lead ions.

$$Pb^{2+} + Na_2SiO_3 \rightarrow PbSiO_3 + 2Na^+$$

To the solution containing lead ions is added, with stirring, an aqueous solution of sodium metasilicate ($Na_2SiO_3 \cdot 9H_2O$, 17 g in 100 ml of water) until there is no further precipitation. If the concentration of the lead salt is known, then 200 ml of sodium metasilicate solution is used for each 0.04 mol of lead ions. If the concentration is not known, it is helpful to allow the precipitate to settle, to withdraw a few milliliters of the supernatant liquid, and to add to it several drops of sodium metasilicate solution to test whether or not precipitation is complete. If no precipitate forms, 10 ml more of sodium metasilicate solution is added and the pH is adjusted to between 7 and 8 by the addition of 2 M aqueous sulfuric acid. For each 100 ml of solution, about 20 ml of 2 M sulfuric acid will be needed. The precipitate is collected by filtration, or the supernatant liquid can be allowed to evaporate in a large evaporating basin in the fume hood. The solid is allowed to dry, packaged and labeled for disposal in a secure landfill, or in accordance with local regulations.

For dilute solutions of lead salts, the sodium metasilicate solution should be added until there is no further precipitation, the pH adjusted to between 7 and 8 by the addition of 2 M sulfuric acid, and the solution allowed to stand overnight before collecting the solid by filtration or allowing the liquid to evaporate.

Solutions of cadmium and antimony salts can be treated similarly to the lead salts, and several other heavy-metal salts can be precipitated as silicates by this procedure. These include iron(II) at pH 12, iron(III) at pH 11, zinc(II) at pH 7 to

7.5, and aluminum(III) at pH 7.5 to 8. Copper(II), nickel(II), manganese(II), and cobalt(II) can be precipitated without adjustment of the pH from that after the addition of the solution of sodium metasilicate. This method for precipitation of heavy metals as insoluble silicates is particularly applicable to such wastes from high school laboratories.

SOLUTIONS CONTAINING MERCURY IONS

Mercury salts can be precipitated as the highly water-insoluble sulfide using sodium or ammonium sulfide. Appropriate precautions must be taken to prevent inhalation of the highly toxic gas, hydrogen sulfide, formed from the soluble sulfide salts in the presence of acid.

$$Hg^{2+} + Na_2S \rightarrow HgS + 2Na^+$$

The waste mercury salts are dissolved as far as possible in water (100 ml for each 10 g of waste). The pH of the solution is adjusted to 10 with 10% sodium hydroxide solution. In a fume hood, aqueous sodium sulfide solution (20%) is added with stirring until no further precipitation occurs. To check whether precipitation is complete a small sample of supernatant liquid is withdrawn and a few drops of sodium sulfide solution are added. If a precipitate or cloudiness appears, more sodium sulfide solution is required. After the precipitate has settled, the liquid is removed by decantation or the solid collected by filtration. The liquid is washed into the drain and the solid packaged and labeled as mercuric sulfide for appropriate disposal. In the U.S., the wastes must be sent to a thermal treatment facility to recover the mercury. Depending on regulations, in other countries, disposal may be into a secure landfill or by encapsulation in a cement block.

SOLUTIONS CONTAINING CHROMIUM IONS

Acidic solutions of potassium dichromate are often used in reactions where a strong oxidizing agent is required. They also used to be widely employed for cleaning glassware. This practice has now been largely discontinued in favor of other types of cleaning solutions that do not contain chromium ions. Insoluble chromium hydroxide is formed by reduction of the dichromate with sodium thiosulfate solution. The efficiency of the reduction and the formation of the product as an easily handled flocculent precipitate rather than as a gel is dependent upon the pH of the solution. For this reason, the solution is first neutralized, then reacidified with a measured volume of acid.

$$Cr_2O_7^{2-} + 3S_2O_3^{2-} + 2H_2O \rightarrow 2Cr(OH)_3 + 3SO_4^{2-} + 3S$$

The method is illustrated for a specific quantity of acidic dichromate solution. To ensure complete precipitation of the chromium ions in an easily filterable form, the pH must be carefully adjusted. Thus, to acidic dichromate solution (100 ml) is added solid soda ash slowly and with stirring until the solution is neutral to litmus. About 108 g of soda ash will be required. The color of the solution changes from orange to green. It is reacidified to pH 1 by the careful addition of 55 ml of 3 M sulfuric acid. The color of the solution returns to orange. Alternatively, if it is possible to measure the pH of the solution reasonably accurately, dilute (about 5%) sodium hydroxide can be added slowly to the original acidic dichromate solution to bring it to pH 1. While swirling, sodium thiosulfate (40 g of $Na_2S_2O_3 \cdot 5H_2O$) is added. The solution becomes blue colored and cloudy and it is neutralized by the addition of soda ash (10 g). After a few minutes, a blue-gray flocculent precipitate forms. The mixture can be filtered immediately through Celite or allowed to stand for 1 week, when much of the supernatant liquid can be decanted. In the latter case, the remaining liquid is allowed to evaporate or be filtered through Celite. Analysis by atomic absorption spectroscopy showed that the supernatant liquid contains less than 0.5 ppm of chromium. Most local regulations allow this solution to be washed into the drain. The solid residue should be packaged and labeled for appropriate disposal depending on local regulations. Aqueous solutions of other chromium salts such as chromium trioxide can be treated similarly.

METAL CARBONYLS

Metal carbonyls such as iron pentacarbonyl and nickel carbonyl are highly toxic and reactive materials and the latter is a suspect carcinogen. These compounds are destroyed by stirring solutions in the appropriate solvent with bleach. The choice of solvent is very important and the conditions necessary to ensuring complete and smooth conversion to the hydroxide are summarized in Table 2. The reactions should be performed in the fume hood.

TREATMENT OF SPILLS

A practical and versatile spill mix has been developed consisting of a mixture of sodium carbonate to neutralize any acid present, clay cat litter (calcium bentonite) to absorb liquid rapidly, and sand to moderate any reaction.

To treat a spill of a mineral acid or bases such as the hydroxides of sodium, potassium, or calcium, wear nitrile rubber gloves, laboratory coat, and goggles. (Self-contained breathing apparatus may be necessary depending on the nature and size of the spill.) Isolate the area of the spill and cover it with a 1:1:1: mixture by weight of sodium carbonate, clay cat litter (calcium bentonite), and sand. When all of the liquid has been absorbed, scoop the mixture into a plastic pail, and in the fume hood very slowly add the mixture to a pail of cold water. Allow to stand for 24 h. Test the pH of the solution, and neutralize if necessary with sodium carbonate *or dilute hydrochloric acid.* Decant the solution to the drain. Treat the solid residue as normal garbage. Sodium carbonate can be replaced with calcium carbonate both in the spill mix and in the neutralization of the aqueous solution.

Table 2 Conditions of the Reaction of Metal Carbonyls with Bleach

$Fe(CO)_5$	+	NaOCl	→	$Fe(OH)_3$
Iron pentacarbonyl solution in hexane: 5 ml in 200 ml hexane		Bleach 65 ml	Stirred for 30 min under helium	Solid filtered, discard aqueous layer to drain, hexane layer recycled or incinerated
$Fe_2(CO)_9$	+	NaOCl	→	$Fe(OH)_3$
Diiron nonacarbonyl solution in toluene: 1 g in 100 ml toluene		Bleach 50 ml	Stirred for 25 h	Solid filtered, discard aqueous layer to drain, toluene layer recycled or incinerated
$Cr(CO)_6$	+	NaOCl	→	$Cr(OH)_3$
Chromium hexacarbonyl solution in tetrahydrofuran: 1 g in 200 ml of THF		Bleach 30 ml	Stirred for 15 min	Solid filtered, packaged for disposal, liquid incinerated
$Ni(CO)_4$	+	NaOCl	→	$Ni(OH)_2$
Nickel carbonyl solution in tetrahydrofuran: 5 g in 200 ml of THF		Bleach 250 ml	Stirred for 2 h under helium	Solid filtered, packaged for disposal, liquid incinerated

The spill mix is also useful to absorb spills of solvent or other hazardous liquids. The resulting solid residue must be treated as a hazardous waste and disposed appropriately.

For spills of those chemotherapeutic drug solutions that are particularly hazardous to handle, it may be desirable to pour a solution of the appropriate deactivating agent, i.e., acidified potassium permanganate, bleach, or synthetic detergent directly onto the spill. Solutions of potassium permanganate for treating spills are prepared by dissolving 4.7 g of potassium permanganate in 100 ml of 3 M sulfuric acid (17 ml of concentrated sulfuric acid added to 83 ml of water). This solution should be prepared fresh daily. If the purple color of the permanganate fades during treatment of the spill leaving a brown mixture, more acidic potassium permanganate solution should be added. The liquid resulting from treatment of the spill is adsorbed onto suitable material such as paper towels or absorbent cotton; the material is placed in a container, covered with more of the deactivating solution and allowed to stand overnight. If the deactivating solution is acidic potassium permanganate, the solution is carefully neutralized with 10% aqueous sodium hydroxide, decolorized by the slow addition with stirring of solid sodium bisulfite, and decanted to the drain. Where the deactivating solution is bleach or synthetic detergent, the liquid is washed into the drain. In all cases, the residual absorbent material is discarded as normal refuse.

CONCLUSION

There are several general chemical techniques that can be used to reduce, reuse, recycle, or recover waste. Where the waste cannot be further minimized, the remaining material can often be safely and effectively converted to nontoxic and environmentally acceptable products. In combination with the strategies to minimize waste chemicals, the disposal procedures allow for the development of an integrated waste management program. We are continuing to develop, test, and document acceptable waste disposal procedures for individual chemicals including azo dyes and pesticides, the latter with two goals in mind: (1) to allow destruction of residual pesticides in containers and (2) to find efficient ways of removing all pesticide residues from workers' clothing.

ACKNOWLEDGMENTS

The author gratefully acknowledges financial support from the Alberta Environment Trust, Alberta Occupational Health and Safety, and the Alberta Heritage Foundation for Medical Research. Consultants in the project are Dr. Lois Browne, Mr. Gordon Weir, and Ms. Rosemary Bacovsky. The laboratory development and testing of the procedures has been done by Donna Renecker, Patricia McKenzie, Carmen Miller, Katherine Ayer, John Crerar, Paul Cumming, Richard Young, Girard Spytkowski, Mui Chang, and Dr. Roger Klemm, all of whom provided many excellent suggestions as the work progressed.

ENDNOTES

1. Pine, S. H., Chemical management, a method for waste reduction, *J. Chem. Educ.,* 61, A95, 1984.
2. Task Force on RCRA, *Less is Better,* American Chemical Society, Department of Government Relations and Science Policy, Washington, D.C., 1985.
3. Pike, R., Szafran, Z., and Foster, J., *Microscale Laboratory Manual for General Chemistry,* John Wiley & Sons, New York, 1993.
4. Williamson, K., *Macroscale and Microscale Organic Experiments,* D.C. Heath, Lexington, MA, 1989.
5. Phifer, R. W. and McTigue, W.R., *Handbook of Hazardous Waste Management for Small Quantity Generators,* Lewis Publishers, Chelsea, MI, 1988.
6. *Prudent Practices for Handling, Storage and Disposal of Hazardous Chemicals in Laboratories,* National Academy of Sciences, Washington, D.C., 1983. [New edition pending.]
7. Pitt, M. J. and Pitt, E., *Handbook of Laboratory Waste Disposal,* Wiley-Halstead, New York, 1985.
8. Lunn, G. and Sansone, E. B., *Destruction of Hazardous Chemicals in the Laboratory,* Wiley-Interscience, New York, 1990.
9. Armour, M. A., *Hazardous Chemicals Disposal Guide,* CRC Press, Boca Raton, FL, 1991.
10. Armour, M. A., Bacovsky, R. A., Browne, L. M., McKenzie, P. A., and Renecker, D. M., *Potentially Carcinogenic Chemicals, Information and Disposal Guide,* University of Alberta, 1986.

11. Armour, M. A., Browne, L. M., and Weir, G. L., Tested disposal methods for chemical wastes from academic laboratories, *J. Chem. Educ.,* 62, A93, 1985.
12. Armour, M. A., Chemical waste management and disposal, *J. Chem. Educ.*, 65, A64, 1988.
13. Castegnaro, M., Adams, J., Armour, M. A., Barek, J., Benvenuto, J., Confaloneri, C. C., Goff, C., Ludeman, S., Reed, D., Sansone, E. B., and Telling, G., *Laboratory Decontamination and Destruction of Carcinogens in Laboratory Wastes: Some Antineoplastic Agents,* International Agency for Research on Cancer, Lyons, 1985.

Part 5:
What Organizations Can Do

CHAPTER 17

Recruiting Vendors to Achieve Waste Minimization and Pollution Prevention

K. Leigh Leonard

CONTENTS

Overview 296
    Why Use Vendors? 296
    Organization of this Chapter 298
Specific Vendor Services 298
    Improving Efficiency of Off-Site Hazardous Waste Disposal 298
    Off-Site Hazardous Waste Recycling 299
    Off-Site Redistribution of Excess Chemicals 299
    Product Take-Backs by Chemical Manufacturers 301
    On-Site Hazardous Waste Recycling 301
    Chemical Inventory Management 302
        Procurement as an Information Source 303
        Procurement as an Internal Control Point 303
        Other Innovative Procurement Options 304
The Limits of the Procurement Process 304
    Market Factors 304
    Purchasing Leverage 305
    The Inertia of the Market 305
Why Seek Innovation? 306
    Service-Oriented Contractual Requirements 307
    Persuading Internal Constituents 307
Conclusion 308
Acknowledgments 308
Endnotes 308

OVERVIEW

This chapter details how chemical product vendors, hazardous waste services contractors, and other commercial enterprises can enhance an institution's efforts to minimize hazardous waste and evolve toward pollution prevention. It also discusses the vendor's role in promoting efficient and cost-effective hazardous waste management. This is in keeping with the idea, outlined in Chapter 1, that a high level of performance of basic hazardous waste management operations provides a critical foundation for an effective waste minimization program.

Initially, the chapter describes, in generic terms, existing and potential services and products that have bearing on waste minimization and pollution prevention. Later sections provide a general conceptual framework for assessing the best way to utilize vendors, and how much innovation may be expected within a given contractual relationship.

This chapter is targeted primarily toward purchasing agents who deal with the procurement of laboratory chemicals or hazardous waste management services, as well as environmental health & safety staff involved with hazardous waste management at an institutional level. Laboratory staff themselves may be less able to make direct use of the information in this chapter, but it will provide them with an idea of what services they may expect of their vendors.

WHY USE VENDORS?

Assuming that vendors have valuable services to offer in this area, why use them for functions beyond provision of stock and research chemicals and removal and disposal of hazardous wastes?

Staff resources are dear. The economic climate of the past several years has meant staff cutbacks in most public institutions and a reexamination of corporate hierarchies in the private sector. "Quality reinvestment" often results in midlevel technical and operational managers being required to take on additional program areas with a diminishing pool of professional staff. In this atmosphere, it is difficult to apply staff resources to functions that are perceived as ancillary to the institution's primary mission. Some managers find it easier to hold onto and even increase S&E (supplies and expenses) budgets than to hire additional staff. Staff dollars are inflexible and very difficult to increase. S&E money can fuel a waste minimization/pollution prevention program through wise investment in innovative products such as commercial distillation equipment, microscale glassware, or a bar-coding system for chemical inventory management. Obtaining such tools may be a major barrier to departmental staff who are otherwise willing to take ownership of a particular waste minimization/pollution prevention program element. Also, if the contract is structured carefully, the vendor can take care of the training and technical assistance needed to get the program underway.

Vendors can allow existing staff to maintain the quality of the base waste management program while maximizing opportunities for waste minimization/ pollution prevention. While environmental compliance is recognized as an imperative function by most larger institutions today, going *beyond* compliance is usually the first thing sacrificed when resources become scarce. At the time this book was written, regulatory mandates for waste minimization are few (see Chapter 4). Each large quantity generator is required to certify that waste minimization efforts are in place to the degree each generator determines to be "economically practicable." Given EPA's low enforcement profile on this requirement, and their lack of authority to dictate waste minimization techniques for specific processes, this is hardly a pivotal regulatory obligation for generators.*

In a time of scarce resources, it is extremely difficult to garner institutional commitment to a considered, planned approach to waste minimization. Usually, the only justifiable waste minimization activities are those that clearly and unequivocally save the institution money. Very often, the environmental staff do not even have time to collect the basic wastestream data and develop the cost/benefit analysis needed to demonstrate the cost savings of a proposed waste minimization technique. Further, a national survey shows that many institutions engage in laboratory waste minimization programs without any mechanism to track volume reduction and associated cost savings to the institution (see Chapter 2). This certainly makes these programs vulnerable to being cut back as resources become more scarce.

A well-utilized hazardous waste contractor should be able to provide wastestream data in a form that is useful, and help to pinpoint and document reductions in specific hazardous wastestreams being shipped off-site. The contractor can also be required to provide detailed cost information by wastestream type and generator source and projections of how costs might change if significantly smaller volumes are generated. These activities put the environmental health & safety program on solid ground when it comes to defending its investments in waste minimization/pollution prevention activities. They may also free up internal staff to maintain the waste management foundation and work more closely with laboratory staff on specific waste minimization/pollution prevention initiatives.

Holding the expectation that vendors provide services and products supporting waste minimization and pollution prevention engenders a future market that is rich in these services. This is a longer-term benefit of working with vendors to provide innovative services that support waste minimization and pollution prevention goals. Twenty years ago material safety data sheets were virtually unknown. Fifteen years ago, there were few hazardous waste disposal options beyond landfill and incineration. Ten years ago there were few contractors that could perform on-site treatment of compressed gases and shock-sensitive materials using procedures other than detonation or open burning. While major market developments in each of these areas was precipitated by federal regulations, there were "front runners" in each service area that predated the regulations. These companies innovated early because they perceived a customer demand for such services, combined with an anticipated, if uncertain, regulatory horizon. Today's environmental health & safety managers owe it to their future counterparts to be demanding customers, doing their part to foster a more diverse service market that

---

*See Comella, P., "The Scope of EPA's Waste Minimization Authority." (Paper presented at the 206th National Meeting of the American Chemical Society, Chicago, August 1993.)

supports waste minimization and pollution prevention efforts. At the same time, managers must carefully explore such initiatives before undertaking them to ensure that the new service or product is really worth the expenditure of additional budget resources.

ORGANIZATION OF THIS CHAPTER

The following section presents a wide range of specific services and products that enhance or accomplish laboratory waste minimization or pollution prevention efforts. Following this, the chapter describes how to maximize vendor roles through existing and new contracts, using the procurement process itself. The last section is oriented toward understanding the limits of the procurement process by assessing the costs and benefits of contracting for innovative services or products that enhance pollution prevention.

SPECIFIC VENDOR SERVICES

Specific services and products that support or enhance waste minimization/pollution prevention in laboratories are presented in opposite order of the waste minimization hierarchy outlined in Chapter 1. The reason for this is that large-scale off-site services (e.g., fuel blending) are generally much further developed than smaller scale on-site services (e.g., on-site solvent distillation, chemical take-back services). Laboratories can implement first the services that are readily available, as they will involve less investment of staff effort to set up. Time that is made available once these services are in place may be directed toward more difficult projects, where the vendor's role in waste minimization/pollution prevention will have to be leveraged or coaxed. This section will not discuss the probable costs for these services and products. Its purpose is to provide examples of creative uses of vendors to provoke the thoughts of the reader. The issue of costs and benefits is discussed in the last section of this chapter.

IMPROVING EFFICIENCY OF OFF-SITE HAZARDOUS WASTE DISPOSAL

Under a well-crafted contract, the laboratory's hazardous waste contractor can play a key role in preparing to seriously undertake a hazardous waste minimization program. In an ideal relationship, the contractor representative will communicate clearly the most beneficial shipment configurations to aid on-site personnel in appropriate segregation and commingling of hazardous wastes. The representative will also perform accurate cost analyses of various scenarios of waste management, including the trade-off between bulking and lab packing a wastestream. The contractor can also be utilized to perform commingling and lab packing techniques on site, which are time-consuming activities. This allows in-house technical expertise to be directed toward waste minimization/pollution prevention activities.

Perhaps the most beneficial function the hazardous waste contractor can provide is a detailed account of the types, volumes, and sources of hazardous wastes being shipped out. This will save staff time when filling out the hazardous waste annual report. This documentation also forms the foundation of the hazardous waste audit — the first step in developing a waste minimization plan.

OFF-SITE HAZARDOUS WASTE RECYCLING

Off-site reuse, recovery, and recycling are presently not considered pollution prevention by EPA and most state regulatory agencies. However, this does not preclude a philosophical preference (perhaps expressed in policy) for off-site management options that emphasize recycling, reuse, or heat recovery. These options include:

- Blending of high BTU hazardous wastes into fuel for cement kilns
- Metal reclamation from wastes such as mercury salts, fluorescent lights, PCB ballasts, "NiCad" batteries, and silver salts
- Large-scale (drum or tanker volume) organic solvent recycling
- Reconditioning of small and large compressed gas cylinders

Engaging some of these services may engender new liability exposures since many of these companies are relatively new, and generally involve more complex processes (and more residuals) than more traditional disposal technologies (e.g., incineration and landfill). Also, in most states, reclamation facilities are still not required to be permitted as hazardous waste treatment processes, meaning they are often subject to less regulatory review than licensed TSD facilities. Incorporating these services may be an especially difficult policy decision for institutions that have historically preferred incineration to minimize long-term liability.

However, if such services are to survive in the shadow of the traditional waste management industry, and grow more sophisticated and reliable in the future, it is critical to participate in the market now. This is not to say that these facilities should not be subjected to the same scrutiny as traditional TSD facilities. However, those who can pass such scrutiny should be engaged where they can enhance an institution's ability to reuse or recycle hazardous waste.

OFF-SITE REDISTRIBUTION OF EXCESS CHEMICALS

This involves working with companies that are essentially chemical brokers. They accept unused chemicals and make them available to potential users: colleges and universities, private laboratories, and other research institutions. These companies should be distinguished from public sector or nonprofit materials exchanges that function strictly as "matchmakers" and never actually take possession of any chemical products.

There are many serious concerns with commercial chemical brokers. *In fact, past problems have been so severe that there are few commercial brokers left in the U.S. today, and most institutions will not even consider utilizing them.* The purpose of addressing the topic here is to alert readers to the types of problems associated with these services, and to identify what qualities a company would have to possess to prevent these problems and assure a safe and legal exchange of chemicals.

The problems with commercial chemical broker services generally stem from their lack of preparation for dealing with the consequences if the chemicals become hazardous wastes at any point in the process. Brokers generally use a large number of outlets for the chemicals, yet often do not adequately document the specific fate of each chemical. The donor may need such documentation to demonstrate to regulators that chemicals are really being used for their intended purpose and not disposed.

If quality assurance is not a primary goal through every stage of the redistribution process, it is inevitable that deteriorated, outdated, or off-specification chemicals will be distributed to a potential user. The entire lot may well be returned to the broker, posing a significant problem if the broker is not licensed to store hazardous wastes, and is not experienced with their proper transportation or disposal. Worse yet, if the broker cannot take back the chemicals, or is no longer in business, the recipient may contact the donor and insist that the chemicals be removed at the donor's expense. Also, if opened containers are donated, and the broker has no testing program to assure quality and purity, a reagent may turn out to be entirely different from what it is alleged to be, with potentially disastrous consequences for the recipient and considerable liability for the donor.

There is also the problem of material safety data sheets (MSDSs). The federal Occupational Safety and Health Administration requires chemical manufacturers and distributors to provide an MSDS for every hazardous chemical sold or distributed. This means that an institution ought not surrender a hazardous chemical to a broker without providing an MSDS. At many institutions, this may pose a problem because MSDSs are lacking for some percentage of their excess chemical stocks. Indeed, that may be one reason the institution no longer wants certain chemicals.

Finally, RCRA requirements relating to "speculative accumulation" pose a major obstacle to the operation of brokerage operations.[1] Brokers may be cited for storing hazardous waste without a license if they accumulate donated chemicals for more than a year without being able to sell them.

Considering these problems, what would be the ideal features of a safe and legal chemical broker service that carries minimal risk for the donor?

- Strict criteria for accepting chemicals (e.g., up-to-date MSDSs, within the manufacturer's shelf life, no unstable chemicals)
- A quality assurance program, including analytical testing of chemicals that had been opened
- A "no-lemon" policy so recipients could return any chemicals found unsatisfactory for a full refund
- Ability to legally store (e.g., permitted TSD storage facility status) as hazardous wastes any chemicals found not to be suitable for redistribution, and to arrange for their treatment or disposal at permitted TSD facilities
- Liability and environmental impairment insurance coverage similar to what would be required for a hazardous waste contractor

At the time this book is being written, the editors know of no commercial chemical brokers that would meet these criteria. The best recommendation is to set up in-house or intra-institutional chemical exchanges (see Chapter 18) or to utilize only those exchange programs that do not take possession of the chemicals, but merely match potential donors with interested recipients.

PRODUCT TAKE-BACKS BY CHEMICAL MANUFACTURERS

This is a narrow form of off-site redistribution of excess chemicals. Some chemical product manufacturers, usually specialized companies with a high service level, offer to take back unused portions of chemicals they have sold to their customers. This is usually done for a small fee.

Interviews with some of these manufacturers reveal that these chemicals are generally not reused or reprocessed, but are disposed as chemical wastes by the manufacturer. (At least one firm does reprocess the chemicals if possible, and will provide the customer with documentation.) Programs where the chemicals are being disposed do not truly accomplish waste minimization in that no real reuse or recycling takes place. Also, expired or deteriorated chemical products may be viewed as hazardous wastes by some regulators, and the vendor would have to be a licensed hazardous waste facility to receive them. Recognizing this gray area, some laboratory chemical suppliers have teamed up with hazardous waste companies to offer a true disposal service.

Product take-back services are invaluable where compressed gases are concerned because these items are so expensive and difficult to dispose through hazardous waste management companies. This practice is generally allowed by state and federal regulators under a long-standing EPA policy that exempts compressed gas manufacturers from RCRA licensing requirements.[2]

Most compressed gas manufacturers will accept back small and large cylinders so they may recertify and reuse the container. They will often accept these containers even when they still contain unused product, as long as the cylinders are in transportable condition according to Department of Transportation requirements. However, they are unlikely to accept back lecture bottles because these are single trip containers that cannot be reused. There are exceptions, however. The University of Minnesota successfully required, as a condition of their contract, that laboratory gas suppliers take back lecture bottles with unused product remaining in them.[3] The University of Illinois at Urbana/Champaign is looking into the feasibility of having their compressed gas vendor provide small quantities of laboratory gases in small cylinders that the vendor could take back instead of lecture bottles.[4]

ON-SITE HAZARDOUS WASTE RECYCLING

There are a couple of waste recycling products that are especially suited for laboratory wastes. A few manufacturers now provide more appropriate equipment for laboratory solvent distillation. Such products have made it possible to safely and efficiently undertake distillation of certain contaminated solvents on-site.

This may be beneficial where the solvent generation rate is too small to interest off-site solvent recyclers, yet it is large enough that there will be a reasonable payback period on the distillation equipment. A good example of a setting where on-site solvent distillation equipment works are histology labs that generate significant volumes (e.g., around 25 gal/week) of lightly contaminated waste xylene.[5] Solvent distillation is discussed in detail in Chapters 16 and 21.

Equipment is also readily available for silver reclamation from photographic fixers and other aqueous silver solutions. These include fairly sophisticated electroplating equipment and end-of-process steel microfilters that plate out some, but not all, of the silver. Silver flake is collected from the on-site electroplating units and marketed directly. The steel filters must be sent to a reclamation facility, and the laboratory is usually paid fair market value for the silver reclaimed.

CHEMICAL INVENTORY MANAGEMENT

Chemical inventory management is a vital step toward reducing wastestreams consisting of old or unused chemical stocks. However, this is a broad subject which has implications that reach beyond the waste management process. Consequently, it will be treated briefly here, with an emphasis on waste management.

Chemical inventory management begins in the laboratory stockroom. The objective is to manage chemical stocks so that product is used efficiently and not allowed to linger on the shelf beyond its safe and useful shelf life. This is especially critical for chemical products that become dangerous as they, or their containers, deteriorate over time or are subject to certain storage conditions.

Bar-coding systems and computerized databases are commercially available to tighten inventory management and increase the effectiveness of in-house chemical redistribution. However, staff resources are usually the major limitation to getting these programs established. Professors and principal investigators rarely feel it is their role to manage such a program. Lab technicians are normally consumed with their routine functions, such as the preparation of reagent solutions, glassware cleaning, waste management, and the timely procurement of necessary supplies and reagents. It is questionable whether student help can be effective in establishing a bar-coded, computerized chemical inventory without professional staff leadership, since student turnover is high, and when they leave they take their expertise with them. There are many cases of chemical inventory management projects getting underway, only to be abandoned later when students leave or when inside staff don't really have the time they thought they would be able to devote. If there was one area where professional consulting services may be wise investment, this might be it.

Along with inventory management, some institutions exercise controls on purchases, gifts, and donations of chemical products to prevent acquisitions that will not be used up, and will eventually have to be managed as hazardous wastes. Gifts and donations may be prohibited outright, subject to approval, or be required to meet certain criteria — such as being able to use up the donated product within a limited time. Purchasing controls may take many forms, including consideration of less-hazardous substitutes or use of chemicals already in stock, purchasing only the amount needed, or purchasing amounts that can be used up within a limited time.

Carefully crafted contracts for the procurement of chemical products can enhance the chemical inventory management program and any purchasing policies an institution may have. In combination, these management tools maximize the potential of reducing wastes comprised of excess chemical products. Against this background, specific contractual mechanisms to promote waste minimization are discussed below.

Procurement as an Information Source

Simply monitoring the types and volumes of chemicals ordered can provide valuable information. This is a "portal" approach to inventory management — everything that comes through the door is monitored, but access to the door is not restricted.[6] Procurement history can be compared with the existing chemical inventory to course correct overstocking. The information can also be used to red flag large flows of chemicals through a laboratory process, earmarking it for a pollution prevention assessment. The University of Minnesota has recently incorporated a provision in their chemical contracts that requires vendors to provide data on purchases annually.[7] They plan to utilize this data in the future for compliance with anticipated federal reporting requirements.

Using a portal approach to procurement is much less threatening to laboratory staff than outright restrictions on chemical purchasing. Chemical suppliers under contract can detail procurement history back to a designated customer representative on a periodic basis (e.g., quarterly). Naturally, this may not capture special orders made by lab staff from vendors not under contract with the institution.

Procurement as an Internal Control Point

If the institutional environment is conducive, purchasing controls can be put into place. For example, anyone ordering a chemical product might be required to get pre-approval of their purchase or to route the request through an individual who will verify that the chemical product is not already in stock. This approach rarely succeeds in large institutions with multiple labs where purchasing is decentralized, for reasons that are beyond the scope of this chapter to detail. However, where this arrangement does exist, there is greater potential to require the use of a single vendor for chemical purchases. This improves the potential for comprehensively monitoring procurement history, because:

- The vendor will be more willing to agree to perform this service as a contractual requirement if 100% of the institution's business is assured
- The data will not be skewed as much by special orders, since lab staff would not be allowed to utilize other vendors unless the primary contractor cannot provide the desired product

Other Innovative Procurement Options

Some chemical companies are providing innovative services that reduce the potential for accumulating excess chemical stocks. The take-back of unused product has already been discussed. "Package-to-order" vendors provide chemical reagents in smaller quantities, in some cases, exactly the quantity needed. This requires significant effort on the vendor's part because every container that is shipped individually must meet stringent Department of Transportation requirements. Naturally, there is some extra cost in procuring chemicals this way, but that should be weighed against the cost of disposing the excess chemical, and the more indirect costs of managing the chemical in the inventory past the time of its initial use. Other services vendors may offer include reducing excess packaging or providing packaging that meets certain criteria (e.g., resistant to breakage and hence spills, not composed of materials that will cause incinerator ash to fail toxicity tests).

THE LIMITS OF THE PROCUREMENT PROCESS

This section explains general concepts of procurement in the context of attaining both services and products described earlier in this chapter. It is not intended to be a comprehensive discussion of the procurement process, but rather to provide a conceptual framework for assessing what will and won't work when attempting to procure services or products that are out of the mainstream.

This chapter has illustrated how contractors and product vendors may facilitate or set the stage for pollution prevention and waste minimization activities. To this point, it has not addressed the feasibility of obtaining each service or product in today's marketplace. How can staff assess their institution's ability to influence vendors to provide new services or products outside of their core missions?

MARKET FACTORS

The nature of the specific market in which a vendor operates has a large bearing on what may be attainable. These characteristics include degree of competition, the market share of various participants, external forces that affect pricing, and so on. If your contract bid specification includes provisions that run counter to current market forces, it is very unlikely they will be met. Taking the hazardous and toxic waste market as an example, it would have been futile during the mid to late 1980s to seek lower-than-average prices for PCB incineration or a rapid turnaround on shipment requests. At this time, the PCB market was experiencing a capacity shortage exacerbated by the unexpected closure of a major PCB incinerator. In the 1990s, the PCB market has loosened up due to added capacity and a shrinking wastestream making lower prices attainable.

Market research is always important when letting a contract. It is critical, though, when seeking services or products that are innovative, esoteric, or not widely sought. Since pollution prevention and waste minimization are not yet circumscribed by a regulatory structure, and the demand for related services is sporadic, not all sectors of the market described earlier in this chapter are well developed. This means that, when seeking services to support waste minimization and pollution prevention, it may be most fruitful to tie them to a market sector that is very robust. For example, if the service desired is an ongoing chemical inventory, consider making this a requirement of your laboratory chemical contract, in exchange for making it an exclusive contract. The vendor's attraction to having a large, exclusive contract could make it worth their effort to provide the chemical inventory service or at least to provide detailed documentation of items procured under the contract.

PURCHASING LEVERAGE

The term "purchasing leverage" refers to how favorably your institution is perceived by participants in the market. If your institution is large, and expects to do a high-dollar volume of business through a contract, the chance of obtaining lower prices, and unusual types or levels of service, increases. This is known as having much purchasing leverage. For example, some states (e.g., Utah and Wisconsin) have let mandatory hazardous waste contracts that serve the needs of all state agencies and institutions. Not only does this limit long-term liability by limiting the number of TSD sites, it also allows much more demanding contract requirements than if each agency bid the contract separately. Typically, a contracting institution that holds a high degree of purchasing leverage also achieves lower-than-average pricing.

Purchasing leverage can also be associated with the "status" of the entity letting the contract. For example, governments, Fortune 500 companies, and certain nonprofit organizations may have added purchasing leverage in some markets just by being who they are. They are a type of institution that is desirable to have on a customer reference list.

If your institution does not have much purchasing leverage, it may band together with similar institutions for contracting purposes, increasing the dollar volume, geographic scope, and/or the "status" of the contract. For example, in some regions, small quantity hazardous waste generators have joined forces to let a contract uniquely designed to service low volume generators with a high variety of wastestreams.

THE INERTIA OF THE MARKET

Some market sectors are particularly resistant to evolving into a new niche. This is especially true of commodity-oriented markets where the quality of the product has been paramount to associated services. For example, despite a marked customer demand, it is notoriously difficult to get chemical companies to take back unused portions of chemicals or to move into associated services such as detailed chemical procurement records or electronic MSDSs.

This can partially be explained by a commodity-orientation inherent to the culture of chemical companies. Their staff have historically measured corporate achievements in terms of the range of specialty and custom chemicals provided, product grade and purity, competitive pricing, and rapid delivery. Despite customer demands, services ancillary to the product itself are not always understood to be a priority within the company. The lack of standard measures by which to gauge services vs. products discourages commodity-oriented vendors from attempting innovations. There are, however, some promising indications that chemical companies are becoming more willing to address the environmental consequences of their business.

The Chemical Manufacturer's Association (CMA) as recently initiated a program called Responsible Care® in reaction to public concerns about chemical risk and to regulatory programs (e.g., The Community Right-to-Know Law) that have been put into place in response to such concerns. Companies who participate in the Responsible Care® program adopt six codes of management practice including community awareness and emergency response, distribution of chemicals, process safety, pollution prevention, employee health and safety, and product stewardship.

The purpose of Product Stewardship, adopted by CMA in 1992, is to "make health, safety, and environmental protection an integral part of designing, manufacturing, marketing, distributing, using, recycling, and disposing of ... products."[8] In theory, under Product Stewardship, chemical manufacturers would take an active role in managing their product from "cradle to grave." So, their management would take into account externalities from extraction of raw materials through the entire life cycle of the product, including its ultimate disposition. While most participating companies are in the very early stages of implementing Product Stewardship, their involvement bodes well for the future of innovative services that promote pollution prevention.

WHY SEEK INNOVATION?

Why should institutions go out of their way to incorporate services into contracts that promote waste minimization or pollution prevention, but which are not yet readily established in the market? Even if this is determined to be feasible within a particular market sector, making innovative services a mandatory part of a contract will likely increase pricing in the contract.

Markets are dynamic entities. However, to evolve, they need clear signals and real pressures. Vendors will not generally provide a special service because one customer politely requests it. If many entities with significant purchasing leverage make the special service a mandatory or highly desirable element of their contracts, vendors will begin to respond. They will respond especially rapidly if the customer demand coincides with external forces that favor provision of the service.

If no one values a service that enhances pollution prevention efforts highly enough to include it in a contract, the service will likely never be born. By seeking services that are on the envelope of the market, and by being willing, within reason, to pay for them, we ensure a more diverse and capable service sector in the area of waste minimization and pollution prevention for the future.

SERVICE-ORIENTED CONTRACTUAL REQUIREMENTS

For reasons already discussed, it is a risky proposition to contract solely for a service or product that does not yet have a life of its own in the market place. The next best thing is to analyze existing contracts to see where ancillary services might be incorporated that would facilitate pollution prevention or waste minimization. Some opportunities specific to laboratories might include the hazardous waste services contract, chemical reagent suppliers, battery manufacturers, specialty gas suppliers, analytical equipment suppliers, and companies that build and install fume hoods and other types of local exhaust ventilation.

For example, a contract to install new fume hoods might require that fume hoods have an organic vapor emission capture and recovery system, or a renewal of the hazardous waste contract might require the contractor to annually assess the chemical inventory to earmark chemicals that should be used up within the next few months to avoid their deterioration and ensuing disposal as hazardous waste. Battery suppliers might be required to provide "green" batteries or provide some means of product return or reclamation for those that aren't. Specialty gas suppliers might be required to track gases sold to the laboratory, take back partially discharged lecture bottles and cylinders, and provide an accurate inventory periodically to the environmental compliance manager. (Some gas suppliers are already well on the way to providing such a service array.)

Of course, not all of these services are readily being provided now. An institution would not want to jeopardize an entire contract by taking a hard line on a single provision. However, depending on the degree of purchasing leverage, there may be no harm in making such provisions optional in the first few years of the contract with a clear message that at some point in the future the provision will become mandatory.

PERSUADING INTERNAL CONSTITUENTS

We cannot expect vendors to develop new services and capabilities without impacting the cost of the core service or product. This fact, by itself, may radically limit an institution's willingness to experiment with innovative services. This is especially true in government procurements where the low-bid approach is typical, and there is high sensitivity about the use of public funds. However, if internal costs and benefits are assessed as well as the actual contract costs, it may be proven that a modest increase in contract pricing may accrue an internal benefit of such a magnitude that it more than returns the increased contract cost.

For example, an enhancement to the hazardous waste services contract costs an extra $0.50 per pound, a 10% increase. Let's say, for the sake of discussion, that the vendor is now required to annually provide data suitable for the Annual Hazardous Waste Generator Report, and to provide quarterly data summaries of

Table 1 Impact of Incorporating Hypothetical Enhancement into Hazardous Waste Services Contract

	With innovation	Without innovation
Disposal cost	$5.50/lb	$5.00/lb
Waste generation	2000 lb	2500 lb
Total cost	$11,000	$12,500
Net savings	$1500 with innovation	

waste generation by source. As a result, in-house staff have more time available to coordinate waste minimization efforts, and the detailed data help to prioritize waste minimization efforts. In the first year, there are 500 less pounds of waste to be disposed, a reduction of 20% of the total wastestream (see Table 1). It is clear that the innovation was worth the price. In fact there is a net savings of $1500, a 12% reduction.

The analysis may not be so clear cut in every case, and often it is not possible to completely assess the results ahead of time. However, having a rationale for the contracting change will increase the potential to persuade internal constituents that it is likely to be worth taking a chance.

CONCLUSION

This chapter detailed existing and potential services and products that can enhance an institution's waste minimization and pollution prevention efforts. It also outlined concepts of procurement that must be understood to successfully contract for innovative services, and to work effectively with procurement staff to incorporate waste minimization and pollution prevention priorities in existing contracts.

The procurement process can be an important tool in building a laboratory waste minimization or pollution prevention program. If your institution happens to be blessed with significant purchasing leverage, and a procurement staff who have time to undertake market research, perhaps you will be inspired to help shape the future market for waste minimization and pollution prevention services.

ACKNOWLEDGMENTS

The author would like to credit Ms. Ellen James, of the University of Wisconsin System Administration, for many of the procurement concepts outlined in this chapter and Ms. Ann Blakely, Earth Resources Corporation, for ideas about commodity-oriented businesses.

ENDNOTES

1. Personal communication, Mr. Russ Phifer, Environmental Assets, June 17, 1994.
2. Presentation by D. Nickens, Earth Resources Corporation, March 7, 1994, at *Environment '94,* sponsored by the Federation of Environmental Technologists.

3. Communication with Mr. Bruce Backus, May 2, 1994.
4. Communication with Peter Ashbrook, February 21, 1994.
5. See, for example; Weeks, C. S. and Davis, E. M., *Solvent Redistillation as a Waste Minimization Strategy,* presented at the Annual College and University Hazardous Waste Conference, 1988.
6. This term was coined by Ms. Patricia Kandziora, University of Wisconsin System Administration.
7. Mr. Bruce Backus, University of Minnesota, phone conversation, May 2, 1994.
8. Bishop, J., Product Stewardship — the chemical industry defines, accepts responsibility for its wares, *Hazmat World,* July 1992.

CHAPTER 18

Surplus Chemical Exchange: Successes and Potential

Jeff Christensen

CONTENTS

Introduction 311
The University of Arizona Chemical Redistribution Program 312
Keys to a Successful Redistribution Program 314
Overcoming Impediments to Surplus Chemical Exchange.......................... 315
Conclusion 317

INTRODUCTION

The purchase and disposal of chemicals in laboratory quantities are expensive necessities. A 500-g bottle of sodium cyanide costs $20.30 through a major chemical supplier. The disposal price of the same bottle as hazardous waste is between $8.00 and $10.00 per pound including transportation, material, and labor costs. In a time of diminishing budgets, buying and then discarding useful chemical reagents is a waste of valuable resources.

An examination of what a university is disposing of as hazardous waste would reveal that most of the materials treated as wastes are perfectly usable chemicals. A mechanism to remove these materials from the hazardous waste loop and offer them to university laboratories as usable reagents helps reduce purchase and disposal costs. Simultaneously, this helps reduce the amount of hazardous waste generated. Eighty percent of the responding institutions to the National Laboratory Survey discussed in Chapter 2 reported that they have some form of a chemical exchange program to help reduce waste disposal costs and minimize waste generation. These programs range from an informal intradepartment chemical exchange to a formal institution–wide program with published lists of available material.

This chapter examines the components of a successful surplus chemical exchange program and offers suggestions on how to reduce conditions that hinder redistribution efforts.

THE UNIVERSITY OF ARIZONA CHEMICAL REDISTRIBUTION PROGRAM

The University of Arizona (UA) started a surplus chemical redistribution program in 1986. The program has undergone many changes since its inception but now operates with most chemical users on campus participating. An indication of the program's success is the total amount of chemicals redistributed. In calendar year 1993, 241 lb and 52 gal of materials were redistributed. Due to sporadic recordkeeping, statistics for the years 1986 through 1990 are unavailable; however, the total amount redistributed from mid–1991 through 1993 is approximately 1200 lb. The disposal cost alone would have been $2,200.00.

The Department of Risk Management & Safety (RM&S) administers the UA's chemical redistribution program. The department accepts, assesses, publicizes, and delivers materials as part of the program.

Before accepting chemicals for redistribution RM&S conducts a brief quality-control assessment on the individual containers. Containers are checked for

- Original container
- Original label
- Expiration date
- Presence of factory seal
- Any visible degradation of material
- No presence of water or caking of powder

If any of these qualifications are not met, the material will not be offered for redistribution. Any potentially explosive materials, such as picric acid or ethyl ether, are automatically excluded from redistribution. Controlled substances and elemental mercury are also excluded from redistribution.

RM&S publishes the list of chemicals available for redistribution quarterly in the department's Laboratory Safety Notes. An example of the list is found in Figure 1. Orders are delivered within 1 or 2 days of request. To reduce multiple trips to deliver one or two containers, several orders are combined to be delivered in one round trip. If requested, RM&S also supplies an MSDS for the material. Special attention is given to the location of the requester. If there are no chemical-using laboratories in the building, e.g., education, RM&S conducts further research to find out why the materials are being requested. If the information obtained still raises doubts, the requested material is not delivered.

The individual order is segregated by DOT hazard class before transport to the requesting laboratory. Containers are placed inside polyethylene bins to provide

Chemical	Size	Quantity
1,1,2–Trichloro–1,2,2–Trifluoroethane	500 ml	1
1,1,2–Trichlorotrifluoroethane	2 l	2
1,2,3,4,5,6–Hexachlorocyclohexane	100 g	1
1,4–Dioxane	2 l	1
2–Ethyl–1–Hexanol	3 kg	1
2–Methoxyethanol	4 l	2
2–Methyl Butane	500 g	3
2,6–Di–tert–butyl–4–methyl–phenol	50 g	1
3–Methyl–1–Butanol	4 l	1
3,5–Dinitrobenzoic Acid	100 g	1
4–(Methylthio)benzaldehyde	50 g	1
4–Tert–butylphenol	100 g	1
5–Sulfosalicylic Acid	125 g	1
a–Amylase	225 g	4
Acetylsalicylic Acid	50 g	1
Acetyl Chloride	500 g	2
Acetyl Chloride	1 kg	1
Acriflavine Hydrochloride	25 g	1
Adipic Acid	75 lb	1
Aluminum Oxide	5 lb	1
Ammonium Molybdate, 4–Hydrate	500 g	1
Ammonium Oxalate	500 g	1
a,p–Dibromoacetophenone	50 g	1
Aura Machine Detergent	3 lb	8
Barium Atomic Absorption Sol.	500 ml	1
Barium Chloride Technical	10 kg	2
Barium Fluoride Ultrapure	25 g	3
Barium Stick	50 g	1
Beta–Chlorolactic Acid	5 g	2
Calcium Phosphate	1 lb	1
Calcium Phosphate Monobasic	1 lb	1
Calcium Sulfate	2.5 kg	1
Carbazole	25 g	2
Carbon Disulfide	500 ml	1

Source: University of Arizona chemical redistribution list, winter 1994, Page 1.

Figure 1 Page from UA's chemical redistribution list.

secondary containment during transport. A copy of the completed Chemical Redistribution Form is taken along with the delivery and functions as the shipping papers.

The list of available chemicals is continually updated. It is RM&S's policy to delete the material from the list if it has been offered for redistribution twice and has not been requested. The deleted material is then disposed of as hazardous waste. The shelves and chemical cabinets dedicated to redistribution chemical storage are inspected daily for leaks or other signs of compromised containers. Materials in failing containers are immediately removed from eligibility for redistribution.

Approximately 30 to 50% of the offered chemicals are requested. Requests vary due to the time of year when the redistribution list is published and the variety of materials offered. Most of the requests are made by research laboratories with teaching laboratories comprising the remainder. Uses of the requested chemicals range from demonstrations at local high schools to electrophoresis experiments.

Risk management was able to save $500.00 by using redistribution chemicals in a water test conducted by the department.

KEYS TO A SUCCESSFUL REDISTRIBUTION PROGRAM

A general knowledge of the laboratory community contributes to the success of a redistribution program. Although it is difficult to predict what chemicals may be useful to the redistribution program, a sense of what will or will not be requested can be developed. If the leading organic chemist on campus has offered some exotic organic reagents for redistribution, there is a great probability that these reagents will not be requested. The laboratory that would have the greatest use for these chemicals has offered them for redistribution, or, if it is known that a researcher uses large quantities of amino acids, the redistribution program administrator could notify anyone offering amino acids for redistribution of the need. The two researchers could then establish direct contact. This method reduces handling and therefore reduces the possibility of an accidental release. It also frees up needed storage space for other redistributable reagents. However, accurate tracking of redistributed chemicals for waste minimization reporting purposes is lost when researchers are involved with direct exchanges.

A redistribution program needs to be treated like a business. The program must offer good service and good product. When a request is made, the order should be delivered as soon as possible. If the requesting laboratory has concerns about the quality of the material after delivery, the problem needs to be resolved quickly. If the requesting laboratory decides that the material is unusable for their purposes, the program coordinator should have the material removed from the laboratory in a timely manner.

A basic understanding of marketing helps the program. RM&S personnel use every opportunity to educate the chemical-using community about the redistribution program. The benefits of waste minimization are emphasized, as well as the "warm, fuzzy" feeling of participating in a program that is helping the environment.

The most important factor contributing to a successful surplus chemical exchange program is publicity. Laboratories need to know that the material is available. RM&S experienced this when the list of available material could not be circulated with the Laboratory Safety Notes. Those interested in chemical redistribution were invited to call RM&S for the current list. There were no calls requesting the list. In contrast, when the list is distributed with the Laboratory Safety Notes, requests for chemicals are made almost immediately.

Another indication of the success of publicity is the frequency of requests after the publication of the current list of available material. Initially there is a flurry of requests that taper off approximately 1 week after publication. However, awareness of the redistribution program and its capabilities leads laboratories to call RM&S at any time to see if a reagent needed for an experiment is available. Requests are also received from laboratories that hear of the program through the grapevine. When this type of request is made, RM&S obtains the necessary information to put the new requester on the mailing list.

As the redistribution program has evolved, organizations having interagency agreements with the UA have contacted RM&S asking about participation in the chemical redistribution program as contributors and requesters. The local community college has also offered chemicals to UA for redistribution. Participation by these outside organizations has been encouraged if certain stipulations are met. First, the material offered by the outside organization is published twice like all other redistribution material. Second, if the material is not requested after the two printings, the outside organization must take the material back for disposal through their hazardous waste program. This ensures that the UA is not absorbing the cost for waste generated by other organizations. High schools or private schools that offer chemicals to the UA in an attempt to avoid paying hazardous waste disposal costs are politely denied participation in the program.

The potential of outside organizations as requesters has not been fully explored by UA. The need for "basic" chemicals is definitely present; however, most of the materials on the redistribution list may be too "exotic" for a community college or high school. Material-handling procedures hinder the potential for non–university entities as requesters. The reagents are considered surplus; therefore, the UA must try to get market value for the surplus property. In other words, the materials must be sold, not given away. If the materials were offered as surplus property, they must be handled by the surplus property office through either sealed or spot bid. In either case, RM&S would lose control over who would ultimately receive the surplus reagents.

Although RM&S considers the UA's Chemical Redistribution Program a success, improvements can be made. The goal is to make the program as user friendly as possible. One area of improvement would decrease the time needed to complete the order form when a request is made by phone. Presently, the RM&S staff member taking the order writes the complete name of the requested material. This is very time consuming. One option under consideration is assigning the chemicals on the redistribution list a stock/part number. The requester could then order the material by that number. Another method offered is use of the fax machine for requesting surplus reagents. In the future, requests by e–mail will also be possible.

OVERCOMING IMPEDIMENTS TO SURPLUS CHEMICAL EXCHANGE

Incentives matter. A successful chemical redistribution program must offer potential users some positive reason to use the program. Conversely, the program must counter any exploitative uses of the system.

Some positive incentives that encourage participation include:

- **First chance at the redistribution list.** Those users who actively participate in the program, either by requesting or by offering, can be given the opportunity to request chemicals from the current list before general circulation. This gives the participant a tangible reward for their use of the program.

- **Public appreciation for use of the program.** If a participant has requested many chemicals or has offered many materials for redistribution, much good will can be obtained by simply showing appreciation for their effort. Everyone likes to receive thanks, especially for participating in an environmentally helpful program. What better way to encourage participation by others than by publicly thanking those who are already participating in the program? A little note in the campus newspaper, staff newspaper, the campus safety newsletter, or a letter to the head of the user's department goes a long way in promoting the program. A future issue of UA's Laboratory Safety Notes will be dedicated to waste minimization, and the various benefits of the chemical redistribution program will be highlighted. This issue will give RM&S an opportunity to give a blanket "thank you" to all those who have participated.
- **Service.** Participation in the Chemical Redistribution Program can be rewarded by extra service. A special waste pickup can be arranged or the participant's laboratory can be moved to the top of the lab cleanout list as a form of recognition for using the program. Help with wastewater problems or an unusual disposal problem can also be provided. In fact, anything that goes beyond the normal service of the department that administers the Chemical Redistribution Program provides incentive for participants to continue in the program or will attract new users to the program.

Unfortunately, there are those who try to exploit the system. The greatest adverse effect of the Chemical Redistribution Program is a laboratory's attempt to use the program as a means to dispose of chemicals clearly unsuitable for redistribution. This problem may be created by those administering the program. At the UA, RM&S's enthusiasm for the Redistribution Program is well known. Laboratories know that, if material is offered for redistribution, RM&S personnel will respond quickly and remove the material from the laboratory. Often, when RM&S personnel arrive to pick up the chemicals, they find unknowns, outdated material, and chemicals that are obviously hazardous waste. Waste chemicals may be buried under good chemicals in the hope that RM&S will take the box and examine its contents at the RM&S storage compound.

To prevent abuse of the program, RM&S personnel now do the quality-control procedures in the laboratory. The individual containers are inspected *very* carefully, no matter the number of containers. The prolonged presence of RM&S personnel in the laboratory can create a very uncomfortable situation for lab personnel. Usually it is uncomfortable enough to cause the laboratory personnel to admit that not all containers offered are redistributable. They then offer to remove the waste chemicals and call RM&S after they have done so. There have been no repeat offenders after a detailed in-lab-quality control check.

Two limitations to surplus chemical redistribution were discussed in Chapter 7. The factor with the most potential for hindering the growth of a redistribution program is the attitude toward shared items. Chemicals offered in a redistribution program are, in a sense, owned by everyone. This fits the definition of a commons. Whether consciously or unconsciously, the idea of shared ownership leads to the destruction of the commons. In a redistribution program, the commons is ruined by actions that lessen the purity or integrity of the chemicals offered. Prior owners who think that they can get rid of a reagent that they have contaminated through their own mishandling threaten the existence of the program.

The only way to combat this attitude is through education. Chemical users need to know the consequences of mishandling chemicals. They need to know the costs of hazardous waste disposal. Users need to be taught that by their actions the entire university community loses. Other laboratories lose the opportunity to reduce their chemical purchase budget and the university loses resources due to increased hazardous waste generation and disposal.

This approach dwells on negative incentives. People respond to positive incentives more readily. The administrators of a redistribution program need to explain how everyone wins with proper chemical handling. The original user may not recover the original expense but he will have reduced costs if he properly uses all of the chemical. Another researcher benefits through obtaining usable material for practically no expense. The university benefits through general reduced chemical purchase savings and reduced hazardous waste generation and disposal.

CONCLUSION

A chemical redistribution program is a successful method of waste minimization. A viable program is user friendly and provides positive experiences for those participating in the program. This is accomplished by providing good service, a good product, good publicity, and continual improvement of the program. If these principles are kept in mind, the program will attract more participants, thus contributing to the university community.

CHAPTER **19**

Cost Savings and Volume Reduction by Commingling Wastes

Peter C. Ashbrook

CONTENTS

Introduction 319
What is Commingling? 320
The Economics of Commingling 320
Role of Commingling in Waste Minimization 321
Potential Safety Hazards 321
Commingling in the Lab 322
Commingling at a Central Facility 323
Commingling at an Off-Site Facility 325
Summary and Conclusions 325

INTRODUCTION

For years most of us have thrown all of our trash into a single trash receptacle. The solid waste disposal industry developed to provide efficient collection and disposal of this trash. This system is easy for us, the generators of the waste, and has been cheap because of the availability of landfills.

In the past few years, as landfills have become more scarce, many communities have decided that the traditional solution to solid waste disposal must be changed for a number of reasons. Foremost among these reasons are cost and the responsible use of resources. As a result, many of us are being asked to recycle newspapers, cans, glass, plastic, and other items; not only that, we are being asked to separate these materials from the trash that cannot be recycled. By segregating these materials according to the available disposal options, we are trying to make full use of these resources.

The solid waste example is applicable to laboratory wastes. For years, the standard disposal method was to put all the wastes down the drain, or perhaps into the trash. As with households, the rationale was that the small amounts of wastes coming from laboratories are not in large enough quantity to be a hazard to anyone or any thing. We found out that this rationale turned out to be false distressingly often. Therefore, laboratory personnel started collecting their wastes by having a separate waste container. A single waste container is convenient for the laboratory personnel, but may not be so convenient for anyone who actually handles the waste and disposes of it.

The hazardous waste disposal industry has developed to efficiently dispose of large quantities of wastes. The smallest container efficiently handled is usually a 55-gal drum, which is a much larger container than most laboratories would want to work with. There are clear cost advantages to commingling wastes for disposal. However, in light of the above discussion, it should also be clear that there are times when commingling is not the best solution.

This chapter will present information to help the laboratory worker determine when to segregate wastes and when to commingle them.

WHAT IS COMMINGLING?

Simply put, commingling is the practice of combining wastes from small containers into larger containers prior to disposal. When discussing hazardous wastes from laboratories, the larger container usually ranges in size from 5 to 55 gal.

Commingling cannot be used to convert a hazardous waste to a nonhazardous waste. The regulations are clear that dilution cannot be used as a method of treatment.

THE ECONOMICS OF COMMINGLING

In most cases, commingling appears to be highly beneficial in cutting disposal costs. Chemical wastes in laboratory quantities (typically 4 l or less) are usually disposed in labpacks at a cost of $10 to 30/kg. Bulk wastes created from commingling can usually be disposed at a cost of $0.50 to $2.50/kg. These figures include labor for packaging and labeling wastes, supplies, transportation, and disposal.

An example of the cost savings possible with solvents illustrates the point. Using the cost for incinerating a typical labpack ($20/kg) and the cost for a incinerating a typical drum of bulk solvents ($1/kg), one calculates a gross savings of $3800 for one 200-l drum when commingling is employed.

Balanced against this cost savings are equipment needs to commingle wastes safely, potential safety hazards when incompatibles are mixed, greater safety hazards due to larger quantities of materials in storage, and the greater consequences when containers leak. In addition, significant costs for having the resulting waste analyzed by an independent laboratory may offset much of the savings.

ROLE OF COMMINGLING IN WASTE MINIMIZATION

From a cost point of view, the potential of commingling is attractive. However, from a waste minimization point of view, there are a number of strategies that should be pursued prior to commingling. When pursued as a top priority, commingling is likely to preclude more desirable waste management strategies such as reuse or reclamation. Commingling is almost always predicated on the assumption that the waste will be disposed. Enlightened waste management and waste minimization strategies, when properly applied, attempt to prevent a waste from being generated in the first place. The widespread use of commingling can divert attention from these waste minimization goals.

The one exception of commingling being based on the assumption that the waste will be disposed is when wastes are being accumulated so that a sufficiently large quantity will make reclamation possible. These cases usually involve combining identical chemicals. For example, a solvent reclamation company may be interested in reclaiming certain chlorinated solvents, such as trichloroethane, but only if there is a minimum amount, such as a 55-gal drum.

Once possibilities for reuse or reclamation have been exhausted, commingling may be an appropriate waste management technique. In these cases, the waste minimization benefit comes primarily from solid waste minimization. When laboratory chemical wastes are not commingled, they are most likely to be incinerated in labpacks. At best, disposing of wastes in labpacks requires four times as many drums as disposing of wastes in bulk form. If these wastes are incinerated, the labpacks will produce more ash and require more resources (drums, packaging material, energy for transportation) than the commingled wastes. In the case of commingling, the smaller containers can be washed so that the glass or metal can be recycled.

POTENTIAL SAFETY HAZARDS

Before pursuing commingling, one must be aware of the potential hazards. Among these potential hazards are

- Possibility of fire or explosion
- Exposure to toxic vapors
- Leaks and spills from larger containers
- Creation of difficult- or impossible-to-dispose-of wastes

Perhaps the greatest danger of commingling is the fact that it is so often uneventful. If commingling is uneventful over a period of time, those doing the commingling may become complacent about the possible hazards. Commingling is inherently hazardous and those commingling chemicals must constantly be aware of potential hazards.

The immediate hazard of improper commingling is the creation of a fire or explosion due to mixing of incompatible chemicals. Prior to any commingling, the chemicals should be reviewed for compatibility. Screening tests of the materials to assure that they are what they are labeled can be helpful. Even with these precautions it is prudent to mix a small amount of the wastes together to verify that there is no unexpected reaction. Then, if all goes as expected, the remainder of the waste chemicals may be mixed.

Even when chemicals are compatible, those doing the commingling must be concerned about exposure to toxic vapors and splashes and spills. Therefore, proper personal protective equipment must be worn including at a minimum appropriate gloves, eye protection, and respiratory protection. The use of a walk-in fume hood or other device to remove toxic vapors is also essential.

Once the commingling is complete, there are still potential hazards. One must be concerned about delayed reactions. For this reason, it is prudent to make provisions for release of excessive pressure buildup. One way to do this with drums is to leave a bung cracked open; however, government inspectors may decide that this practice violates requirements for keeping the container closed at all times except when materials are being added or removed. Leaks and spills are also of concern. A leaking 55-gal drum presents a much greater hazard than a leaking 1-gal container. Containers used to hold commingled wastes must be provided with secondary containment in the event of spills, just as one would provide containment for smaller containers. If commingling is done using 55-gal drums, it would be prudent to have several 85-gal drums on hand in the event one of the 55-gal drums springs a leak. Emergency procedures should be developed to address the possibility of a spill of commingled wastes. The kinds and amounts of spill cleanup materials required will probably be different from those required for smaller spills.

Though not a safety hazard, one must also be careful that one does not create a drum that is impossible to dispose. The most common example is with mercury. At the time this was written, there were few disposal companies in the U.S. that could even accept wastes containing mercury at levels above 260 ppm. Even when a company could be found, the disposal cost was very high.

Finally, keeping good records of the wastes that were commingled is important to help characterize the resulting waste. Even with good records, it is prudent to have a representative sample analyzed by a reputable laboratory.

COMMINGLING IN THE LAB

Each laboratory should evaluate all the wastes produced and make sure that commingling is performed in conformance with the procedures of the institution. If there is only one laboratory at the institution, this review will be straightforward. At larger institutions, the lab manager should consult with the hazardous waste management staff to determine appropriate procedures.

As a rule of thumb, commingling in the lab should only be done for identical wastes. The reason for this guideline is that one does not want to preclude more desirable waste minimization opportunities by mixing different wastes together prematurely. For example, mixing compatible solvents together will probably save money on disposal costs compared with disposal in labpacks. However, such mixing will likely eliminate the possibility of recovering any of the solvents by distillation.

Containers used for commingling in the lab should be no larger than 5 gal (20 l). Most labs are not equipped to deal with leaks from drums. There are too many things that can go wrong for drums to be viable in labs. Even 5-gal containers may be too large. The largest container used is a function of what provisions the lab can make for potential spills, how frequently wastes are removed from the lab, and what provisions the institutional hazardous waste management staff or its disposal contractor have for handling these larger containers. Many universities have found that poly containers work best because of resistance to corrosion and light weight. These containers may be emptied at a central waste management facility and returned to the laboratory.

When commingling wastes, laboratory workers must keep good records about the materials mixed. Good records will allow for better management of the wastes after they are removed from the lab and will allow waste handlers to minimize safety hazards. Some regulatory agencies and some disposal companies may require a listing of all hazardous waste codes for the wastes that are commingled into a drum. Good records also will minimize the need for outside analytical work. With the cost of a full-scale TCLP analysis well over $1000 to determine whether a waste exhibits hazardous waste characteristics, such information has clear economic benefit as well.

Solvents are probably the most common wastes commingled in laboratories. However, any compatible wastestreams generated on a continuing basis would be suitable. For solvents, some laboratories try to segregate halogenated solvents from nonhalogenated solvents.

COMMINGLING AT A CENTRAL FACILITY

If your laboratory is part of a larger facility, your institution probably has a central waste management building. This is the case at many universities and the larger private sector research laboratories. The first and most important purpose of a central storage facility is to provide storage space for wastes. At a minimum, such space allows for wastes to be removed from laboratories on a timely basis and makes for efficient use of the time of the hazardous waste disposal contractor.

Most people working at central storage facilities soon realize that there are many simple practices that can significantly reduce disposal costs. Commingling, of course, is one such practice. Workers at a central facility must exercise more caution than laboratory workers when commingling because the wastes tend to be in larger quantities than in labs and because they have less knowledge than the lab workers about the wastes they are handling.

The simplest kind of commingling is to combine identical wastes from partially filled containers into a single container. For example, one might have six 500-kg bottles of lead acetate, each of which contains no more than 50 g. These could all be combined into one of the containers. Since disposal costs are usually based on the number of containers rather than the actual amount of wastes in them, this technique can produce significant savings if one has been disposing of many partially filled containers.

On a larger scale, the commingling of different solvents is more common. As mentioned above for laboratories, some institutions find it makes economic sense to separate halogenated solvents from nonhalogenated solvents. If the institution has very many labs, commingling would usually be performed using 55-gal drums. However, in some cases commingling into smaller containers may be worthwhile. If halogenated solvents are commingled, poly or poly-lined containers should be used rather than metal ones. The reason for this is that halogenated solvents may dissociate into hydrochloric acid if water is present. If the dissociation occurs in enough quantity and the drum is stored for very long, the acid can eat through metal causing a leak.

Most common solvents are suitable for commingling. Solvents likely to polymerize or cause polymerization should not be commingled except possibly under highly controlled conditions. Amines and thiols are generally not commingled because of odors. One should also avoid adding wastes containing heavy metals to solvent wastes, although small amounts should not create a problem. All containers with flammable liquids must be grounded to minimize the chance of fires through static electricity. If poly drums are used, the liquid can be grounded by placing a steel wire into the liquid and grounding the wire.

Certain solvents may be present in sufficient quantity to commingle separately. We found that our Physical Plant's paint shop uses xylene to clean their street painting equipment. Because we receive a substantial amount of slightly contaminated xylene from histology labs and other sources, we combine xylene into separate containers rather than mixing it with other solvents. This used xylene is taken to the paint shop for their use before being returned to us for final disposal. With a little imagination, other uses of slightly contaminated solvents may be found at your institution. If you do reuse slightly contaminated solvents, make sure you comply with OSHA Right-to-Know information and labeling requirements. Though not usually considered a solvent, used oil is often commingled out separately. In most communities used oil can be shipped for recycling at a much lower cost than disposal as a waste solvent.

Larger institutions may also wish to consider commingling compatible aqueous wastes. Common wastes in these categories include certain acid solutions, solutions with heavy metals, or developers. As noted above, you may wish to exclude wastes containing mercury to keep the level from exceeding 260 ppm. Any wastes containing corrosives should be combined into poly or poly-lined drums to prevent the corrosion problems described above. One simple safety precaution when commingling aqueous wastes is to check the pH of every waste container. This will prevent acid wastes from being inadvertently combined with waste bases.

Other waste types we have commingled include used oil, waste and excess paint, and used silica gel. Additional possibilities include mixing organic solids with solvents or certain soluble inorganic solids into aqueous wastes. Prior to doing any commingling, you should check with your contractor to determine if the expected waste will be acceptable for disposal. The disposal contractor will frequently request a laboratory analysis of commingled wastes — this cost must be factored into whether the commingling activity is worth pursuing. In the case of used oil, we analyze each drum for PCBs prior to shipping it off for recycling.

Even if you choose not to commingle your own wastes with your own staff, on-site commingling still may be an option. Some contractors will commingle your wastes at your site. Most of these contractors will even provide a portable hood for vapor control. Although this method may not save your institution as much money as if you did it yourself, it does have the advantages of being done under your supervision and it may be easier to dispose of the wastes because of the contractor's experience. If you do hire a contractor to perform on-site commingling, be sure to provide adequate oversight to make sure they are giving adequate attention to safety and regulatory concerns.

COMMINGLING AT AN OFF-SITE FACILITY

In recent years, hazardous waste contractors have discovered that there is a large market for disposal of laboratory wastes. As competition for this business has increased, some contractors have looked at ways to reduce the cost of disposal of such wastes. Several companies have developed sites to which wastes are shipped in labpacks. At these sites, the labpacks are unpacked and compatible wastes from many different generators are commingled to take advantage of bulk disposal rates. Some of these cost savings are passed on to each generator. Thus, these companies can offer generators lower disposal prices than companies that merely ship labpacks for disposal.

Off-site commingling facilities will do many of the activities described in the previous section. This has the advantage of relieving the generator of providing adequate staff and resources to provide such services on-site. On the other hand, the activities are outside of the control of the generator. In addition, generators will probably find that they obtain greater savings when doing the commingling themselves.

SUMMARY AND CONCLUSIONS

The primary waste minimization benefit of commingling is a reduction of materials used to package waste for disposal. Commingling, if performed safely and properly, can substantially reduce hazardous waste disposal costs compared with disposal of waste in labpacks. Before pursuing any commingling project, one must consider potential safety hazards and develop contingency plans in the event of spills or leakage from the larger containers. In most cases, the preferred point of commingling is at the institution's central storage facility rather than the laboratory.

Part 6:
Case Studies and Applications

CHAPTER 20

Applications for Waste Solvent Recovery Using Spinning Band Distillation

John A. Mangravite, Roger R. Roark, Jr., Roger R. Roark, Sr., and Paul VanTriest

CONTENTS

Introduction 330
Methods and Materials 331
    Water Analysis 331
    Gas Chromatographic Analysis 331
    High-Performance Liquid Chromatographic Analysis 332
    Ultraviolet Analysis 332
    Fluorescence Analysis 332
    Refractive Index 332
    Boiling Points 333
    Infrared Analysis 333
    Equipment 333
    The 8300 Distillation Unit 333
    The 8400 Distillation Unit 333
    The 8600 Distillation Unit 336
    The Pureform® 2000 Recovery System 337
Results 337
Hospital Wastes 340
    Xylene 341
    Xylene Substitutes 346
    Ethanol and Alcohol Mixtures 347
    Formalin 347
High-Performance Liquid Chromatographic Wastes 351
    Acetonitrile 352

| Methanol 356
| Isopropanol 356
| Nonpolar Solvents 361
| Tetrahydrofuran (THF) 362
| Hexafluoroisopropanol (HFIP) 362
| Environmental Analyses Waste 366
| Dichloromethane (DCM) 366
| Freon 367
| Routine Laboratory Waste Solvents 369
| Acetone 369
| Toluene 369
| Others 371
| Practical Aspects of Solvent Recycling 371
| State and Federal Regulations 371
| Safety Factors 374
| Cost Benefits 374
| Conclusion 375
| Endnotes 375

INTRODUCTION

In Chapter 15, we presented a detailed discussion of the theoretical aspects of distillation and the characteristics of spinning band distillation units. The purpose of this chapter is to describe the details of those processes and solvent systems in which this equipment has been used successfully for waste minimization and solvent recovery. Since there are a myriad of solvent waste problems, some of which are unique to an individual laboratory, we have organized this chapter into four general areas:

1. Hospital wastes
2. High-performance liquid chromatographic wastes
3. Environmental analyses wastes
4. Wastes from routine laboratory procedures

Our primary purpose for working in this area has been to design equipment that can be operated in a simple and efficient manner, with the highest regard for safety, and to develop procedures for recovering and reusing the waste solvent within the laboratory setting. This is not to say that the solvent recovery equipment must be physically located in the laboratory in which the waste is generated (in fact it is advantageous to have a separate facility for carrying out the recovery process), but the recovery process should be carried out on-site. Depending upon the locale, a permit may be required for on-site recovery. There are some compelling arguments for this approach. The waste mixture can be carefully controlled and inventoried, thereby reducing or eliminating problems of recovery caused by the mixing of different wastes. In addition, the cost and safety problems associated with the transfer of hazardous materials to another facility are eliminated.

Another aspect of primary concern for a laboratory engaged in solvent recovery is the quality of the recovered product. If such material is to be reused, some assessment of its purity must be made. While many laboratories have the proper analytical instrumentation for determining the properties of the recovered solvent, others do not. This chapter will not only describe the successful solvent recovery situations, but will also provide the analytical data demonstrating the purity of the recovered material. The following section describes the analytical techniques and equipment we have used for this purpose.

METHODS AND MATERIALS

We have conducted a variety of analytical tests on the solvents recovered from the wastes described in this chapter, and have compared the results with those obtained from the commercially pure solvent. All the analyses are common procedures employed in assessing the quality of a given material and have various sensitivities dependent upon the type of impurity being analyzed.

WATER ANALYSIS

Water content was routinely determined using two methods:

1. Karl Fischer titration — Samples were titrated with a Hydranal® composite reagent[1] using a Brinkmann Instruments, Inc. 633 Automatic Karl Fischer Titrator with a sensitivity from 10 ppm to 100% water.
2. Gas chromatography — A gas chromatographic technique[2] was used to verify the results of the Karl Fischer experiment. In this procedure, samples were analyzed using a Hewlett-Packard 5730A Thermal Conductivity Detector (TCD) gas chromatograph fitted with a 6′ × 1/8″ SS column packed with Chromosorb 101. This method was particulary accurate in the 0.01 to 50% water range.

GAS CHROMATOGRAPHIC ANALYSIS

Different gas chromatographic procedures were employed which were dependent upon the type of material being analyzed.

1. FID detection — All samples were routinely analyzed for trace organic impurities using a Hewlett-Packard 5890 gas chromatograph equipped with a flame ionization detector (FID) and an all-purpose Ultra2 capillary column (30 m × 0.31 μ).
2. ECD detection — A Hewlett-Packard 5880 gas chromatograph equipped with an electron capture detector (ECD) was used for trace analysis of halogenated compounds and in solvents used in pesticide analyses.

3. GC-MSD — In some cases, samples were analyzed for trace impurities using a Hewlett-Packard 5880 gas chromatograph equipped with a 5970 mass selective detector (MSD) fitted with an Ultra2 capillary column (30 m × 0.25 µ). This allowed for identification by mass spectrometry of unknown components.

HIGH-PERFORMANCE LIQUID CHROMATOGRAPHIC ANALYSIS

In those cases where a very high purity grade of solvent was the desired outcome, HPLC analysis was performed using a Perkin-Elmer Series 4 liquid chromatograph equipped with autosampler and UV detector. Columns varied with the analysis; however, they were usually an Alltech normal phase silica column or a DuPont Zorbax reverse phase C-8 or C-18 column. Two procedures were employed:

1. Static HPLC analysis (baseline studies) — A sample of the recovered solvent (100%) was allowed to flow at a high flow rate ($\approx$4 ml/min) for 30 min and the baseline compared to that obtained with a commercial HPLC solvent.
2. Dynamic HPLC analysis — Samples of the recovered solvent and a commercial HPLC-grade material were compared after 60-min on-column enrichment of a 30:70 solvent/water mixture at a flow rate of 3 ml/min through a C-18 reverse-phase column for acetonitrile or methanol or a silica column for nonpolar solvents. Impurities that might be present would accumulate on the column and were subsequently stripped off using a step gradient of 70:30 solvent/water for 15 min followed by a step gradient to 100% pure solvent. This technique provides a very sensitive analysis of trace impurities in high-purity solvents.

ULTRAVIOLET ANALYSIS

Ultraviolet analysis was carried out on a Cary 219 UV-visible spectrophotometer using 1-cm quartz cells. Two procedures were employed:

1. Direct comparison of a recovered solvent with the waste solvent and a comparison with commercial HPLC-grade solvent.
2. A tenfold preconcentrated sample, prepared by flushing down 100 ml of the sample to 10 ml using an unheated rotary evaporator. This procedure concentrates impurities for a more sensitive analysis.

FLUORESCENCE ANALYSIS

Fluorescence spectra were recorded on a Perkin-Elmer MPF-44A spectrophotometer.

REFRACTIVE INDEX

Refractive indices were obtained on a Bausch and Lomb refractometer and are uncorrected.

BOILING POINTS

Boiling points were obtained directly from the head temperature and are uncorrected.

INFRARED ANALYSIS

Infrared (IR) analysis was obtained using NaCl or quartz cells and a Perkin-Elmer 1600 FT-IR spectrophotometer.

EQUIPMENT

All of the solvent recovery experiments, except one, described in this chapter were carried out on spinning band distillation units manufactured by B/R Instrument Corp., Pasadena, MD. The lone exception was the recovery of formalin in which a nonspinning band type was used. A description of this equipment is included in this section. It should be mentioned that there is standard laboratory equipment used in our experiments that will not be discussed in detail. These include recirculating baths, fume sensors, spill containment systems, in-cabinet fire extinguishers, and storage containers. All of this equipment is designed to maximize the safety of a recycling program and is strongly recommended.

THE 8300 DISTILLATION UNIT

The 8300 spinning band distillation unit is pictured in Figure 1. This system is contained in a cabinet that measures 54" × 19" × 19". The unit is controlled by a microprocessor that is operated at 115VAC 50/60Hz 10A. In Table 1, the operational characteristics of this unit are presented. This system is routinely used in situations in which small volumes of waste are generated. Typically, laboratories of small hospitals that generate fewer than 50 l of recyclable solvents such as xylene and alcohols can use this system to carry out two 5-l runs per day. This system has also found use in small colleges and universities to recycle solvents generated in teaching laboratories.[3] The unit is operated at 2200 rpm, the operational speed of greatest efficiency. At this speed less than 1 h of equilibration is necessary for all but the most difficult separations. Figure 2 shows the efficiency as a function of equilibration time at two different boilup rates. The test mixture was methylcyclohexane/toluene (Δ boiling point = 10°C).

THE 8400 DISTILLATION UNIT

The 8400 system is shown in Figure 3. This system is a larger unit capable of recycling volumes up to 22 l. It comes in a cabinet that measures 72" × 19" × 19" with a microprocessor operating at 115VAC 50/60Hz 15A. Table 2 describes the operational characteristics of this system. These data show that the 8400 is capable

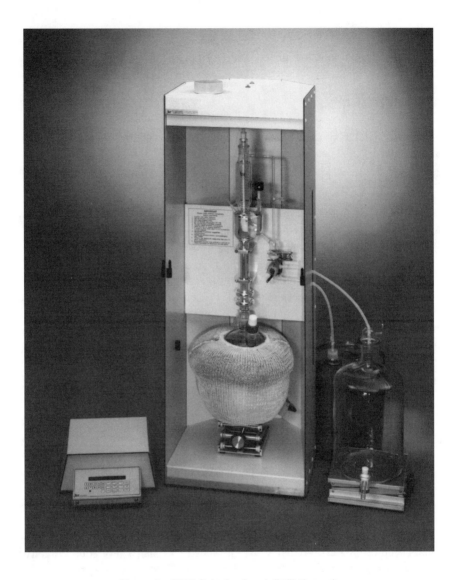

Figure 1 8300 Spinning band distillation unit.

Table 1 Operational Characteristics of the Model 8300

Pot size	5 l
Throughput	250–3000 ml/h
Holdup	1–10 ml
Motor speed	2200 rpm (Teflon band)
Pressure drop	Not measured
Vacuum capability	Not available
Efficiency	2–18 plates
Boiling point separations	to 20°C

WASTE SOLVENT RECOVERY USING SPINNING BAND DISTILLATION 335

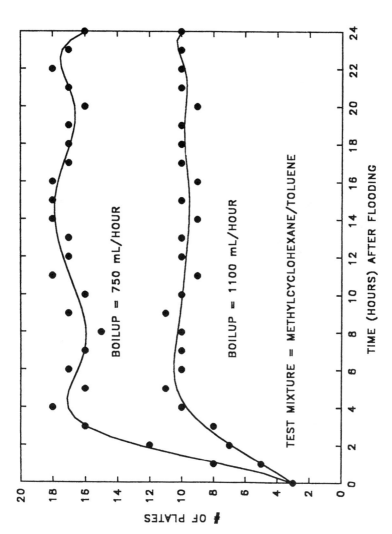

Figure 2 Efficiency vs. equilibration time for the 8300 still.

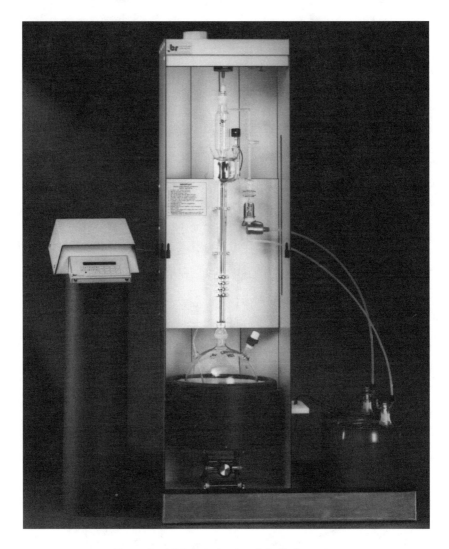

Figure 3 8400 Spinning band distillation unit.

of recycling a much wider variety of solvents and larger volumes. It has greater efficiency and vacuum capability. Figures 4 and 5 present results describing the efficiency of the unit with regards to equilibration time and band speed. For use as a solvent recovery still, the unit is operated at 2200 rpm.

THE 8600 DISTILLATION UNIT

The model 8600 system shown in Figure 6 is the most versatile of spinning band distillation systems. It is capable of very high efficiency which allows for a wide variety of solvent recycling applications. High throughput separations of mixtures with a 5°C boiling point separation can be achieved in this unit. It is

Table 2 Operational Characteristics of the 8400

Pot size	12–22 l
Throughput	250–5000 ml/h
Holdup	1–10 ml
Motor speed	2200 rpm (Teflon band)
Pressure drop	0.5–2 Torr
Vacuum capability	Available
Efficiency	5–30 plates
Boiling point separations	to 10°C

Table 3 Operational Characteristics of the 8600

Pot size	12–22 l
Throughput	250–5000 ml/h
Holdup	2–15 ml
Motor speed	2200 rpm (Teflon band)
Pressure drop	0.5–2 Torr
Vacuum capability	to 0.05 Torr
Efficiency	5–50 plates
Boiling point separations	to 5°C

Table 4 Hospital Wastes

Solvent	Waste source
Xylene	Tissue processing
Xylene substitutes	Tissue processing
Ethanol and alcohols	Staining
Formalin	Tissue fixation

contained in a cabinet measuring $84'' \times 19'' \times 19''$ and is controlled by a microprocessor with the same electrical requirements as the 8400 unit. Table 3 details the operational characteristics of the system that is normally run at 2200 rpm motor speed. Figures 7 and 8 describe the results of efficiency studies carried out with this unit.

THE PUREFORM® 2000 RECOVERY SYSTEM

Several years ago, B/R Instrument Corp. developed a nonspinning band distillation unit that has found wide use for solvent recovery and waste minimization from aqueous solutions. This still is pictured in Figure 9. The unit is totally automatic including the filling of the pot flask. By far, the most important application of this distillation system has been the recovery of formaldehyde from used formalin solutions. We describe the work carried out with this unit later in this chapter.

RESULTS

As mentioned in the Introduction, we have organized this chapter so that four general areas of solvent recovery will be discussed. Tables 4 to 7 present a tabulation of the different solvent wastes that will be discussed.

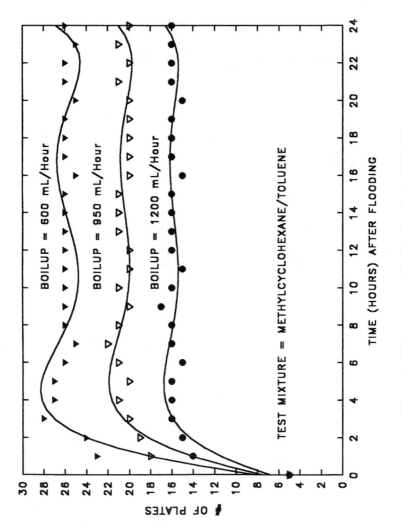

Figure 4 Efficiency vs. equilibration time for the 8400 still.

WASTE SOLVENT RECOVERY USING SPINNING BAND DISTILLATION 339

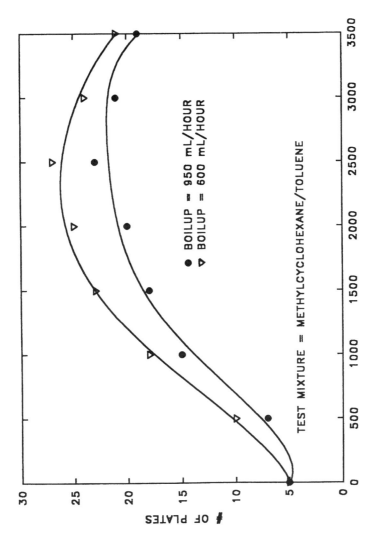

Figure 5 Efficiency vs. band speed for the 8400 still. Equilibration Time = 24 hours.

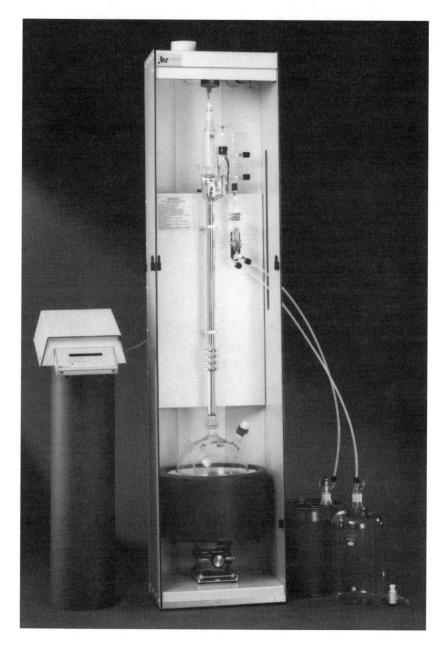

Figure 6 8600 Spinning band distillation unit.

HOSPITAL WASTES

There are many procedures in anatomical pathology laboratories located in hospitals and private facilities in which large quantities of organic solvents are generated as waste material. Tissue processing and staining procedures use large

Table 5 High-Performance Liquid Chromatography Wastes

Solvent	Waste source
Acetonitrile	Reverse phase
Methanol	Reverse phase
Isopropanol	Normal phase
Nonpolar solvents	Normal phase
Tetrahydrofuran	Size exclusion
HFIP	Size exclusion

Table 6 Environmental Analysis Wastes

Solvent	Waste source
Dichloromethane (DCM)	Pesticide analysis
Freon	Extraction

Table 7 Routine Laboratory Waste Solvents

Solvent	Waste source
Acetone	Cleaning
Toluene	Degreasing
Others	

amounts of xylene and xylene substitutes and ethanol or alcohol mixtures. Formalin is used in tissue fixation. All of these situations are quite amenable to solvent recovery.

XYLENE

Xylene wastes generally contain water and alcohol impurities that must be removed before the xylene can be reused. The percentages of these impurities vary greatly from laboratory to laboratory. We have carried out a survey of xylene wastes from over 200 hospital laboratories that are currently using B/R Instrument Corp.'s spinning band distillation equipment for xylene recycling and who participate in a gas-chromatographic quality-control program in which up to six samples are analyzed three times a year in our Easton, MD facility. The results of this survey were dramatic. The percent alcohol contamination ranged from 0.6 to over 64% with the average waste being 14.9%. In all cases, an excellent-quality xylene was recovered in high yield using either an 8300 or 8400 distillation unit. A typical distillation was programmed to operate automatically using the following experimental parameters:

Equilibration time	=	0
Reflux ratio	=	0
Boilup	=	2000 ml/H
Forecut	=	Ambient to 135°C
Heart cut	=	135–142°C

Figure 10 shows the FID gas chromatogram of the starting waste, a sample of pure histological grade xylene, and the first and heart cuts of a 10-l sample run on

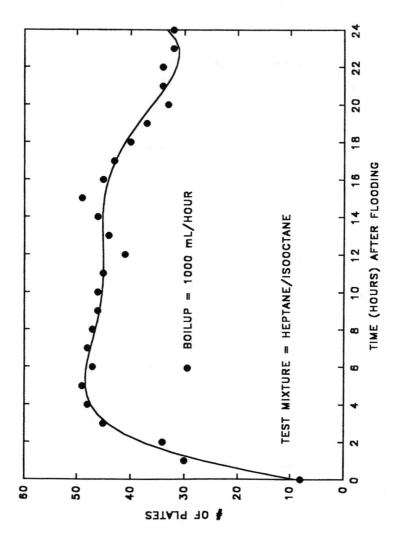

Figure 7 Efficiency vs. equilibration time for the 8600 still.

WASTE SOLVENT RECOVERY USING SPINNING BAND DISTILLATION 343

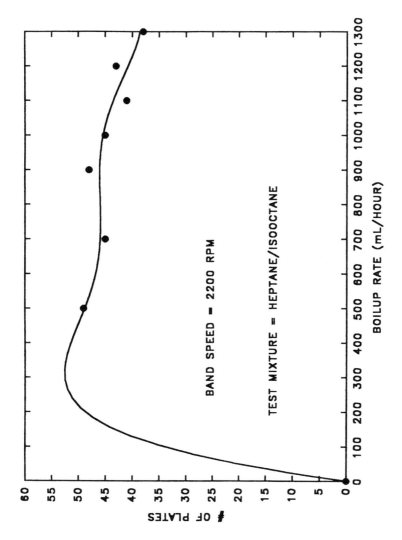

Figure 8 Efficiency vs. boilup rate for the 8600 still. Equilibration Time = 2 hours.

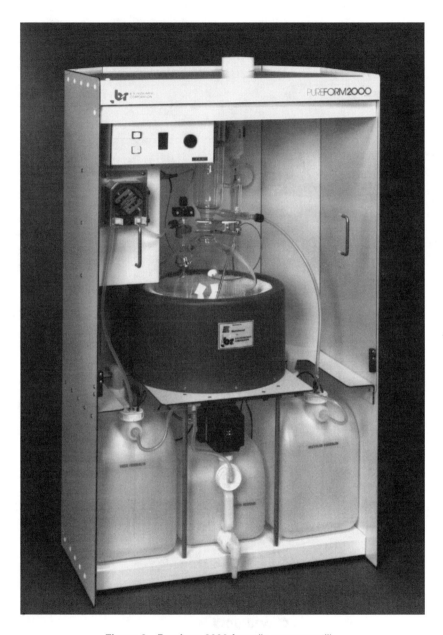

Figure 9 Pureform 2000 formalin recovery still.

an 8400 unit. The initial waste was 85:15 xylene/ethanol. It should be noted that histological grade xylene contains all three isomers (ortho, meta, and para) and some ethylbenzene. The percentage of ethylbenzene varies widely from manufacturer to manufacturer. As is evident from these chromatograms, the heart cut ($\approx$8000 ml) is identical to an unused xylene. This amounts to a recovery of about 94% of reusable solvent. The total time of the run was 5 h.

WASTE SOLVENT RECOVERY USING SPINNING BAND DISTILLATION 345

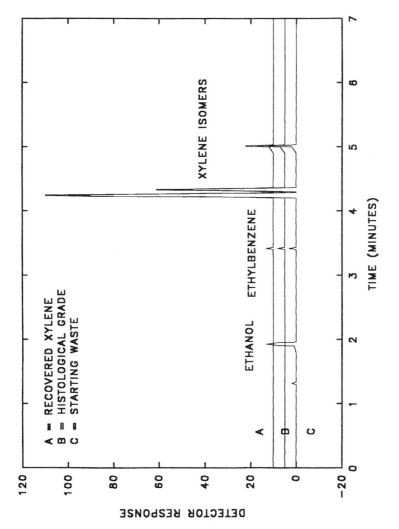

Figure 10 FID gas chromatograms of xylene samples. Recovered xylene from 8400 distillation.

Table 8 Recovery of 85:15 Xylene/Ethanol

Volume (ml)	8400 Distillation (% xylene)	Rotary evaporator (% xylene)	Atomizing plate technology (% xylene)
0–1000	17.4	43.9	47.1
1000–2000	69.4	63.0	60.2
2000–3000	99.8	93.4	84.4
3000–4000	100	98.8	94.4
4000–5000	100	99.0	97.0
5000–6000	100	99.9	98.0
6000–7000	100	100	98.4
7000–8000	100	100	99.4
8000–9000	100	100	100
9000–10000	100	100	100
Time of run (h)	5	20	4
% Recovery of pure xylene	94	59	24

Recently, we have carried out and reported on[4] a comparison study in which an 85:15 xylene/ethanol waste was recycled using rotary evaporation[5] and atomizing plate technology and the results compared to the spinning band distillation described above. Table 8 presents a summary of the results of this study. Ten-liter pot charges were analyzed every 1000 ml for % xylene and the total % purity of the combined last 8000 ml compared. In addition, run times and % recovery of a pure xylene were tabulated. It is evident from these results that spinning band distillation gives a much higher recovery of a pure product than the other techniques and the recovery is carried out in a reasonably short time. A typical laboratory can process 20 l/day (100 l/week) using one 8400 distillation unit if two runs are performed a day. Under such conditions, the cost of this unit can be easily recouped within a year (assuming a price of $10/l for purchased xylene and a disposal cost of $5/l). The ease of use and safety of this unit coupled with the economical and ecological advantages of recycling xylene should prompt all laboratories generating xylene waste to embrace this technique. For smaller hospitals, the 8300 distillation unit capable of 4-l pot charges would meet most requirements. Because of the lower capacity, payback would take somewhat longer (up to 2 years).

XYLENE SUBSTITUTES

In recent years, a variety of xylene substitutes have been developed in attempt to replace this highly flammable and toxic solvent. In general, these substitutes fall into four classes and are marketed under various trade names such as Histoclear, Histosolv, Clear-Rite 3, Hemo-D, Americlear, etc. The chemical components are one of the following: limonene (a naturally occurring terpene that is a by-product of the citrus industry), aliphatic hydrocarbon mixtures with lower toxicity than aromatics, some high-boiling aromatic hydrocarbon mixtures which have lower volatility than xylene but are just as toxic, and mineral oil mixtures. In practically all cases, we can successfully recycle these materials with a slight modification in the distillation parameters to provide for the higher boiling points of these mixtures. In fact, many hospital labs that have switched to these substitutes are using 8300 and 8400 units for recycling. The one instance in which recycling using our spinning band units has not been successful is with Microclear which is a mineral oil having too high a boiling point to be distilled using the Teflon spinning band employed in our units.

ETHANOL AND ALCOHOL MIXTURES

Clinical laboratories in both health care facilities and research institutions have a very high demand for alcohol solvents. Both the cytology and histology laboratories utilize large amounts of these solvents (as ethanol, 2-propanol, or a reagent alcohol mixture composed of ethanol, methanol, and 2-propanol) as both a fixative and in staining and slide preparation procedures. Under normal histological use, ethanol is contaminated with stains used in slide preparation, methanol, organic tissue, and varying amounts of water. When such a mixture is distilled in an 8300 or 8400 distillation unit, a 95:5 ethanol/water azeotrope is obtained. A normal protocol for such a distillation involves the following:

Equilibration time	=	1 h
Reflux ratio	=	1:1–2:1
Boilup	=	1000–2000 ml/h
First cut	=	77–82°C for reagent alcohol
	=	75–78°C for ethanol

Gibbs[6] has obtained throughputs of 2.4 l/h when the reflux valve is left wide open, while 1.1 l/h was obtained at a 2:1 reflux ratio. Figure 11 shows the distillation curve for the batch distillation of an 85:15 ethanol/water waste containing nonvolatile stains. Figure 12 shows the gas chromatographic analysis of starting waste and recovered azeotrope. A 95:5 ethanol/water standard is also shown. These data demonstrate the high recovery (93%) and the high-purity of the recycled solvent. This recovered 95:5 azeotrope is perfectly suitable for reuse in histological procedures or any other procedure requiring a high-purity ethanol solvent. If reagent alcohol is used in the clinical procedures, recycling produces a high recovery of a 90:5:5 mixture of ethanol/methanol/2-propanol with a small percentage of water. Figure 13 shows the gas chromatogram of a recycled reagent alcohol.

Gibbs[6] has carried out a complete environmental risk assessment in the large-scale recovery of both alcohol and xylene. Using standard EPA methods for collection and analysis of samples, he found that levels of solvent emissions during the recovery process were significantly below any health risk standard and magnitudes below any fire risk.

FORMALIN

In 1987, OSHA revised its ruling on the toxic material formaldehyde and promoted it to one of a group of chemicals regarded as potential carcinogens. While the ruling focused on this carcinogenic aspect, it also emphasized in its new standards the toxic and allergic effects of this chemical. The standards for moni-

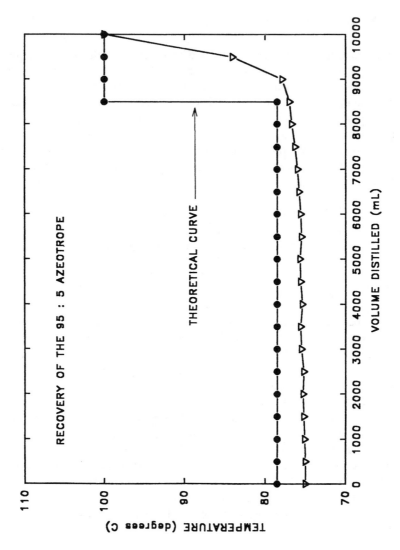

Figure 11 Distillation curve for the 8400 distillation of 85:15 ethanol/water.

WASTE SOLVENT RECOVERY USING SPINNING BAND DISTILLATION 349

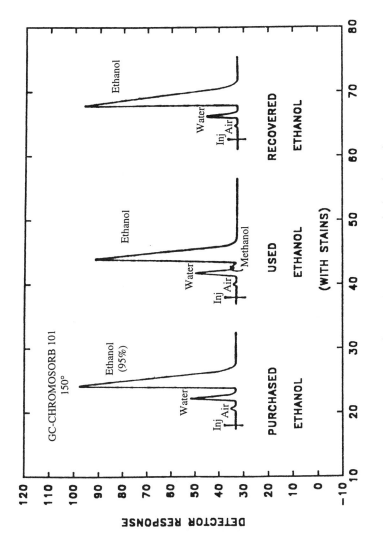

Figure 12 TCD gas chromatograms of ethanol samples.

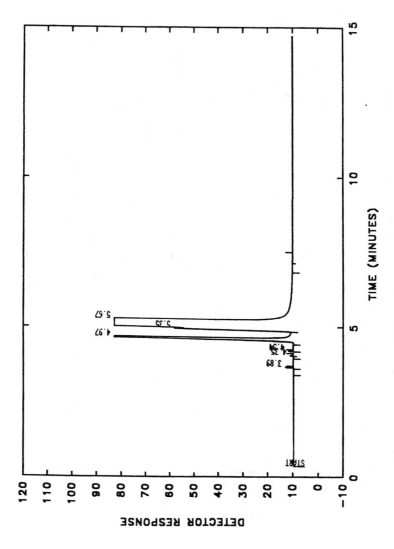

Figure 13 TCD gas chromatogram of reagent alcohol.

Table 9 Summary of Data for Formalin Recovery

Hospital	% Formaldehyde		
	Starting material	Recovered distillate	Bottoms
North Arundel General	3.27	3.76	0.14
GBMC	3.33	3.83	0
St. Agnes	3.31	3.28	0
Mt. Sinai	4.04	4.28	0.81
St. Francis	3.23	3.47	0

toring became quite stringent and the costs of disposal increased. Since recycling now became an attractive alternative to disposal, we initiated a series of experiments to demonstrate the feasibility of distillation as a method of recovering and reusing used formalin solutions. This effort culminated in the development of a non-spinning band system, the Pureform 2000®, for this purpose.

Table 9 summarizes a series of experiments that were conducted on a wide variety of formalin waste solutions from different hospitals near Baltimore. It is obvious from these results that the recycled material was of relatively constant concentration and was recovered in high yield. Further experimentation with the cooperation of Anatech Ltd.[7] allowed us to develop a complete protocol for recycling formalin solutions for reuse and to provide for proper disposal of the waste material. This procedure involves placing up to 19 l of filtered or unfiltered formalin waste in a filling jug of the Pureform 2000® commencing the automatic procedure. The still pot is automatically filled with 9.5 l of waste and condenser cooling water is started along with the pot heater. After distillation commences, material is removed until the pot temperature rises to 108°C. When this temperature is reached, heating is terminated and the pot material allowed to cool to below 52°C. At this time, the pot material is dumped to waste and then refilled with a second 9.5-l batch of used formalin. In most of the many laboratories that are using this system, a recovery of about 80% is achieved and this material is consistently about 10% formalin (4% formaldehyde). This distillate is then buffered using a kit developed by Anatech Ltd. The still bottoms which comprise about 20% of the pot charge contain about 0 to 3% formalin which can be destroyed using a second kit designed by the aforementioned company. This material can then be easily discarded. It should be noted that air monitoring of this almost totally automatic procedure showed no detectable formaldehyde escaping from the system.

We are currently investigating other solvent recovery and waste minimization uses for this unit which should prove universal when aqueous-based wastes are involved. We will report on these at a later time.

HIGH-PERFORMANCE LIQUID CHROMATOGRAPHIC WASTES

Our initial efforts in the investigation of in-house laboratory recycling of used solvents was in the recovery of high-purity solvents, especially from liquid

Table 10 Analytical Data for Acetonitrile Recovery

Analysis	HPLC grade	Recovered pure solvent	Recovered azeotrope
Boiling point(°C)	81–82	82	77–78
Refractive index	1.3466	1.3465	
Water (%)	0.07	0.08	15.7
GC purity (%)	99.97	99.98	99.7
UV cutoff (nm)	190	190	190

chromatographic wastes. We were prompted to study this area because of the high initial costs of these solvents and the ethical and financial burden of disposal. Our original work[8] involved the recovery of acetonitrile from an HPLC mobile phase containing this solvent along with an aqueous buffer. We have subsequently extended this work to the recovery of mobile phases from both normal- and reverse-phase HPLC procedures and from size-exclusion experiments.

ACETONITRILE

The acetonitrile wastes that were used in our initial studies were either an acetonitrile-4% aqueous ammonia solution or a 30:70 acetonitrile/acetate buffer mixture obtained from theophylline or tricyclic antidepressant analysis[9] or analysis of opiates.[10] These solvent mixtures were equilibrated for 2 h in an 8400 distillation unit and distilled at a throughput of 300 to 500 ml/h. The lower throughput than in our other recovery procedures produced a recovered material of better quality than that produced at faster recovery rates. The first 10% was discarded in all distillations and a pure acetonitrile (boiling point = 81 to 82°C) was recovered when the pot charge was 96:4 acetonitrile/buffer. When the second mixture was recycled, the heart cut was obtained at 77 to 78°C and contained the expected[11] azeotrope containing 15.7% water. Solvent purity was determined from boiling point, refractive index, Karl Fischer titration, UV and fluorescence analysis, and chromatographic data. The solvent specifications are shown in Table 10. It is obvious from this data that the recovered solvents contained the same gross properties when compared to a commercial high-purity HPLC sample. In addition to these results, we conducted two experiments that provided a better assessment of solvent purity. In the first experiment, samples of recovered acetonitrile were preconcentrated by reducing 100-ml portions to 10 ml by flash evaporation at room temperature. The tenfold preconcentrates were then spectrophotometrically compared to samples of unused HPLC-grade acetonitrile prepared in the same manner. A comparison of the UV spectra is seen in Figure 14. While the spectra of the undiluted samples show nearly identical absorbance, the preconcentrated spectra show a higher absorbance in the 210- to 225-nm region for the unused acetonitrile sample. A second type of experiment that also demonstrated the high purity of the recovered solvent was a dynamic HPLC analysis conducted in the following manner: samples of acetonitrile (either recovered or unused HPLC

WASTE SOLVENT RECOVERY USING SPINNING BAND DISTILLATION

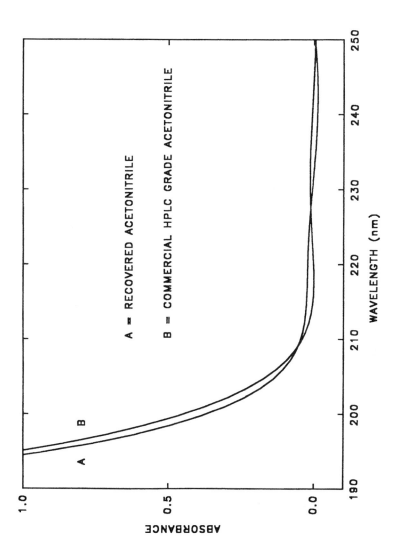

Figure 14 UV spectra of preconcentrated samples of acetonitrile.

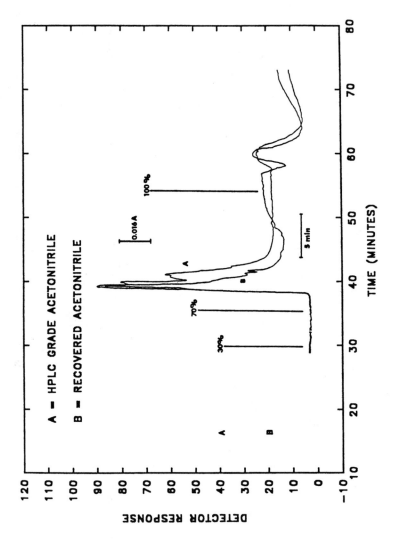

Figure 15 Dynamic HPLC analysis of acetonitrile samples.

Table 11 Recovery of Acetonitrile/Water Azeotrope from Different Wastes at High Throughput

Waste	Recovered Acetonitrile		
	UV cutoff (nm)	% Water	% Recovery
35% ACN 65% buffer	191	15.6	96
20% ACN 78% buffer 1% TCE/1%TFA	190	14.9	93
30% ACN 67% buffer 3% 2-propanol	192	15.3	46
67% ACN 22.5% water 0.5% TEA	191	15.6	94

grade) were compared after 60 min of on-column enrichment using acetonitrile/water, 30:70, passed over a C-18 reverse-phase column at 3 ml/min. Impurities accumulating on the column were then stripped off by using a step gradient of acetonitrile/water, 70:30, for 15 min, followed by a step gradient to 100% acetonitrile. The comparison results are shown in Figure 15. The lower level of impurities in the recovered solvent is amply demonstrated in this chromatogram. The chromatography was monitored using a UV detector at 210 ηm.

In recent years, we have conducted a large number of experiments on a wide variety of acetonitrile wastes using an 8600 distillation unit which because of its higher efficiency (greater number of plates) has allowed us to increase the throughputs (up to 800 ml/h) in this recovery. We have obtained consistent results in this recovery as demonstrated by the data in Table 11. This table presents a summary of four experiments in which acetonitrile is present in concentrations of less than 70%. As can be seen, a high recovery of azeotrope is obtained with excellent UV quality. We are presently attempting to develop procedures in which the water can be removed from the azeotrope after distillation by an economical and safe drying procedure. Until such a procedure is developed, the chromatographer can use the azeotrope by programming the chromatograph to add the proper makeup solvent for a gradient run.

It should be noted that, when acetonitrile HPLC wastes containing methanol and water are distilled, a much more complex situation is seen. The presence of a constant boiling mixture containing all three components mitigates against the separation. It turns out, however, that regardless of the composition of the starting mixture if one collects material up to 70°C, the composition of the distillate is nearly constant. Table 12 tabulates data from distillations conducted with different starting mixtures. In all cases, a mixture approximating 44:52:4, acetonitrile/methanol/water is obtained. The purity of these distillates was greater than 99.9% by GC analysis and was of sufficient purity to be reused providing the necessary buffers and make-up solvents were added. While this recovery is certainly much more involved, it might be justified under certain circumstances.

Table 12 Data for the Distillation of Acetonitrile/Methanol/Water Mixtures

Starting compositon (ACN/MeOH/H_2O)	Composition of distillate (ACN/MeOH/H_2O)
30:30:40	46:50:4
40:40:20	44:52:4
25:25:50	45:51:4
25:25:50	42:54:4

Table 13 Analytical Data for Methanol Recovery

Analysis	HPLC grade	Recovered pure solvent
Boiling point (°C)	64–65	64–65
Refractive index	1.3311	1.3310
Water (%)	0.09	0.17
GC purity (%)	99.80	99.80
UV cutoff (nm)	205	205

METHANOL

Aqueous methanol mobile phases are commonly used in reversed-phase HPLC analyses including analysis of aromatics,[11] pigments,[12] phenols,[13] and weak acid drugs.[14] In our initial studies,[15] used samples were distilled in an 8400 at throughputs of about 300 ml/h. Currently, we carry out methanol recoveries in an 8600 distillation unit at throughputs close to 800 ml/h. Figure 16 shows the distillation curves for several different aqueous methanol mixtures. In this display, % water is plotted as a function of volume distilled. These graphs demonstrate the difficulty of completely removing all the water from the distillate at high throughputs. Even when longer equilibration times and slower takeoffs are employed, some water is present. This has been noted elsewhere.[17] Table 13 displays the analytical data for methanol recovery, while, more importantly, the preconcentrated UV data and the dynamic HPLC runs are shown in Figures 17 and 18. These data emphatically demonstrate the high degree of purity of the recovered material. While the cost of high-purity methanol is not that of acetonitrile, the ease of recovery should mitigate against its casual disposal.

ISOPROPANOL

Isopropanol is employed to a much lesser extent in HPLC analyses than methanol and acetonitrile; however, it is the solvent of choice in certain procedures.[18] We have recovered isopropanol from used aqueous mobile phases utilizing the same distillation conditions as those employed with the other HPLC mobile phases. As in the case of acetonitrile, distillation of aqueous isopropanol solutions leads to the recovery of an azeotrope with about 13% water. The purity of the recovered material is identical to that of an HPLC-grade solvent. Figure 19 shows the distillation curve for a run on an 8600 unit at a throughput of 800 ml/h.

WASTE SOLVENT RECOVERY USING SPINNING BAND DISTILLATION 357

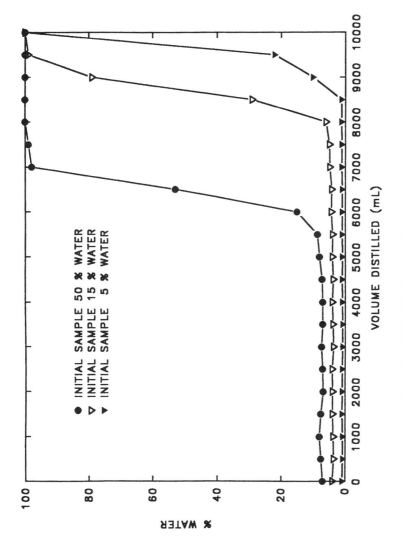

Figure 16   8600 distillation of methanol/water mixtures.

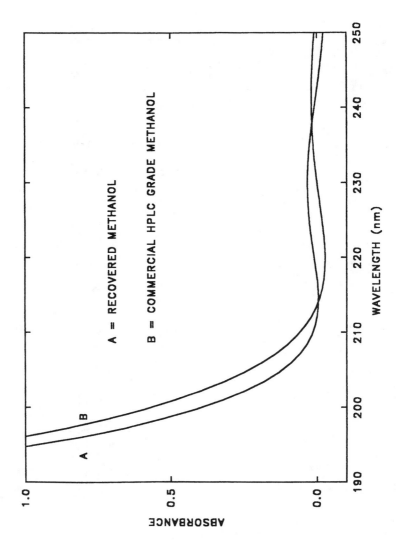

Figure 17 UV spectra of preconcentrated samples of methanol.

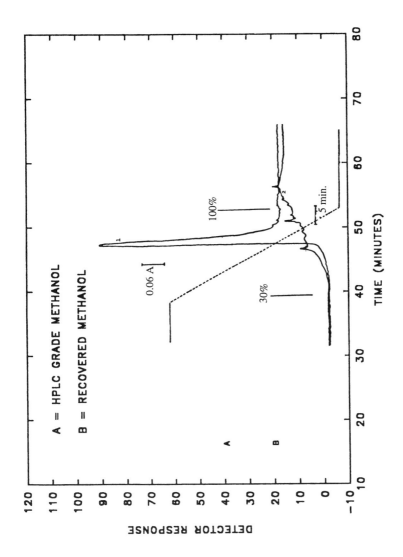

Figure 18 Dynamic HPLC analysis of methanol samples.

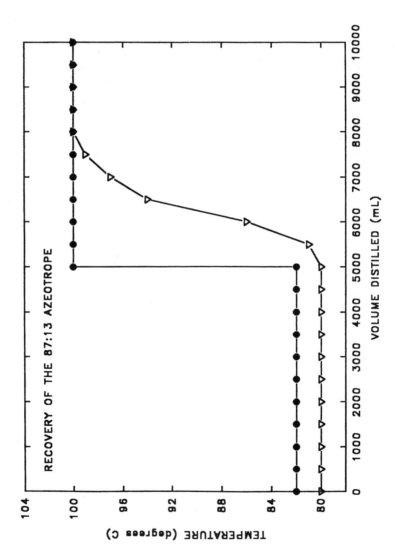

Figure 19 Distillation curve for the 8600 distillation of 50:50 isopropanol/water.

Table 14 Analytical Data for Hexane Recovery

Analysis	HPLC grade	Recovered pure solvent
Boiling point (°C)	68–69	68–69
Refractive index	1.3755	1.3751
Water (%)	0.01	0.01
GC purity (%)	99.9	99.9
UV cutoff (nm)	195	195

Table 15 Analytical Data for Cyclohexane Recovery

Analysis	HPLC grade	Recovered pure solvent
Boiling point (°C)	80–81	81–81.5
Refractive index	1.4258	1.4257
Water (%)	0.02	0.01
GC purity (%)	99.9	99.8
UV cutoff (nm)	200	202

Table 16 Analytical Data for Heptane Recovery

Analysis	HPLC grade	Recovered pure solvent
Boiling point (°C)	98–99	98–99
Refractive index	1.3878	1.3872
Water (%)	0.01	0.01
GC purity (%)	99.9	99.9
UV cutoff (nm)	200	199

Table 17 Analytical Data for Isooctane Recovery

Analysis	HPLC grade	Recovered pure solvent
Boiling point (°C)	99–100	100
Refractive index	1.3915	1.3918
Water (%)	0.03	0.04
GC purity (%)	99.7	99.8
UV cutoff (nm)	215	215

NONPOLAR SOLVENTS

When normal-phase HPLC analyses are employed, nonpolar solvents are used as the mobile phase. Hexane is used in the analysis of hydroxylated aromatics[19] and azo dyes[20] on a porous silica normal phase column, while cyclohexane is the solvent of choice when alkylamino metabolites are analyzed by ion-pair chromatography.[21] Similiarly, heptane and isooctane are routinely used in HPLC procedures[22] and the latter along with hexane is important in pesticide analyses.[23] While there are certain mobile phase mixtures (such as Hexane, THF, DCM combinations) that do not lend themselves to recycling, in many cases, we have been able to recover a high purity solvent. Tables 14–17 present analytical data for these nonpolar solvents recovered from used mobile phases. We will spend further discussion on Hexane recovery in the next section.

Table 18 Analytical Data for Tetrahydrofuran Recovery

Analysis	HPLC grade	ACS grade	Recovered pure solvent
Boiling point (°C)	65.6	65.5–66.7	65.5
Refractive index	1.4001	1.4095	1.4001
Water (%)	0.004	0.101	0.005
GC purity (%)	99.9	98.2	99.8
Fluorescence (nm)			
320	1%	5%	1%
360	1%	26%	1%
400	1.5%	54%	1.5%
440	1%	15%	0.5%

TETRAHYDROFURAN (THF)

THF is an extremely useful solvent for natural and synthetic resins and is incorporated as a mobile phase in size exclusion chromatography. The following data were obtained from distillation in an 8600 distillation unit of 8 l of THF waste from an analysis of polystyrene polymers.[24] Distillation was conducted at a throughput of 900 ml/h after an equilibration time of 1 h. The distillation curve is shown in Figure 20. Table 18 compares the recovered solvent to two grades of commercial THF (unstabilized HPLC grade and an ACS grade). In addition, the UV spectra are shown in Figure 21. The results in Table 18 show the high quality of recovered solvent, although GC analysis indicated that an impurity not found in HPLC grade was present. This impurity was suspected to be a peroxide from a known[25] THF oxidation reaction. The UV spectra confirmed this suspicion. On allowing the recovered, almost anhydrous, solvent to stand in the presence of air, the UV absorbance increased with a concomitant decrease in GC purity. After 3 days, the absorbance of the solvent had increased to such a level that it would no longer be useful for HPLC procedures. The same behavior was observed when a freshly opened bottle of HPLC-grade THF was exposed to air. When the same experiment was carried out with the ACS-grade solvent, the rate of UV deterioration was considerably slower, presumably due to the presence of a larger water content in this solvent. As noted in Chapter 15, a simple chemical test for the presence of peroxide is available and removal of this dangerous contaminant can be achieved with copper(I) chloride treatment. Anhydrous, peroxide-free THF can be stored for longer period of times if stored under a nitrogen atmosphere; otherwise, it must be used immediately after recycling.

HEXAFLUOROISOPROPANOL (HFIP)

HFIP has become a very important solvent for use in size-exclusion chromatography and other procedures. The initial price of this solvent is extremely high and its disposal is both environmentally difficult and costly. Figure 22 shows the capillary gas chromatograms obtained using a mass selective detector (MSD) for an HFIP waste containing toluene, DMF, and several other volatiles obtained from a polymer analysis. Also included in this figure is the chromatogram for a

WASTE SOLVENT RECOVERY USING SPINNING BAND DISTILLATION 363

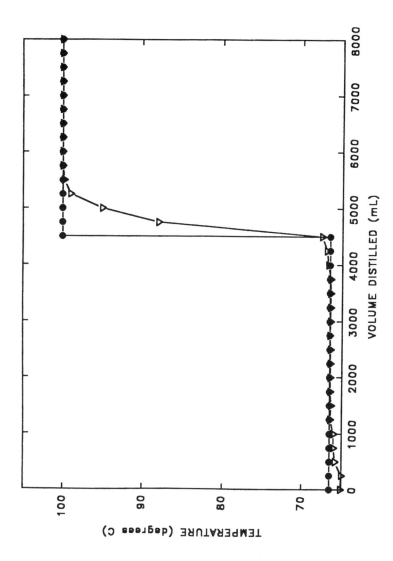

Figure 20 Distillation curve for the 8600 distillation of 55:45 THF/water.

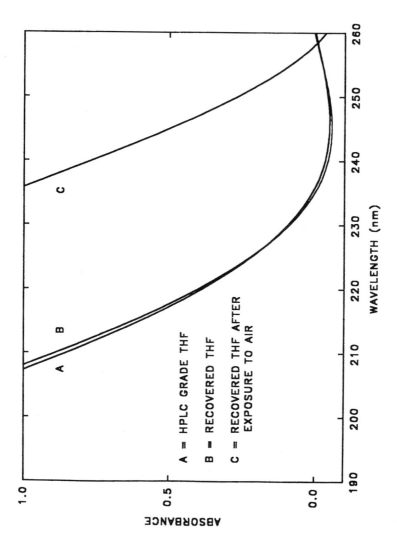

Figure 21   UV spectra of tetrahydrofuran (THF) samples.

WASTE SOLVENT RECOVERY USING SPINNING BAND DISTILLATION 365

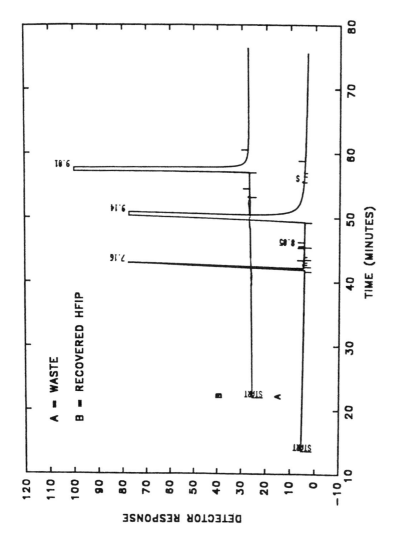

Figure 22 Capillary GC-MS chromatograms of HIFP samples.

recovered HFIP obtained from an 8400 distillation of 2 l of this waste at a throughput of 400 ml/h. This recovered material represents the heart cut obtained with a boiling range of 57.6 to 58.2°C. Approximately 3700 ml was obtained (about 90% recovery) of material with purity greater than 99.5%. There are presently a number of laboratories employing spinning band distillation in the recovery of this expensive solvent.

ENVIRONMENTAL ANALYSES WASTE

One of the most important areas in which solvent recycling has become extremely significant is in the environmental analysis laboratories. There are numerous facilities that are currently evaluating and analyzing large numbers of samples. Most of the standard EPA methods require large volumes of such solvents as dichloromethane (DCM), 1,1,2-trichlorotrifluoroethane (freon), hexane, and isooctane. In addition, a solvent such as DCM is used often in biochemical applications such as peptide sequencing procedures. In this section, we report on some of our work in this area and mention some work that is currently being carried out by others.

DICHLOROMETHANE (DCM)

Our initial studies of DCM involved the recovery from a peptide-sequencing waste containing 55% DCM, 35% dimethylformamide (DMF), and 10% methanol, water, and nonvolatiles. We carried out a recovery using an 8400 distillation unit at a reflux ratio of 1:1, resulting in a throughput of about 1000 ml/h. Three fractions were obtained:

Fraction # 1	38–40°C	93% DCM/7% methanol
Fraction # 2	40–148°C	DCM/water/methanol/DMF
Fraction # 3	148–152°C	100% DMF

It was necessary to treat fraction #1 with anhydrous calcium chloride (20 g/l) to remove methanol and a trace of water from the azeotrope. The purity of this recovered DCM was greater than 99.9%. It should be noted that, as the third fraction was being collected, considerable decomposition was seen in the pot due to thermal degradation of DMF. If DMF recovery is desired, the system should be shut down before distillation of DMF occurs and a vacuum applied to distill this material below its decomposition temperature.

A second situation involved a DCM/methanol (50:50) waste obtained from extraction. As noted above, direct distillation would result in a 93:7 DCM/methanol azeotrope. Instead, we pretreated the mixture as follows: 8 l of the waste was extracted with 2 l of water and the bottom layer (after separation into two layers) removed and dried with anhydrous calcium chloride (20 g/l). After 20 min drying time, the DCM was decanted into an 8400 and distilled at a throughput of 1000 ml/h. The distillation curve is shown in Figure 23. Included in this figure is

Table 19 Recovery of Pesticide-Spiked DCM

Experiment No.	Reflux ratio	Throughput (ml/h)	Carryover (ng/ml)
1	4:1	524	48.1
2	3:1	536	25.6
3	2:1	510	9.5
4	2:1	685	25.2

the distillation curve for direct distillation of a 50:50 mixture showing the recovery of the azeotrope. A recovery of about 80% of a greater than 99.9% pure DCM was obtained.

A final situation is that in which DCM is obtained as a waste material from pesticide analysis. A number of studies have been carried out in which spiked samples of pesticide-grade DCM was distilled directly in an 8600 distillation unit under different conditions. Table 19 presents a summary of data obtained from experiments in which DCM was spiked with lindane, 4,4'-DDD, and 4,4'-DDE at levels of 62,000 ng/ml each. Four experiments at different throughputs and reflux ratio were performed. The distillate was analyzed for pesticide contamination by GC-ECD and the results reported as carryover of the pesticides. While some contamination was found, it is obvious that considerable purification has taken place with this sample with artificially high pesticide contamination. Recent work[26] with actual wastes with much lower pesticide contamination has shown a considerable improvement in this recovery process. The recovery protocol will be reported on at a later time.

FREON

Another important and environmentally sensitive solvent that is used in EPA methods and other extraction procedures is 1,1,2-trichlorotrifluoroethane (freon). The major impurity in used freon samples are hydrocarbons. Our initial studies, which resulted in the use of this technique by a variety of labs recycling low hydrocarbon content waste freon, involved the recovery of used freon obtained from test procedures for oil and grease and a method for analyzing petroleum hydrocarbons in water. An 8600 distillation unit was employed and the normal protocol called for a 15-min to 1-h equilibration and a reflux ratio of 5:1. For a 10-l batch run, it is suggested that the first 5 to 10% of distillate be discarded before obtaining a heart cut in the 46 to 49°C boiling range. Analysis of the distillate was performed using an FT-IR technique in which a sample was placed in a 10-cm quartz IR cell and the spectrum scanned in the 3000 to 2900 cm^{-1} range with an absorption measurement carried out at 2929 cm^{-1}. Table 20 presents data for a series of runs carried out by an independent laboratory in which the hydrocarbon content of the heart cut is displayed. It is obvious that the recovered freon is of excellent quality and can be reused. Experiments are currently being conducted in which freon samples with higher concentrations of hydrocarbons are being recycled. It appears that it will be necessary to carry out a double distillation (a redistillation of the heart cut) to obtain satisfactory reusable material. Pretreatment of the waste is also a possibility.

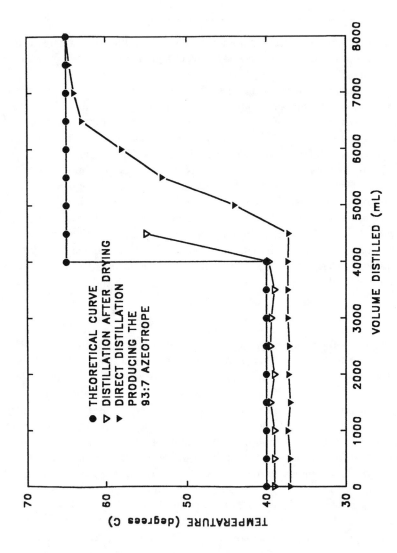

Figure 23 Distillation curve for the 8400 distillation of 50:50 DCM/methanol.

Table 20 Hydrocarbon Content of Recovered Freon

Run No.	Boiling point (°C)	Reflux ratio	Discarded Forecut (ml)	Hydrocarbon Content (mg/l)
91231	47–49	10	300	0.0006
91232	47–49	10	300	0.0066
91233	47–49	5	230	0.0103
91234	47–49	3	450	0.0099
91235	47–49	5	400	0.0083
91240	47–49	6	450	0.0033
91248	46–49	2	600	0.0122
91250	46–49	5	500	0.0001

ROUTINE LABORATORY WASTE SOLVENTS

Besides the situations described in preceding sections, there are many other situations in industrial and academic laboratories in which large quantities of ordinary organic solvents are generated as waste materials. In most cases, if the used solvents are managed correctly (i.e., different wastes are segregated from each other), rapid recycling can be achieved in a spinning band distillation unit to obtain recovered material of sufficient purity for reuse. If the solvent is produced in some routine procedure or analysis, then it is advisable to incorporate the recycling as part of the protocol. In this section, we discuss some of the routine solvents that we have successfully recycled.

ACETONE

Acetone is one of the most common solvents found in a laboratory and it is routinely used for glassware cleaning and degreasing of precision parts. In addition, it is used as an extraction solvent or as a delusterant solvent for cellulose fibers.[27] While it is moderately toxic and flammable, in most cases, the water miscibility of the material allows it to be disposed of down the drain as long as no other hazardous component is present or no local sewage ordinance prohibits this procedure. In any event, an attractive alternative is to collect the used acetone and recycle it under high throughput conditions. The major contamination is water and usually nonvolatiles. Figure 24 shows the distillation curve of a 70:30 acetone/water waste that was distilled at a throughput of 2000 ml/h. Excellent recovery of a better than 99.8% acetone of demonstrably reagent grade was obtained. It is obvious that on a routine basis large quantities of sufficiently pure acetone can easily be recovered.

TOLUENE

We have carried out dozens of experiments in which the major objective was to recover used toluene. These have included the recovery of this solvent from relatively simple mixtures such as waste generated from degreasing procedures in petroleum laboratories to complex solutions containing mixtures of toluene/MEK/

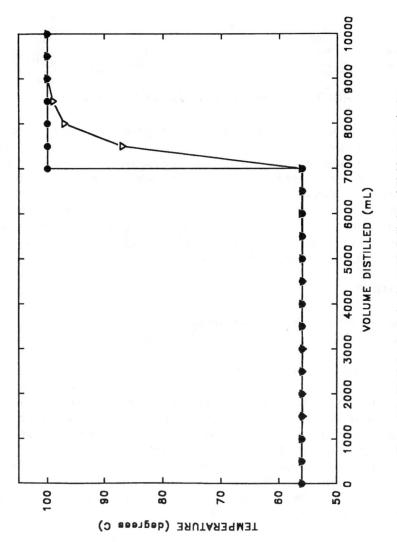

Figure 24 Distillation curve for the 8400 distillation of 70:30 acetone/water.

acetone. All of these have been successful as long as the proper experimental conditions are employed. For example, the latter recovery, because of the complexity of the waste, required a much lower throughput and a fairly high reflux ratio, 10:1, to obtain satisfactory results.

One of the more interesting recoveries involving toluene (and xylene) was the recovery of this solvent from liquid scintillation waste.[28,29] Figures 25 and 26 show distillation curves of low-level radioactive waste material obtained from the University of Alabama Medical Center which was distilled on a spinning band distillation unit. This toluene-based waste (33 or 50% toluene) contained tritiated steroids along with aqueous detergents and fluors. In these curves, the distillation progress is followed by plotting the radioactivity (counts/min/ml), as measured in a liquid scintillation counter vs. volume distilled. It is seen that immediately the radioactivity decreases to background level. All of the initial radioactivity will be concentrated in the pot, thereby minimizing the volume needed to be disposed. Gibbs[6] followed up our initial work by examining the potential environmental risk associated with the recovery process. He conducted environmental air sampling that demonstrated that no radioactivity was being emitted during the recovery process. In addition, he developed a simple and risk-free procedure for decontaminating the very small amount of radioactivity that remained in the distillation unit after use.

OTHERS

There are numerous other situations in which solvent recycling is attractive. We have reported[3] on situations in which active laboratories have incorporated spinning band distillation into their routine procedures for recovery of used liquid wastes. For example, at West Chester University, one of us initiated a program in which over 15 different solvents generated as waste in ordinary teaching laboratories were recycled. The annual savings in cost and disposal in this relatively small (10,000 students) university amounts to over $3000. The economical and environmental benefits are obvious.

PRACTICAL ASPECTS OF SOLVENT RECYCLING

It is clear from the above discussion that a large number of solvent wastes can be recycled to produce a material of such high quality that it can be reused. This includes the high-purity solvents that can tolerate only very low quantities of impurities. Besides this technical aspect of recycling, there are a number of practical reasons for a laboratory to enter into this endeavor.

STATE AND FEDERAL REGULATIONS

The management of hazardous waste has become an important aspect of laboratory operations. The Resource Conservation and Recovery Act (RCRA) is

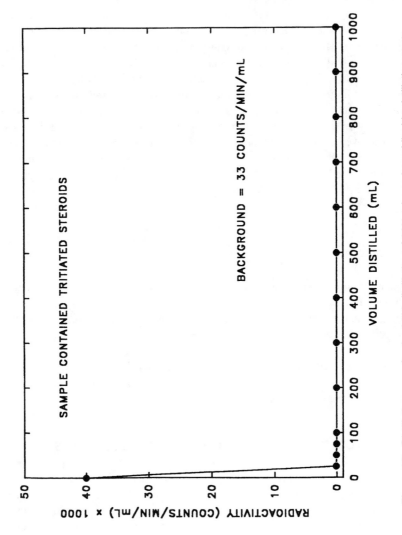

Figure 25 Distillation curve for the 8400 distillation of 33% toluene-based liquid scintillation waste.

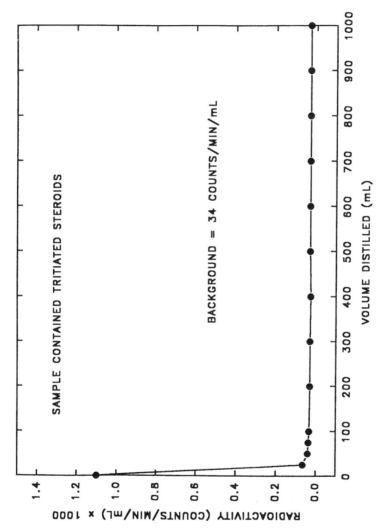

Figure 26 Distillation curve for the 8400 distillation of 50% toluene-based liquid scintillation waste.

designed to reduce hazardous wastes by tracking and regulating wastes from cradle to grave. The generator retains the liability of the waste material until it is either destroyed, rendered nonhazardous, or beneficially recycled to a reusable product. Under this regulation, solvent recycling is accepted as a very viable alternative to waste removal.

SAFETY FACTORS

Because of the flammability and toxicity of organic solvents, one of the overriding concerns with regards to solvent recycling is the safety of the process. The characteristics and safety record of the spinning band units used in this work are outstanding. The electronics of the microprocessor are UL listed to UL 1262 by EL laboratories (an OSHA nationally recognized testing laboratory) and the microprocessor incorporates a number of safety features designed to stop operation of the unit in the event of some malfunction. This includes shutdown if any problems are encountered with the heating mantle in which the heating element is protected by an Al liner, cooling water malfunction, or any other potential difficulties. The units can be vented at 100 to 150 cfm at about 5 ft from the solvent container. In addition, there is attached within the cabinet an automatic fire extinguisher and a Sentinel 125 Automatic Vapor Sensor with sensitivities as high as 0.0237 ppm formaldehyde (0.624 ppm xylene). All in all, these units are now the most attractive distillation systems for the safe, automatic, and economical recovery of used solvents.

There are other safety aspects to a solvent recycling program that must be recognized. The involved laboratory must keep accurate records of the solvent waste and the recycled material and must ensure that different waste materials are carefully segregated from each other. The laboratory personnel must be made aware of the flammability and toxicity of the material involved and must ascertain that no adverse chemical reactivity may result from the distillation process. While this is not much of a problem with ordinary solvents (except the case of peroxide-forming solvents such as THF which we have discussed), there is the possibility that some unusual cases may arise. In such an event, it is essential that the operator determine the characteristics of that solvent.

COST BENEFITS

With the rising costs for solvent purchasing and waste removal, there is a major economic benefit in recycling. Tables 21 and 22 show several cost/benefit payback analyses for some common, relatively inexpensive, solvents. These quantities are about those generated in an average-sized anatomic pathology laboratory. The expensive high-purity solvents would lead to even greater savings. In these calculations, we have used typical recovery rates for reusable high purity material. The recovered solvent can be recycled numerous times, although there is the expected overall loss of material due to handling. With the initial cost of implementing a solvent recovery program including purchasing of equipment

Table 21 Payback Analysis for Alcohol, Xylene, and Xylene Substitutes

Solvent	Alcohol	Xylene	Xylene substitutes
Weekly usage (gal)	15	1	7
Unit cost ($)	7.00	11.00	13.00
Weekly cost ($)	105	11.00	91.00
Recovery (%)	60	95	95
Cost reduction ($)	63.00	10.42	87.98
Disposal cost ($)	150.00	10.00	70.00
Total ($)	213.00	20.42	157.98
Weekly total ($)		392.40	
Annual total ($)		20,405.00	

Table 22 Payback Analysis for Formalin

Weekly usage (gal)	14
Unit cost ($)	3.00
Weekly cost ($)	42.00
Recovery (%)	85
Cost reduction ($)	35.70
Disposal cost ($)	434.00
Weekly total ($)	479.70
Annual total ($)	23,944.00

and other factors, the economic benefit is very attractive. Many laboratories can recoup their initial investment within 2 years.

CONCLUSION

In Chapter 15 we discussed the operational characteristics of spinning band distillation systems and described the theory and limitations of distillation as a waste minimization process. In this chapter, we have discussed the practical aspects of solvent recycling. To be sure, there are a number of solvent wastestreams that are not amenable to recovery by direct distillation; however, in most cases, laboratory recycling is possible. The overall benefit of a solvent recovery program is tremendous, both in the savings involved in purchased material and in its disposal and in the positive environmental impact of such a program. It is expected that as the cost of these solvents spiral upward, more and more laboratories will routinely incorporate recycling into their procedures. We hope this chapter has helped in this effort.

ENDNOTES

1. Crescent Chemical Co., Inc., 1324 Motor Pkway., Hauppage, NY 11788.
2. Hogan, J. M., Engel, R. A., and Stevenson, H. F., A gas chromatograhic method for water analysis, *Anal. Chem.*, 42, 249, 1970.
3. Mangravite, J. A., Recovery of waste solvents, in *Library of Environmental Science Technology*, Chereminisoff, P. N., Ed., Gulf, Houston, 1989, chap. 2.

4. Warbritton, A. R. and Mangravite, J. A., Solvent Recycling in the Histology Lab: A Comparison of Fractional Distillation and Flash Evaporation Techniques, National Meeting of the National Society for Histotechnology, Monterey, CA, September 1992.
5. Brinkmann Instruments, Inc., Buchi Rotavapor 176 Ex operated at 10 mmHg and 100°C, Cantiague Road, Westbury, NY 11590.
6. Gibbs, L. M., Environmental risk factors in solvent recycling, *Am. Clin. Prod. Rev.*, November 1983.
7. Anatech Ltd., 1020 Harts Lake Road, Battle Creek, MI 49015.
8. Katusz, R. M., Bellew, L., Mangravite, J. A., and Foery, R. F., Recovery of HPLC grade acetonitrile by spinning band distillation, *J. Chrom.*, 213, 331, 1981.
9. Kabra, P. and Marton, L. J., An HPLC method for antidepressant drugs, *Clin.Chem.*, 28, 687, 1982.
10. Wu, C. Y. and Wittick, J. J., Analysis of opium derivatives, *Anal. Chem.*, 49, 359, 1977.
11. Horsley, L. H., Azeotropic data, in *Advances in Chemistry Series Number 6*, ACS Publications, Washington, D.C., 1952, 6.
12. Schmidt, J. A., Henry, R. A., Williams, R. C., and Dieckman, J. F., Liquid chromatographic analysis of aromatics, *J. Chrom. Sci.*, 9, 645, 1971.
13. Kirkland, J. J., Pigment analysis, *Chromatographia*, 8, 661, 1975.
14. Court, W. A., HPLC analysis of phenolic derivatives, *J. Chrom.*, 130, 287, 1977.
15. Wheals, B. B., Reverse phase analysis of weak acid drugs, *J. Chrom.*, 122, 85, 1976.
16. Foery, R. F., Mangravite, J. A., and Katusz, R. M., HPLC Solvent Recovery by Spinning Band Distillation, 32nd Pittsburgh Conf. on Analytical Chemistry, Atlantic City, NJ, March 1981, 460.
17. Evers, E. C. and Knox, A. G., Methanol distillation, *J. Am. Chem. Soc.*, 73, 1739, 1951.
18. Kirkland, J. J. and Antile, P. E., HPLC mobile phases, *J. Chrom. Sci.*, 15, 137, 1977.
19. Kirkland, J. J., Analysis of hydroxylated amines, *J. Chrom. Sci.*, 10, 593, 1976.
20. Passarelli, R. J. and Jacobs, E. S., HPLC analysis of azo dyes, *J. Chrom. Sci.*, 13, 153, 1975.
21. Borg, K. O., Gabrielsson, M., and Jonsson, T. E., Ion pair chromatography, *Acta Pharm. Suec.*, 11, 313, 1974.
22. Dolphin, R. J., Hoogeveen, L. P. J., and Wilmont, F. W., HPLC analyses, *J. Chrom.*, 122, 259, 1976.
23. Cochran, W. P. and Chau, A. S. Y., Pesticide analysis, in *Advances in Chemistry, Series Number 20*, ACS Publications, Washington, D.C., 1971.
24. Kirkland, J. J. and Antle, P. E., THF as a solvent for HPLC analysis, *J. Chrom. Sci.*, 15, 137, 1977.
25. Hamstead, A. C. and VanDelinder, L. S., Peroxide formation in THF, *J. Chem. Eng. Data*, 5, 383, 1960.
26. Kelley, K. P., private communication.
27. Hawley, G. G., Acetone uses, in *The Condensed Chemical Dictionary*, 8th ed., Van Nostrand Reinhold, New York, 1971.
28. Mangravite, J. A., Gallis, D., and Foery, R. F., Recovery of toluene and xylene from liquid scintillation waste, *Am. Lab.*, 15, 24, 1983.
29. Mangravite, J. A., Volume reduction from low level radioactive waste material, *Atom*, 1987, 383.

CHAPTER 21

The Implementation of Waste Minimization Strategies in a Biotechnology Research and Development Laboratory

Richard A. Senn

CONTENTS

Introduction and Background 377
    Why Waste Minimization? 378
    How We Got Started 379
    Company Background 380
    Chemical Use 380
    Wastestreams 381
    Waste Minimization Audit 382
Source Reduction Waste Minimization Opportunities 382
    Product Substitution 382
    Waste Segregation 384
    Recycling Waste Minimization Opportunities 385
    Management Controls 386
    Waste Minimization Program Benefits 387
Endnotes 388

INTRODUCTION AND BACKGROUND

To many hazardous waste professionals, waste minimization may seem the most appropriate to manufacturing facilities where process modifications and other waste minimization strategies can significantly minimize the waste generated from a particular manufacturing process or in a major academic institution where significant quantities of individual laboratory wastestreams are produced. However, because of the complexity of the research laboratory, there are also many waste minimization opportunities in an R&D environment. Waste production savings for individual wastestreams may not be significant by themselves, but combining improvements from a number of minor waste minimization efforts can have a major overall impact on a research lab's hazardous waste production. In addition, waste production savings can be significant because the large number of low volume wastestreams generated by laboratories are more costly to manage than bulk shipments. Lab pack shipments are more labor intensive, require a higher degree of technical expertise and involve more paperwork. Therefore, the per pound waste disposal savings for a lab pack shipment will generally be greater than that for a bulk wastestream. The guiding principles of waste minimization are valid regardless of the size of the institution. The major goal of this chapter is to show that waste minimization is a valuable environmental strategy for research laboratories and that implementation of a waste minimization program fosters a win-win situation where the environmental impact of an institution's hazardous waste production is minimized while still providing cost savings and other corporate benefits.

Before discussing the implementation of a waste minimization program at Agracetus, it is necessary to define what is included in the definition of waste minimization. For the purposes of the Agracetus program, waste minimization includes both source reduction and recycling. It is important to point out the inclusion of recycling as a waste minimization technique. Some environmental professionals do not include recycling as a valid waste minimization strategy because one is only reusing an existing wastestream.

WHY WASTE MINIMIZATION?

To begin a successful waste minimization program, it is essential to establish program goals. What is the purpose in establishing a corporate waste minimization program? Perhaps it is to minimize future generator liability through waste reduction. Maybe the primary goal is the reduction of disposal costs. Every institution has its unique needs. The important point is to understand these needs before undertaking the program. Otherwise, it becomes difficult to evaluate the program's effectiveness.

There were four primary goals in establishing a waste minimization program at Agracetus. First, like every organization implementing a waste minimization program, reduction of hazardous waste disposal costs was a primary goal. At the time the waste minimization program began, Agracetus was defined as a small quantity generator (<1000 kg/month) under Wisconsin hazardous waste regulations. Therefore, the magnitude of hazardous waste production was not large. However, because 90% of the waste production consisted of expensive lab pack shipments of research chemicals, there were still significant opportunities for cost savings.

The second goal was to change our generator status from small quantity generator to a very small quantity generator. In Wisconsin there are definite regulatory advantages to this change in generator status. A very small quantity generator is defined as a generator who generates less than 100 kg (220 lb) of hazardous waste in a month and does not accumulate quantities of hazardous waste greater than 1000 kg. This status does not require that the generator meet the manifest requirements of NR 615.08 and the notification requirements of NR 600.05. The very small quantity generator also does not have to submit an annual hazardous waste production report to the Wisconsin Department of Natural Resources, as do small and large quantity generators. Successfully changing the generator status meant a decreased regulatory responsibility.

A third goal was to minimize the time and resource commitment dedicated to hazardous waste management by the environmental health and safety department. At Agracetus, hazardous waste management is only one of the many regulatory responsibilities of the environmental health and safety department. Other responsibilities include radiation safety, chemical and environmental safety, biosafety, employee health, worker's compensation, OSHA compliance, and regulatory and management responsibilities for an animal facility. With this myriad of responsibilities, finding ways to reduce time commitment in one responsibility of environmental health and safety means that there is more time to devote to other priorities.

Finally, waste minimization was one of the areas in environmental health and safety where a more proactive approach to environmental management was an achievable goal. It is an important component of corporate image to concentrate on ways of eliminating hazardous waste before it is produced, instead of developing new treatment or disposal methods for waste that is already generated. This proactive approach can lead to very positive public relations for the institution.

HOW WE GOT STARTED

Agracetus' waste minimization program began as a combination of strategies developed by staff and recommendations provided through the University of Wisconsin (UW)-Extension's Technical Assistance Program. One of the most difficult aspects of implementing waste minimization in a small- to medium-sized laboratory facility is having the internal expertise to implement a program. In a facility this size, there is generally not a hazardous waste specialist who has enough familiarity with hazardous waste management to understand all the potential methods for waste minimization. Hazardous waste management may be part of the responsibilities of the environmental health and safety manager, or a part-time duty for a scientist or the facilities manager. Obviously, this variety of personnel with hazardous waste management responsibilities can lead to this function not receiving appropriate attention. Therefore, developing a waste minimization program may not be a high priority because of lack of time or understanding by internal resources.

It became evident that Agracetus fell into this category, where there were not sufficient internal resources to develop an in-depth waste minimization program.

Therefore, although some waste minimization strategies were developed internally, the major portion of the program was implemented with outside technical assistance. The University of Wisconsin Extension, Department of Engineering Professional Development had received a grant to implement a pilot waste minimization technical assistance program in Wisconsin. Agracetus was one of seven companies in Wisconsin to take part in this pilot program. The technical assistance provided by UW-Extension consisted of evaluating the Agracetus facility, performing a waste audit, providing a written report of audit results, and making recommendations to reduce or minimize the generation of hazardous wastes. The results of this audit were the focal point in developing a series of policies and procedures to minimize waste generation in the Agracetus laboratories, greenhouses, and facilities department.

COMPANY BACKGROUND

To appropriately understand how this waste minimization project helped Agracetus minimize hazardous waste generation, it is helpful to first provide background information about the company, its chemical usage, and existing wastestreams. Agracetus is a 65,000-ft^2 biotechnology research and development facility, consisting of 30,000 ft^2 of laboratory space, 10,000 ft^2 of greenhouses, a 3,300 ft^2 animal facility and support facilities. The research focus at Agracetus is to develop genetically improved products in both the agricultural and biomedical fields. In agricultural biotechnology, Agracetus is developing genetically improved products through genetic engineering. This program includes the development of genetically improved cotton plants and a plant bioreactor program that utilizes plants to produce industrial enzymes, and pharmaceuticals. In biomedical research, Agracetus is developing human gene therapy products and vaccines. One project involves developing an AIDS vaccine.

CHEMICAL USE

In the process of developing these agricultural and biomedical products, it is necessary to use many chemicals. The majority of these chemicals are used during extraction and purification processes in the molecular biology labs. For example, laboratory staff use phenol/chloroform solutions during DNA extractions. Other frequently used hazardous chemicals include acrylamide, ethidium bromide, and 2-mercaptoethanol. Agracetus maintains a chemical inventory database of all chemicals that enter the facility. This database shows that over 1000 different chemicals are used in the research and development laboratories.

However, chemical use is not confined to just the research laboratories. The greenhouses and facilities departments and the animal facility also use hazardous chemicals. To control insect problems, a variety of pesticides are sprayed in the greenhouses every week. This creates pesticide rinsewater and bulk unused or outdated pesticides. The facilities department also utilizes a variety of hazardous chemicals in maintenance and remodeling projects. These include paints and paint thinners, caustic acids and bases, floor strippers, and varnishes, and cleaning chemicals. In the animal facility, a variety of disinfectants and cleaning products are used to keep the facility germ free. These include mineral acids, phenols, iodines, and halogens.

WASTESTREAMS

Use of a diverse set of chemicals also produces a diverse set of wastes. Major wastestreams include organic and inorganic research chemicals, chlorinated organic solvents, nonchlorinated organic chemicals, contaminated labware, pesticide rinsewater, and used oil. The largest wastestream at Agracetus is organic and inorganic research chemicals. This consists of a variety of unused or outdated chemicals. Many of these research chemicals become outdated before their expiration date. Therefore, they become waste material. Although the quantites are generally less than 100 g, disposal is extremely expensive because they represent such a wide variety of chemical classifications — flammable liquids, organic acids and bases, mineral acids and bases, water and air reactives, poisons, carcinogens, mutagens, drugs, and pesticides. Because of this diversity, many of these chemicals cannot be mixed as wastes and this significantly increases the complexity and cost of their disposal.

Much easier to handle, but still a significant wastestream problem in research laboratories, are chlorinated and nonchlorinated organic solvents. Phenol/chloroform solutions, used in DNA extractions in the molecular biology labs, are the major component of this chlorinated wastestream. Nonchlorinated organic solvents include toluene, alcohols, mercaptoethanol, formaldehyde, and other reseach organic liquids.

The third major laboratory wastestream is contaminated labware. These are gloves, bench coverings, weighing material, and other solid labware that are contaminated with chemicals that should not be included as part of the normal solid wastestream. This includes chemicals such as ethidium bromide, phenol, and acrylamide. Contaminated labware is collected in refuse containers separate from the normal trash. If handled correctly, contaminated labware is nonhazardous. However, if it is inadvertently mixed with hazardous waste there is the potential to significantly increase the volume of hazardous waste generated because the contaminated labware would then also be considered hazardous.

The greenhouses generate the largest volume of liquid hazardous waste through pesticide rinsewater generated during spraying operations. The ten greenhouses are sprayed frequently to control insect problems. Greenhouse staff are certified pesticide applicators, qualified to spray most pesticides. They mix the pesticide solutions and use either backpack or portable spraying units for application. After spraying is completed, the sprayers are rinsed out to remove any residues of pesticide concentrate. This residue solution can still have hazardous levels of pesticides. In Wisconsin the regulatory agency for pesticides is the Department of Agriculture. If the pesticide concentration in the wash water is less than 10 ppm, the wash water is considered nonhazardous. If the pesticide concentration exceeds

10 ppm, it is considered a hazardous waste and is regulated by the Wisconsin Department of Natural Resources under NR 600 like many other chemical wastes.

The largest wastestream from the facilities department is used oil. This is generated from equipment and machinery used in the facilities, greenhouse, and laboratory departments and includes crankcase, transmission, and other engine oils. This waste oil is collected in 5-gal carboys and transferred to 55-gal drums for recycling. Although the majority of this wastestream is nonhazardous, the used oil can become contaminated with chemicals that would define the oil as a hazardous waste.

WASTE MINIMIZATION AUDIT

Once waste minimization goals were established and technical assistance from UW-Extension had been obtained, the next step was to perform a waste minimization audit. Before one can determine if there are effective ways to minimize hazardous waste, it is necessary to know what chemicals are used in the facility and current disposal methods. To better understand current chemical usage and waste-handling practices, chemical purchasing information for a whole year was collected and analyzed. To evaluate this massive amount of information most effectively, it was necessary to input data into a DBase III database with specific data fields: chemical name, quantity purchased, project making request, supplier, date of purchase order, date chemical received, and expiration date.

Chemical purchasing information from the Dbase III database was then compared to disposal records obtained from manifests and lab pack lists. The purpose of this comparison was to evaluate potential candidates for waste minimization. Chemicals used in the largest quantities or that offered limited disposal options were identified as the best candidates for waste minimization.

SOURCE REDUCTION WASTE MINIMIZATION OPPORTUNITIES

PRODUCT SUBSTITUTION

Evaluation of information from the waste minimization audit helped to identify a number of potential candidates for waste minimization. The first source reduction technique looked at was product substitution. Because of the diversity of chemicals used during research, there is the opportunity to minimize many hazardous wastestreams by substituting a less-hazardous chemical for one that presents more difficult disposal options. One of the best examples of this strategy is the use of "No Chromix" to replace dichromates in sulfuric acid. For many years, dichromates have been added to sulfuric acid to create a very strong oxidizer which is used to clean laboratory glassware. However, chromium is one of the heavy metals that is highly toxic to the bacteria that help degrade organic wastes in sewage treatment. At chromium concentrations of 0.5 mg/l anaerobic sewage treatment becomes gradually less effective. And, because toxic metals are usually considered toxic above 1 to 3 ppm, the U.S. Public Health Service Drinking Water Standard limits chromium to 0.05 mg/l. Therefore, either disposal of chromate acid cleaning solutions has to be monitored very carefully to conform with these sewer disposal requirements or chromate cleaning solutions must be collected to dispose of as hazardous waste or treated in-house to separate the chromates. "No Chromix," a strong cleaning solution with no toxic metal components, eliminates the chromate waste disposal problem. "No Chromix" solutions can be neutralized and disposed of in the sanitary sewer system.

Another product substitution involved replacing mercury thermometers with red liquid thermometers whenever possible. Biotechnology research laboratories have many uses for thermometers, including water baths, incubators, and drying ovens. Unfortunately, the use of so many thermometers presents the potential for many accidents. If these accidents involve mercury thermometers, there are also hazardous waste disposal implications. Although mercury can be successfully recycled in many laboratory applications, recycling is not practical with broken mercury thermometers. The mercury is in small quantities and is generallly difficult to remove from the water bath or other affected equipment. Amalgamating the mercury with a mercury spill kit is often the only effective way to remove it. By utilizing the purchasing department to substitute alcohol for mercury thermometers, it is possible to gradually phase out most uses of mercury thermometers in the laboratories through product substitution and avoid the waste disposal problems associated with mercury. However, occasionally research needs dictate the use of mercury thermometers.

In addition to the use of many different chemicals, most biotechnology laboratories also use radioisotopes in their research. At Agracetus these radioisotopes include phosphorous 32, phosphorous 33, cadmium 115, chromium 51, iodine 125, sulfur 35, carbon 14, and tritium. The effectiveness of different research experiments is usually determined by dissolving these radioisotope extracts in a solution that causes the radioisotope to fluoresce when bombarded by radiation from a radium-sealed source in a scintillation counter. Traditionally, scintillation cocktails have been toluene or xylene based. These scintillation cocktails present an increasingly difficult and expensive disposal alternative, because the solution is a mixed waste that is both radioactive and hazardous. Most hazardous waste disposal companies will not accept radioactive wastes and the few radioactive waste disposal facilities that will accept this type of waste charge a premium for this service. Therefore, Agracetus began using "Opti-Fluor," a biodegradable pseudocumene scintillation cocktail from Packard Instruments that has a flashpoint above 150°F and is therefore not classified as a flammable liquid like toluene- or xylene-based scintillation fluids. Depending on the requirements of the institution's radioisotope license, the "Opti-Fluor" or other similar products can either be disposed of in the sanitary sewer system or incinerated through a commercial radioactive waste broker as nonhazardous radioactive waste.

Opportunities for product substitution in a R&D facility are not just confined to the laboratories. Product substitution opportunities were also found in the Agracetus greenhouses, substituting toxic organophosphate and carbamate pesticides with less-toxic alternatives. For example, "Margasan-O" is a new naturally occurring biodegradable and environmental friendly pesticide extracted from the neem tree in India. Ground neem seeds are commonly added to stored grain in India to keep out insects and neem leaves are placed in books or stored in clothing to repel moths. Research by the USDA and other agencies showed that neem seed could control over 80 major insect species. Agracetus was one of the greenhouses used to test market a specific application of this pesticide. Neem was found to be an effective replacement for Dursban, Malathion, and other organophosphate insecticides used to control thrips. Neem's major advantage is that it degrades rapidly under ordinary environmental conditions and therefore does not create the hazardous waste disposal problems inherent with organophosphate pesticide rinsewater and bulk solids. Other pesticide product substitutions used include replacing organophosphate and carbamate pesticides with less toxic synthetic pyrethroids.

Product substitution opportunities were also found in the facilities department. Dichloromethane was used as an effective degreaser to remove grease and dirt from tools and machinery parts in the facilities department. However, dichloromethane has safety and disposal concerns because it has been shown to be a potential carcinogen. After studying different product substitution alternatives, it was decided to use the parts washer service provided by Safety Kleen. They utilize a proprietary cleaning solution that is less toxic than dichloromethane. Safety Kleen picks up the dirty parts washer solution and recycles it, avoiding the cost of continually repurchasing and disposing of parts washer solvents.

WASTE SEGREGATION

Waste segregation is also a useful waste minimization technique in R&D laboratories. Developing waste segregation policies can avoid mixing hazardous and nonhazardous wastestreams. Without these policies, additional hazardous waste production and disposal costs can be incurred. Before implementing waste segregation policies at Agracetus, nonhazardous solid waste occasionally became contaminated with chemical waste. As mentioned previously, one solid wastestream at Agracetus is chemically contaminated labware. This consists of gloves, bench top covers, and other materials contaminated with ethidium bromide, acrylamide, and other very toxic chemicals. Although the concentration of chemicals is fairly low, it is advantageous to remove this material from the normal trash to avoid custodial staff handling these materials. Chemically contaminated labware is segregated from normal trash and shipped to a special waste landfill. If this chemically contaminated wastestream is not carefully monitored, it can become a hazardous wastestream.

In one case broken mercury thermometers were inadvertently mixed with chemically contaminated labware. When the contents of the drum were analyzed by the disposal contractor, the mercury concentration exceeded the regulatory limit. Consequently, the whole drum had to be disposed of as hazardous waste instead of nonhazardous chemically contaminated labware. Frequent occurrences of contaminating labware with hazardous chemicals will have a significant adverse effect on the volume of hazardous waste production.

Waste segregation in a research lab is the most valuable in separating waste research chemicals into different categories based upon chemical composition. Many unused or unwanted research chemicals in biotechnology laboratories are not hazardous by RCRA definition. Segregation of laboratory chemicals before including them in lab pack shipments can avoid including nonhazardous chemicals in a hazardous waste shipment. In a large research institution with many different chemicals in use, this segregation into chemical classifications could become very complex. For source reduction at Agracetus it was sufficient to segregate waste research chemicals into three categories: inorganic chemicals (salts and nutrients), inorganic chemicals-toxic metals, and organic chemicals. Segregating inorganic and organic chemicals became the most important waste reduction technique for the labs. Many inorganic chemicals in biotechnology labs are used in the preparation of media solutions for plant growth experiments. These chemicals are often nonhazardous and, because wastes are generally in small quantities, they can either be landfilled or dissolved in water and disposed of in the sanitary sewer system. These disposal options are also valid for proteins and other nontoxic organic chemicals used in the research labs. To evaluate which chemicals could be considered nonhazardous, several information sources were used. The primary source of disposal information was the Agracetus material safety data sheet (MSDS) file. Agracetus maintains MSDS on all chemicals used in the facility. These MSDS often can provide the necessary information to make a disposal decision. If this source was insufficient, the disposal information in the Sigma and Aldrich catalogs and Prudent Practices for the Handling, Storage, Disposal of Hazardous Chemicals in Laboratories were used as reference sources. If all these information sources could not provide satisfactory disposal information, the manufacturer's phone number was obtained through purchasing records and a manufacturer's technical representative was consulted for disposal recommendations.

Waste segregation was also utilized in the collection of used oil. Equipment and machinery at Agracetus that uses different engine oils has to be routinely changed as part of a preventive maintenance program. Most of these oils are free from chemical contamination and therefore can be recycled as part of a used oil program. However, some of the oils can become chemically contaminated. This includes ultracentrifuges that may become contaminated with ethidium bromide, vacuum pumps contaminated with radioisotopes. Segregating these contaminated oils from other waste oils avoids potentially treating all of the used oil as hazardous waste.

RECYCLING WASTE MINIMIZATION OPPORTUNITIES

The waste minimization audit also identified two important recycling opportunities. One recommendation was to recycle unwanted or unused chemicals between the labs. Because research projects often change focus and scope, there are frequently chemicals that do not get used. When one lab no longer has need for a chemical, another lab may find a use for it. Since it is research chemicals that create large and expensive lab pack shipments, a chemical recycling program can be one of the most effective strategies in minimizing hazardous waste production in a laboratory facility. In one case at Agracetus, one of the research labs was moved to Columbia, MD, to consolidate research efforts. This reseach lab left many chemicals behind. By inventorying these chemicals and distributing the list to research staff, one third of the chemicals were taken by other labs for use in their research programs. This simple recycling initiative meant a savings in disposal costs of several thousand dollars. In addition, recycling these chemicals meant they did not have to be repurchased. Avoided purchasing costs may equal or exceed the savings from avoided waste disposal costs.

Recycling recommendations also helped reduce the volume of pesticide rinsewater produced in the greenhouse. Before a waste minimization program was implemented, pesticide rinsate from sprayers was collected in 55-gal drums for disposal. This produced several hundred gallons of rinsate and unused concentrate solution annually that had to be disposed of through a waste contractor. To minimize pesticide wastewater production, it took only a little extra time to spray unused concentrate back on the plants that were originally sprayed. This simple management practice significantly reduced the pesticide concentration in the rinsate and resulted in obtaining regulatory approval to dispose of the less concentrated solution in the sanitary sewer system, as long as the volume did not exceed 10 gal/week.

MANAGEMENT CONTROLS

An effective waste minimization program is not limited to just the source reduction and recycling of chemical wastestreams. Developing management practices that reduce the potential for purchasing excess quantities of chemicals or chemicals with difficult or expensive disposal options is just as valuable. Through the recommendations of the technical assistance program and internal evaluation, several management practices were developed that have been instrumental in minimizing the volume of hazardous waste generated at Agracetus.

The most effective management practice actually predates the development of a formal waste minimization program. When new research laboratories were built in 1984, a decision was made to centralize the purchasing, preparation, and storage of most chemicals. Instead of having individual laboratories order and retain their own supply of general-use chemicals, commonly used chemicals were transferred to a central media lab. This management practice minimized duplication in the ordering of commonly used chemicals because the media labs, not individual research labs, were responsible for monitoring and replenishing chemical supplies.

The centralization of flammable liquids and acids also helped reduce waste generation. A flammable liquid storage room built in 1984 contains eight chemical storage cabinets, seven for flammable liquids and one for acids. All commonly used flammable liquids, organic chemicals, scintillation cocktail fluids, acids, and bases are stored in these cabinets. Media lab staff are also responsible for monitoring and replenishing chemicals stored in this room. Although there are still many research chemicals that individual labs must order, centralization of commonly used solid and liquid chemicals in the media lab and the flammable liquid storage room has had a major impact on minimizing the tendency to purchase excess or duplicative quantities of commonly used chemicals.

For those chemicals that are not part of centralized chemical storage and purchasing, management control can be exercised by knowing what chemicals are entering the facility. This goal is achieved at Agracetus by maintaining a database of all commercially purchased chemicals. With this database information, new chemicals can be evaluated for potential disposal problems and management practices implemented that will limit the chemical's waste potential. For example, ethyl ether disposal can become expensive if partially used containers accumulate. Through oxidation, potentially explosive peroxides can form in these containers. If this occurs, it requires a highly trained and expensive chemical treatment team to stabilize the ether before it can be safely handled. After several cans of ether-containing peroxides were identified, a policy was established to minimize the recurrence of this disposal problem. All incoming purchase orders of ether and other peroxide-forming flammable liquids then had a policy attached requiring researchers to date and dispose of all ether containers after 6 months. The more accurate and complete the chemical inventory database is, the more effective a chemical database will be to identify chemicals with difficult and expensive disposal options.

WASTE MINIMIZATION PROGRAM BENEFITS

Developing a waste minimization program should provide benefits that compensate for the time and expense in implementing the program. There were several major benefits that justified developing a waste minimization program at Agracetus. The most tangible benefit was a decrease in hazardous waste production and lower disposal costs. Before implementing a waste minimization program, hazardous waste production exceeded 5000 kg/year at Agracetus. After implementation, hazardous waste production was reduced to less than 500 kg/year. Therefore, Agracetus successfully attained two of its waste minimization program goals: reduction in waste disposal costs and changing generator status from small quantity to very small quantity.

Another potential benefit of the waste minimization program was to decrease the amount of time spent on hazardous waste management by the environmental health and safety department. However, as implementation developed, this became a very difficult concept to measure. Setting up a waste minimization program is very time consuming. This was offset by a number of cost savings: reduced purchasing time through recycling programs; decreased waste management time expenditure by spraying pesticide concentrate back on the crops; spreading the involvement in waste minimization to many people outside the environmental health and safety department and decreased probability of dealing with time consuming problem wastes like peroxides in ether. Conceptually, over a sufficient period of time, the time savings developed in waste minimization should outweigh the time expenditure in its development. However, to accurately measure this tradeoff in time expenditures is very difficult.

A less tangible benefit to the Agracetus waste minimization program, but certainly no less valuable, was the minimization of corporate risk. Since hazardous waste disposal is a cradle to grave responsibility, minimizing generator liability is an important benefit of a waste minimization program: waste that is not generated cannot become a corporate liability. The most potentially damaging component of generator liability is Superfund liability. The more waste generated, the higher the degree of potential liability. Source reduction minimizes the legal and settlement costs of becoming a potentially responsible party (PRP) in a Superfund litigation. Waste minimization became an important strategy in avoiding future Superfund involvement.

An even less tangible benefit from a waste minimization program is positive public relations. Business and industry is increasingly interested in being recognized as a good corporate citizen and taking a proactive approach to environmental management. Waste minimization is one of the best examples of a proactive environmental management strategy. Not only can waste minimization reduce hazardous waste disposal costs, it can also generate good public relations.

The development of a waste minimization program at Agracetus provided significant tangible benefits in decreased disposal costs, decreased hazardous waste production, a change in generator status and a reorganization of hazardous waste time expenditures. Adding the intangible benefits of minimizing corporate risk and positive public relations provided additional justification for the concept that the environmental impact of an institution's hazardous waste production can be minimized while still providing cost savings and other corporate benefits.

ENDNOTES

1. Larew, H., Agricultural Research, August 1981, 13–14.
2. Committee on Hazardous Substances in the Laboratory, *Prudent Practices for the Handling, Storage, Disposal of Hazardous Chemicals in Laboratories,* National Academy of Sciences, Washington, D.C., 1983. [New edition pending.]
3. UW-Madison Department of Engineering Professional Development, Waste Minimization Audit, 1987.

CHAPTER 22

Waste Minimization and Management at 3M R&D/Laboratories

Linda J. Tanner

CONTENTS

Introduction 390
R&D — A $930 Million Effort at 3M 390
Hierachy and Cardinal Rules 391
Compliance Monitoring 391
Waste Stream Profile Program 392
Minimizing R&D/Laboratory Waste 392
    Modifying Experimental Procedure 392
    Minimizing Chemical Purchases 393
    Restocking/Redistributing Unused Chemicals 394
    Recycling "Waste" Materials 394
    Reclaiming Elemental Mercury 395
    Recycling Empty Containers 395
        Plastic and Glass Containers 395
        Container Recycling Preparation 396
        Five-Gallon Metal Pails and 55-Gallon Metal Drums 396
    Cleaning R&D/Laboratory Benches, Hoods, and Equipment 397
    Energy Reductions — "Green Lights Program" 397
Recycling and/or Disposing of R&D/Laboratory Waste 398
    Planning for Waste Disposal 398
    Becoming Familiar with Toxic Properties of Materials 399
    Converting to Less Hazardous Waste 399
    Disposing of R&D/Laboratory Chemicals under TSCA 399
    Battery Disposal Guidelines 400
    Disposal to the Sewer System 400
    Disposal to the Atmosphere 401

Conclusion 401
Endnotes 402

INTRODUCTION

At 3M, waste minimization is part of a corporate-wide commitment to an idea that makes both environmental and financial sense. In 1975, 3M adopted its Corporate Environmental Policy which states that all 3M operations must strive to prevent pollution at the source and to develop products and processes with minimal environmental impact. The focal point of this commitment has become 3M's voluntary Pollution Prevention Pays (3P) program. Over the past 17 years, 3P has prevented nearly one billion pounds of pollutants from entering the environment while saving the company over a half billion dollars. In 1990, 3M pledged to reduce worldwide plant environmental releases 90% and the generation of wastes by 50% by the year 2000. 3M was also among the first to sign the International Chamber of Commerce Business Charter for Sustainable Development.

R&D — A $930 MILLION EFFORT AT 3M

At many companies, R&D/laboratories would be excluded from such efforts. With the increasing pressures of a global economy, R&D efforts are becoming the decisive element in securing a company's position in the marketplace. It is critical for the success of R&D that researchers remain flexible and able to respond creatively to both external and internal customer needs. Those needs include meeting existing and anticipated environmental regulations, improving product performance, as well as assisting with zero defect, cost reduction, total quality, and time compression efforts.

In a sense, it is the "job" of researchers to make waste. To enable them to engage in their important work unencumbered by restrictive regulations, researchers' activities are often excepted or exempted from "Right-to-Know" and pollution prevention laws. Given the type of work performed in R&D/laboratories, it is not usually practical to attempt to quantify waste minimization efforts. An experiment intended to take a day or week may be terminated after an hour due to unsuitable results. Conversely, an experiment with an anticipated short duration may stretch on for weeks as the unexpected develops, and researchers strive to determine the answers. In addition, specific waste minimization research projects are often proprietary in nature and could significantly impact the future creation of a commercially superior product or process.

Although research and development is critical to 3M's success, 3M R&D/laboratories share the company's emphasis on pollution prevention as the key component of a responsible waste management strategy. Experience at 3M has shown that this voluntary approach to waste minimization is effective even without restrictive environmental regulations. This chapter will deal with site-wide R&D/laboratory waste minimization techniques which could be adopted by any R&D or laboratory facility.

HIERARCHY AND CARDINAL RULES

All 3M R&D/laboratories follow this hazardous waste minimization preferred hierarchy:

Waste Minimization Hierarchy

1. Reduce waste at the source (generation prevention).
2. Reuse or recycle.
3. Use appropriate treatment for remaining waste.

3M R&D/laboratories are guided by four "cardinal rules" of waste management which are observed at all times.

3M's Four Cardinal Rules

1. Hazardous waste must be packaged and labeled as such immediately.
2. Never sewer capturable amounts of hazardous materials.
3. Never allow evaporation for the purpose of disposal.
4. Always follow the 3M Compactor/Dumpster Policy.*

COMPLIANCE MONITORING

To insure that every person in every 3M R&D/laboratory is following corporate waste minimization and management guidelines, R&D/laboratory management, in conjunction with Environmental Engineering & Pollution Control (EE&PC), provide awareness and education programs to all staff.

Each R&D/laboratory has a safety and health committee which conducts monthly safety inspections. A specific checklist is followed to guarantee a thorough inspection. If a potential problem is identified, the committee leaves a corrective action note for the responsible party.

EE&PC performs unannounced inspections of R&D/laboratories and formal environmental audits every 2 years. Whenever potential problems are identified, written corrective action notices are delivered to those responsible.

---

* The 3M Compactor/Dumpster Policy presents guidance on what is and is not acceptable for disposal in compactors and dumpsters. Those acceptable items include noncontaminated paper, wood, corrugated, plastic, and metal; clean empty containers (5 gal or less); small pressurized (nonaerosol) containers; small volumes of bagged loose materials and other landfillable materials. Unacceptable items include hazardous or contaminated materials; liquids; items capable of causing injury; containers with hazardous warning labels; gas containers; large volumes of landfillable materials. 3M makes special arrangements for the waste management of unacceptable materials including recycling materials through the Resource Recovery Department.

WASTE STREAM PROFILE PROGRAM

To ensure the accurate identification, preparation, and shipment of waste, the Waste Stream Profile Program was developed by the 3M Waste Management staff. For this program, a "wastestream" is defined as any solid, containerized liquid, or containerized gas waste. Wastestreams include both disposal and recycling processes.

For each R&D/laboratory wastestream, a 3M Waste Stream Profile Form is filled out and sent to the corporate EE&PC office. Instructions for filling out this form are printed on the reverse side of the form and are contained in the *3M Waste Management Program* manual. Each 3M facility Transportation/Waste Coordinator has a copy of the manual.

The Waste Stream Profile Form requires the 3M employee to work with their coordinator to:

1. Name the wastestream and the process that produces the waste.
2. Identify the waste contents with chemical name, chemical formula, or trade name.
3. Use a percentage range to quantify each of the materials in the wastestream (even for items such as rags).
4. Describe the physical state: liquid, solid, or gas.

This profile is evaluated and classified by the EE&PC Waste Management staff, and an identification number is assigned to that unique wastestream. The R&D/laboratory reporting the wastestream is then supplied with a 3M Waste Stream Profile Reference sheet. This computerized sheet includes all of the information needed to label, prepare, and eventually ship the wastestream.

The 3M R&D/laboratory Waste Minimization Guidelines, which are followed in all 3M R&D and laboratory facilities, are presented in the next section of this chapter. The guidelines were developed to help researchers meet 3M environmental, health, and safety standards. The guidelines appear in the *3M Guide to Laboratory Practices*, a project of the 3M Technical Council — Health, Safety and Environment Committee. The manual was developed by the Education/Training Subcommittee made up of members from Industrial Hygiene, Safety, and Environmental staff groups, as well as laboratory personnel.

MINIMIZING R&D/LABORATORY WASTE

MODIFYING EXPERIMENTAL PROCEDURE

The creative, innovative nature of R&D/laboratory work makes it difficult to establish a uniform approach to every waste minimization situation that may occur in a manufacturing setting. In the R&D/laboratory, the challenge is to create within each employee a personal resolve to consider waste reduction continually in all phases of his or her work. At 3M, management support, formal and informal encouragement, policies, and 3P awards all serve to inspire the resolution to integrate waste reduction strategies into all operations.

As part of this process, researchers are beginning to implement the principles of life cycle analysis (LCA) in the design of 3M products, processes, and packaging. Among other ingredients, an LCA involves the evaluation of the environmental and human health concerns of a product from its initial design phase through raw material acquisition, manufacture, transportation, storage, use, and ultimate disposal. Life cycle concerns can impact laboratory practices, as well as the design of products and processes.

At the bench level, one simple approach to minimize waste involves reducing the size of reaction flasks, except of course, in scale-up experiments. At this time, most 3M R&D/laboratories have completely eliminated 22-l and even 1-l reaction flasks during initial research phases. The majority of experiments now occur in 100-ml flasks. 3M researchers also spend much of their time doing water-based experiments and developing pollution-free products.

3M Waste Reduction Strategies for R&D/Laboratories

- Source Reduction — prevent generation
- Substitute nonhazardous materials
- Source segregation — hazardous vs. nonhazardous
- Recycle/reuse
- Standardize operating instructions
- Properly label containers
- Improve loss prevention
- Inventory materials at least once a year
- Set goals, establish baseline
- Eliminate unnecessary chemicals
- Assess packaging
- Date materials — use older first
- Restock unused/partially used chemicals
- Optimize purchasing
- Purchase chemicals shipped in cylinders only from vendors who accept returns of empties
- Purchase more efficient equipment
- Use "just in time" material delivery
- Use silver recovery units
- Reduce leakage, spills, and air emissions
- Convert "waste" into product/resources
- Clean glass with detergents — not sulfuric or chromic acid
- Scale down experiments/procedures (use smaller reaction flasks)
- Complete reactions to eliminate hazardous waste
- Use more sensitive analytical equipment

MINIMIZING CHEMICAL PURCHASES

Smaller volume chemical purchases are favored for a number of reasons. Breakage is substantially less likely in smaller packages. In addition, smaller containers are emptied faster so that decomposition of reactive compounds is less likely to occur. Also, smaller containers are easier to handle, lessening the risk of accidental exposure.

Purchasing chemicals in larger quantities often dictates a need for other equipment, such as a variety of transfer containers, funnels, pumps, and labels. Additional labor to subdivide the larger quantities into smaller containers and additional personal protective equipment may also be needed. These tenets are presented in the American Chemical Society's publication, *Less is Better* (see Endnote 1).

It is important to note that for some chemicals the waste disposal price is considerably higher than the original purchase price. Therefore, the cost of handling and disposal must be considered when figuring the cost of chemicals. Unfortunately, manufacturers, vendors, and distributors do not always make small quantities available.

RESTOCKING/REDISTRIBUTING UNUSED CHEMICALS

3M maintains a Laboratory Supplies Stockroom at 3M Center in St. Paul that accepts usable surplus chemicals from the company's various R&D/laboratory facilities. A list of surplus chemicals must be sent to the Laboratory Supplies Stockroom for preliminary approval before shipping. The Laboratory Supplies Stockroom will accept most unused chemical portions if the containers are neat and clean with labels in good condition. In turn, these containers are restocked and reissued at a reduced cost. The new computer system factors the original cost of the chemical and adds a storage and service charge. The system also tracks the amount of chemical remaining in the containers (1/4, 1/2, 3/4, full).

3M does *not* use or distribute reclaimed laboratory chemicals from or to outside resources because of quality-control and liability concerns.

RECYCLING "WASTE" MATERIALS

3M has programs for recycling many waste materials generated by R&D/laboratories. The R&D/laboratory recycling program is managed through the 3M Environmental Engineering and Pollution Control (EE&PC) and the Resource Recovery Departments. The 3M Center campus of R&D/laboratory facilities employs a full-time recycling coordinator to implement, pilot, and monitor the various programs. The following are some examples of reclamation and recycling programs.

- Drums and pails
- Office paper
- Solvents
- Mercury
- Lead-acid batteries: car, forklift, truck batteries
- Other metals: steel, aluminum, razor blades, etc.
- Glass and plastic containers
- Cardboard

- Precious metals: silver, gold, etc.
- Plastic scrap, film scrap

RECLAIMING ELEMENTAL MERCURY

3M R&D/laboratory policy is to save for reclamation all collected elemental mercury, including spilled mercury that is reasonably clean.

At the 3M Center in St. Paul, the mercury is packaged in thick-walled, high-density polyethylene bottles. The bottles are labeled "Metallic Mercury for Recycling" and sent to the Laboratory Supplies Stockroom. Nationwide, 3M R&D/laboratories package, label, and ship elemental mercury according to the 3M Waste Stream Profile Reference Sheet.

The Laboratory Supplies Stockroom also accepts the following mercury-contaminated or mercury-containing materials from 3M Center:

1. Articles that contain mercury (mercury batteries, thermostats, thermometers — padded and not broken, electric switches, appliances, medical or scientific instruments).
2. Mercury-contaminated materials, such as broken mercury thermometers.
3. Contaminated spill clean-up materials, such as gloves, rags, and sorbents.
4. Mercury-containing compounds.

Each type of material must be packaged and labeled separately prior to being sent to the Laboratory Supplies Stockroom.

Outside 3M Center, R&D/laboratories must establish a Waste Stream Profile and package, label, and ship the mercury-contaminated or containing material according to the Waste Stream Profile Reference Sheet. The waste mercury vendor services can be quite expensive. At this time they are approximately $1800 to $2000 per drum.

To reduce the potential for mercury spills and waste, R&D/laboratories are encouraged to use alternatives to mercury-containing items, such as nonmercury thermometers.

RECYCLING EMPTY CONTAINERS

All 3M Center buildings have recycling programs for empty 5-gal metal pails, 55-gal drums, some plastic containers, as well as clear, brown, and green glass containers. At this time, it appears that 3M is one of the few companies that recycles (clean) chemical reagent bottles.

Plastic and Glass Containers

At the Center, plastic containers that are recycled include #1 PET "necked" and #2 HDPE "necked" containers such as beverage, detergent, and chemical bottles. Chemical reagent (solvent) bottles and jars and food-type bottles and jars are examples of types of glass being recycled.

3M Center Recycling

	1991	1992	1993
Glass	26,000 lb	49,600 lb	47,600 lb
5-gal pails	1,500 pails	5,400 pails	5,000 pails
55-gal drums	2,300 drums	1,870 drums	2,110 drums
Paper	2,750,000 lb	2,840,000 lb	2,785,730 lb
Corrugated	790,000 lb	970,000 lb	1,074,000 lb
Plastics (HDPE and PET)	1,400 lb	4,100 lb	5,000 lb
Phone books	15,400 lb	17,500 lb	17,500 lb
Aluminum (donated to United Way)	1,050,000 cans	864,000 cans	788,570 cans
Solvents	54,000 gal	44,300 gal	52,700 gal
Precious metals	25,000 troy oz	28,900 troy oz	51,300 troy oz
Cafeteria fats and oils	1,800 gal	2,000 gal	2,400 gal
Waste oils (tank cleaning materials)	30,200 gal	5,700 gal	1,600 gal
Compactor waste	3,500 tons	3,600 tons	3,400 tons

(Additional Programs under development for Construction and Confidential Wastes.)

Container Recycling Preparation

Glass and plastic containers are required to be empty and clean of residue before recycling, according to the following guidelines:

1. Capturable amounts of hazardous chemicals are *never* sewered. Capturable quantities of hazardous materials are put directly into an appropriate hazardous waste container.
2. Empty containers that contain only water-soluble residues are *triple* rinsed by the researchers with water.
3. If the reagent is not water soluble, an appropriate water-soluble solvent, such as acetone or ethanol, is used prior to the water rinse. The organic solvent rinses are then placed in the proper hazardous waste container. All subsequent water rinses may go to the sanitary sewer.
4. Chemical containers are *not* placed in laboratory hoods for the purpose of evaporating any remaining chemical residues.
5. Labels need not be removed, but must be obliterated (Department of Transportation). Polyvinyl coating (PVC) labels, if present, must be removed by the researchers with a razor knife.
6. Cleaned containers are collected by the custodial staff once each week. The triple-rinsed containers are then sold to a recycler where they are melted and reused.

Five-Gallon Metal Pails and 55-Gallon Metal Drums

At 3M Center, when ordering metal pails and drums, priority is given to vendors who provide liners and/or take back their clean empty containers. Pails and drums can be reused to containerize waste bound for the corporate incinerator, if they meet requirements set by the Department of Transportation (DOT) and the Environmental Protection Agency (EPA).

Containers that are *not* to be reused or recycled are pails and drums that have contained materials classified as

Poison	Flammable solid	"P"-listed waste (acute hazardous waste)
Radioactive	Etiologic/infectious	Organic peroxide

These containers are incinerated at the corporate incinerator.

"Clean" pails, those free of residue, are shipped to a shipping dock area for collection. These "clean" pails go to a non-3M recycler, which refurbishes the pails or on to a company that purchases them for scrap metal.

"Contaminated" pails and drums are shipped out for cleaning and processing by a non-3M recycler. These are prepared for shipping as follows:

1. Pails must be empty, meaning *no pourable liquid* and *less than 1 in.* of residue. The original container labels and markings should be maintained.
2. "Empty Drum Certification" labels are placed on the pail — *not* on the cover.
3. Prepared pails are brought to a shipping dock for collection.
4. Pails with residues should be sealed, with tightened covers and bungs, and shipped as though full using the Waste Stream Profile Reference Sheet.

CLEANING R&D/LABORATORY BENCHES, HOODS, AND EQUIPMENT

When cleaning up after an experiment, it has been routine practice to use solvents, such as acetone, toluene, or methyl ethyl ketone (2-butanone). In an effort to minimize the unnecessary use of certain solvents in the laboratory, 3M has undertaken pilot projects in two R&D/laboratory facilities. EE&PC provided the researchers with several cleaning alternatives along with basic recommendations on their use.

Alternative Laboratory Cleaners

Alconox (lab soap) and water
3M Brand Citrus-Based Natural Cleaner — degreases, cleans
Simple Green — cleans, degreases
Zep Big Orange — degreases, removes adhesives
Zep E.S.P. — degreases, cleans
Zep X-489 — degreases, works with crosslinked and noncrosslinked materials
Zep 55 — may replace 1,1,1-trichloroethane
Zep X-3388 — hot/cold degreaser
For product information call: 3M: 1-800-362-3456
                               Zep: 1-800-366-3395

ENERGY REDUCTIONS — "GREEN LIGHTS PROGRAM"

When considering ways to reduce waste in the laboratory, the conservation of energy is often overlooked. Yet outmoded, wasteful lighting fixtures contribute a great deal to the cost of operating a large R&D/laboratory complex. At 3M, all facilities are working to reduce energy consumption as part of a global corporate program.

In 1990, the EPA and industrial partners, including 3M, instituted the "Green Lights" program in an effort to reduce energy consumption in large commercial buildings and R&D/laboratories. By switching to energy-efficient lighting fixtures, less electricity would be needed from power plants. With less combustion for power generation, the levels of carbon dioxide, sulfur dioxide, and nitrous oxide would be reduced. Thus would occur the double benefits of lower energy costs to industry and lower emissions to the environment. The 3M Center R&D/laboratory facilities were part of the company's involvement with the program.

To date, the number of light fixtures replaced at 3M Center has totaled over 150,000. The total cost of the program was approximately $4 million. The local electrical utility contributed a rebate of roughly $450,000 for upgrading to more efficient fixtures. The annual savings of over $700,000 per year will continue long after the payback period of less than 5 years.

RECYCLING AND/OR DISPOSING OF R&D/LABORATORY WASTE

Even under the most comprehensive waste minimization programs, R&D/laboratories will continue to generate wastes. The remainder of this chapter presents guidelines for the management of wastes in 3M R&D/laboratory facilities.

PLANNING FOR WASTE DISPOSAL

The 3M Waste Management Program encourages employees to make the disposal of products, by-products, and scrap an integral part of experiment planning and the hazard review process. Hazard reviews are required under the Occupational Safety and Health Administration (OSHA) Laboratory Standards. These reviews afford researchers the opportunity to identify and characterize potential wastes before they are generated. With this information, researchers can plan ways to eliminate or manage the wastes. At 3M, disposal issues may be discussed with any of several resources, including supervisors, Transportation/Waste Coordinators, EE&PC environmental facility contacts, or Transportation/Regulated Materials contacts.

3M Hazard Reviews

Initiate Hazard Reviews whenever
- starting a project or new project phase:
  a new project, process, or project phase
  the restart of an idle project or process
  the scale-up of a project or process
- there is a significant change to a project:
  a change in raw materials (where a new hazard is introduced into the workplace)
  an equipment or instrumentation modification (and a new hazard is introduced into the workplace)
  a change in personnel

- significant new hazards are recognized:
    a change in toxicological data (and a new hazard is recognized)
    an existing hazard is discovered or expected
- there is an incident
    any unexpected event involving real or potential damage to people, property, or the environment

BECOMING FAMILIAR WITH TOXIC PROPERTIES OF MATERIALS

In compliance with OSHA standards covering occupational exposures to hazardous chemicals in laboratories, 3M has developed and implemented a Chemical Hygiene Plan. The plan sets forth procedures, control measures, protective equipment, and work practices to help researchers protect themselves from the health and safety hazards presented by hazardous chemicals used in the R&D/laboratories.

3M R&D/laboratory personnel must become familiar with the toxic properties of the materials used at their facilities, including the toxic properties of the possible reaction products. This information, which is needed to fill out the 3M Waste Stream Profile, is often included in the MSDS. When handling potentially hazardous waste, researchers are also expected to:

1. Account for trace elements or impurities in possible reaction products, referring to the MSDS and the manufacturer's technical data sheets.
2. Treat with respect reaction products whose toxic properties are unknown. These uncertain hazards are to be noted on the 3M Waste Stream Profile.
3. Never mix incompatible wastes. There are many resources for information on potentially incompatible materials such as *Safe Storage of Laboratory Chemicals and Hazardous Laboratory Chemicals Disposal Guide* (see Endnotes 2 and 15).

CONVERTING TO LESS HAZARDOUS WASTE

When it can be done in a practical and safe way, researchers are urged to convert *small quantities* of especially hazardous materials to less-hazardous materials as part of the design of the experiment or reaction. This should be done before the materials become waste. For instance, strong carcinogens can be oxidized in solution. Another example would be to convert highly reactive substances to less-reactive substances. Reactions can be moderated by dilution, cooling, or the slow addition of the proper neutralizing agents. If the heat of the reaction is a potential problem, it can be diluted before neutralizing. A good source for information on this activity is *Destruction of Hazardous Chemicals in the Laboratory* (see Endnote 10).

DISPOSING OF R&D/LABORATORY CHEMICALS UNDER TSCA

The Toxic Substances Control Act (TSCA) has some requirements governing the use and disposal of new experimental materials produced solely for R&D

purposes. With regards to the disposition of laboratory chemicals, the *3M Guide to Laboratory Practices* directs employees to maintain compliance with TSCA regulations and to dispose of all R&D/laboratory chemicals in accordance with individual Waste Stream Profiles and Reference Sheets. Researchers are advised to consult with their Transportation/Waste Coordinator.

BATTERY DISPOSAL GUIDELINES

3M has published guidelines for the recycling and disposal of batteries at all U.S. operations, including R&D/laboratory facilities. The guidelines contain specifics on packaging, labeling, and shipping requirements. All batteries, except alkaline and carbon zinc, must have a Waste Stream Profile. Batteries containing mercury, nickel-cadmium, and lithium should be collected separately from other waste and managed as hazardous waste.

- Alkaline and carbon zinc batteries may be disposed of in compactors or dumpsters.
- Lithium batteries must be separated from other batteries whenever possible. They should be packaged in plastic bags, separate from other batteries to avoid charging. Mercury, silver, and nickel-cadmium batteries can be packaged together in the same pail as long as the researcher and/or shipping and receiving personnel separate them by type and seal them in plastic bags. Batteries should be placed in a heavy plastic liner placed in a 5-gal 17H pail. The sealed container should be labeled according to the Waste Stream Profile and sent to the corporate waste handling center. There the batteries are further sorted for recycling or hazardous waste incineration or landfilling, as is appropriate.
- All lead-acid batteries are recycled through the 3M Resource Recovery Department.

DISPOSAL TO THE SEWER SYSTEM

Aside from recycling, incineration at the corporate hazardous waste incinerator is the preferred chemical waste disposal method at 3M. Disposal of nonhazardous materials to the sewer is permitted in circumstances where it can be determined that waste materials will not mix with other materials to create hazardous conditions and where such disposal activities comply with local ordinances.

Employees are expected to consult with a supervisor or a Transportation/Waste coordinator concerning applicable local and state regulations, as well as 3M policy, regarding sewer disposal. In addition to EE&PC environmental facility contacts, Transportation/Waste Coordinators are available to answer questions about these regulations and policies.

In general, the following guidelines must be observed:

A. It is *not acceptable* to sewer capturable amounts of *hazardous materials,* including materials defined by the Resource Conservation and Recovery Act (RCRA) as "hazardous waste," by the Department of Transportation (DOT) as "regulated materials," or by Occupational Safety and Health (OSHA) as a "health hazard." These materials are recycled or disposed of as corporate incinerator waste. The

only exception to this is the water rinsate from "empty" containers that held residual water-soluble hazardous materials.
B. Chemicals that are foul-smelling or irritating to the eyes also must *not be sewered.*
C. *Nonhazardous* materials that are water-soluble, nonhalogenated, biodegradable substances in quantities of less than 1 gal (final solution) are acceptable for sewering, as long as:
  1. The materials are within a specific pH range. This range varies slightly for each 3M facility, depending upon their site wastewater permits. It is therefore advisable to consult with the facility Transportation/Waste Coordinator.
  2. Any particles in a water suspension or water mixture are dilute enough to be easily pourable.
  3. Any appropriate photochemical processing chemicals containing silver have been processed through a silver recovery unit.
  4. Recent new technologies indicate that 10% buffered or unbuffered formalin can be sewered if it is treated with a neutralizing solution, such as ALDE-X Aldehyde Management System or VYTAC 10 F Neutralizing Solution (which may need to be treated with water or a dilute NaOH solution to meet sewering pH requirements).

DISPOSAL TO THE ATMOSPHERE

Evaporating liquids or discharging hazardous vapors, gases, fumes, and dusts to the atmosphere are *not* considered appropriate disposal methods at 3M. Hazardous substances discharged into the atmosphere can damage the environment as well as the R&D/laboratory ventilation system. They can be recirculated into the building ventilation system or endanger workers who may be on the roof.

3M policy prohibits the use of R&D/laboratory hoods as disposal systems. These devices are designed as backup safety equipment to contain vapors, gases, fumes, or dust when an experiment fails. They are not to be used to evaporate materials from open chemical containers. Other R&D/laboratory guidelines include

1. Keep all chemical containers closed in the R&D/laboratory hood except during reactions or actual transfer. (Hoods should not be used for chemical storage!)
2. Keep hazardous waste containers closed unless actively adding or removing waste.
3. *Do not* leave unattended open "waste" containers with uncovered funnels inserted.
4. Evaporate water-only residues out of triple-rinsed chemical containers. *Do not* evaporate chemical residues.

CONCLUSION

The R&D/laboratory waste minimization guidelines presented in this chapter undergo continuous reviews and updates to meet the changing needs of the company. They are a part of the total quality environmental management process at 3M. Every operating unit views the generation of waste as an unproductive activity that should be avoided. Pursuing the philosophy of pollution prevention is one avenue taken by 3M as it moves toward sustainable development in the next century.

ENDNOTES

1. American Chemical Society Task Force on RCRA, *Less is Better, Laboratory Chemical Management for Waste Reduction*, American Chemical Society, Washington, D.C., 1985.
2. Armour, M. A., *Hazardous Laboratory Chemicals Disposal Guide*, CRC Press, Boca Raton, FL, 1991.
3. Bergstrom, W. and Howells, M., Hazardous Waste Reduction for Chemistry Instructional Laboratories, Minnesota Technical Assistance Program, Minneapolis.
4. Boortz, M. J., Hazardous waste minimization and management at an R&D facility, *Environmental Progress*, 9(1), 30, 1990.
5. Bringer, R. P. and Benforado, D. M., Pollution prevention as corporate policy: a look at the 3M experience, *Environmental Professional*, 11, 117–126, 1989.
6. Brzybytek, J., Solvent stabilization, *Chromconnection*, September 1991, p. 4.
7. Cohan, D., Wapman, K., Corey Trench, W., and McLearn, M., Beyond Waste Minimization: A Life-Cycle Cost Approach to Managing Chemicals and Materials, presented at the Air & Waste Management Association Annual Meeting, Kansas City, June 1992.
8. Filardi, Raul E., Sr., Waste Minimization in a Research and Development Environment — A Case History, presented at the Air & Waste Management Association Annual Meeting, Kansas City, June, 1992.
9. Fischer, K. E., Developing A Plan For And Instructing Laboratory Personnel In The Proper Disposal of Hazardous Laboratory Waste, Environmental Protection Agency — National Enforcement Investigations Center, Denver.
10. Lunn, G. and Sansone, E. B., *Destruction of Hazardous Chemicals in the Laboratory*, John Wiley & Sons, New York, 1992.
11. 3M, Laboratory Waste, 3M Guide to Laboratory Practices, Section 16, 18, and 19, St. Paul, MN, January 1993 (not commercially available).
12. 3M, 3M Waste Management Program Manual, St. Paul, MN, 1990 (not commercially available).
13. 3M Environmental Engineering & Pollution Control, Product Life-Cycle Assessment; An Approach for the Future, St. Paul, MN, 1991 (not commercially available).
14. Moser, J. H., Reducing Laboratory Wastes, presented at the Central States Water Pollution Control Association, St. Charles, IL, May 16, 1991.
15. Pipitone, D. A., *Safe Storage of Laboratory Chemicals*, 2nd ed., John Wiley & Sons, New York, 1991.
16. Rice, S. C., Incorporating waste minimization into research and process development activities, in *Hazardous Waste Minimization*, Freeman, H., Ed., McGraw-Hill, New York, 1990, p.153.

17. Salisbury, C. L., Wilson, H. O., and Priznar, F. J., Source reduction in the lab, *Environmental Testing Anal.*, March/April 1992, 48.
18. Sands, M. A., Establishing a Waste Minimization Communication Program in a Research and Development Laboratory, Lawrence Livermore National Laboratory, Livermore, CA, March 1992.
19. Ralph Stone and Co, Inc., Waste Audit Study — Research And Educational Institutions, prepared for the California Department of Health Services, Alternative Technology Section, Toxic Substances Control Division, August 1988.
20. U.S. Environmental Protection Agency, Safety, Health and Environmental Management Division, Hazardous Waste Reduction Efforts at EPA Laboratories (draft), Washington, D.C., 1991.
21. University of Illinois at Urbana-Champaign, Division of Environmental Health and Safety, 101 Ways to Reduce Hazardous Waste in the Lab, Urbana, IL, May 20, 1991.
22. Wedin, R. E., Pollution prevention — when less is better, *Today's Chemist Work*, 1992.

CHAPTER 23

Implementation of Waste Minimization Strategies

Bruce D. Backus

CONTENTS

Abstract	406
Scope	406
Introduction	406
The University of Minnesota Waste Minimization Program	407
Waste Minimization Methods	407
Assessment and Evaluation	407
Economic Assessment, Planning, and Administrative Support	408
Purchasing	409
Segregation and Characterization	409
Source Reduction	410
Chemistry Teaching Laboratories	411
Recycling	412
Reclamation	412
Management	413
Incentives	413
Training	413
Low-Level Radioactive (LLR) Wastes	414
Future Plans	414
The Planning, Policy, and Regulatory Framework for Waste Minimization	415
Planning and Policy Framework	415
Some Guiding Principles for Waste Minimization	416
Regulatory Oversight	416
Comments	416
Endnotes	418
Appendix 1	420

ABSTRACT

Strategies for reducing hazardous waste and low-level radioactive (LLR) waste are reviewed. The strategies are based on waste minimization programs implemented by the University of Minnesota statewide system. Strategies and techniques involve assessment and evaluation, cost analysis, segregation and characterization, source reduction (through process modification or product substitution), recycling, reclamation, management of operating practices and inventory, and training. Methods of implementing the waste minimization strategies are reviewed. The Region V United States Environmental Protection Agency (EPA) has formally approved of the University of Minnesota's waste minimization policies in three finalized Part B Hazardous Waste Facility (RCRA) Permits.

SCOPE

This paper looks at the waste minimization of hazardous, or toxic, chemical wastes, low-level radioactive wastes, and mixed wastes (both chemical and low-level radioactive wastes) generated by an educational/research institution. It reviews procedures to reduce solid waste generation, water usage, and energy consumption, but not stormwater run-off and food waste generation.

INTRODUCTION

Waste minimization, waste reduction, and pollution prevention are dominating current environmental discussion. Benefits of waste minimization can include decreased waste treatment costs, decreased future liability costs, improved working conditions/worker safety (employees are exposed to smaller amounts of hazardous materials), improved public relations, and compliance with mandated pollution prevention laws. If your institution is a small research lab, college, or university, there is also the potential benefit of reducing your generator status from a Resource Conservation Recovery Act (RCRA) large quantity generator to a small quantity generator, which would decrease regulatory paperwork, compliance, and cost burdens.

The University of Minnesota waste minimization program has developed gradually since 1980, primarily in response to economic and regulatory compliance concerns. The methods used by the University parallel those suggested in the "Guides to Pollution Prevention: Research and Educational Institutions,"[24] with the exception that the University has traditionally used one-on-one interviews with department administrators and lab managers for gathering information, rather than using questionnaires. As mentioned in the EPA "Guides," there are obstacles in academic settings, not found in industry, that hinder waste minimization. University management structures are often decentralized which diffuses commitment to waste minimization goals. Decentralization makes it difficult to track the movement of chemicals through the system, and educational systems produce small quantities of a great variety of wastes, rather than a few large wastestreams as found in industry.

THE UNIVERSITY OF MINNESOTA WASTE MINIMIZATION PROGRAM

WASTE MINIMIZATION METHODS

The waste minimization methods listed here, while not all inclusive, provide examples of methods that can be applied to many research hazardous wastestreams. The University of Minnesota waste minimization program, approved by the U.S. EPA Region V, contains the following elements:

- Assessment and evaluation of wastestreams (preliminary audit)
- Economic assessment, planning, and administrative support
- Segregation and characterization
- Source reduction methods (product substitution and process modification/downsizing)
- Recycling or waste exchange
- Reclamation
- Distillation, neutralization, and deactivation
- Management (audit and inventory control)
- Training
- Review and readjustment of pollution prevention goals

ASSESSMENT AND EVALUATION

Waste chemical streams at the University of Minnesota were analyzed for volume, regularity of production, and ease of reduction. The larger, easier to reduce wastestreams were tackled first. Examples of the "easier" to reduce wastestreams are the use of biodegradable scintillation solutions instead of xylene- or toluene-based solutions and the reclamation of silver from waste photofixer. The assessment was performed by the Environmental Protection and Radiation Protection Divisions in the Department of Environmental Health and Safety (DEHS), with input from affected generators. Even though these wastestreams are easier to reduce, it still took several years to educate all the generators as to proper management techniques.

After waste reduction procedures were put in place, an evaluation was made of their effectiveness. In evaluating the effectiveness of the new biodegradable scintillation cocktails, it was found that researchers were reluctant to give up xylene- and toluene-based cocktails out of fear that they would lose resolution in their studies. To address this concern, the Radiation Protection Division had a leading University researcher test the effect of using biodegradable cocktails on resolution in scintillation studies. The researcher determined that there was no loss of resolution and the publication of these results helped persuade many researchers to make the switch to biodegradable materials. For those researchers who refused to make the switch, the University started charging a $60/gal surcharge for the pickup of these materials. This surcharge further eliminated the use of solvent-based cocktails.

ECONOMIC ASSESSMENT, PLANNING, AND ADMINISTRATIVE SUPPORT

The University of Minnesota administration is becoming more cognizant of waste reduction issues. In 1990, the Senior Vice President of Finance and Operations at the University of Minnesota formed a multidepartmental Waste Abatement Advisory Committee whose short-term goal (Fall 1990) was to "develop a plan to cap or decrease waste flow" not just for hazardous and LLR waste, but also solid waste (campus and hospital), bioinfectious waste, small animals, large animals, fly ash, and debris. The committee must provide options not only for reducing the waste, but also the cost of waste handling. Another division has been formed to look at energy conservation in the design of new facilities and remodeling of existing facilities. These two groups must work closely together, as some waste reduction techniques result in the net increase in energy usage. For example, energy costs and power plant emissions must be taken into account whenever using a distillation recovery method on solvent wastes.

[The energy conservation division is replacing existing fluorescent lights and ballasts throughout the University (at a cost of 4.8 million dollars) with new energy-efficient models. The pay-back period on this project is approximately 3 years. Although this replacement will have energy conservation benefits, the PCB ballasts and fluorescent bulbs from these old lights must be handled as hazardous waste. Estimated disposal cost is $900,000 over a 2-year period. The PCB ballasts cannot be recycled, but mercury, aluminum, and potentially glass from the fluorescent light bulbs are reclaimed through a local recycling company.]

Independent of the above committees, the Department of Environmental Health and Safety (DEHS) at the University of Minnesota sought solutions to growth of hazardous and LLR wastestreams for the past decade. Success in reducing waste growth in the past was largely due to direct, cooperative work with generators, rather than by administrative mandate.

A key to receiving administrative support for waste minimization activities in 1989 and 1990 was DEHS's cost analysis for various chemical and LLR wastestreams. This cost analysis is provided as Appendix 1. Due to this analysis, DEHS is able to show the administration the cost per kilogram reduction if waste minimization techniques are applied. These cost savings can be used in traditional financial analyses to determine rates of return and net present value of investments for waste minimization equipment or alternative nontoxic substitutes.

Overall planning for waste minimization is now performed jointly by the Waste Abatement Advisory Committee and DEHS. DEHS targets two specific wastestreams a year for an in-depth study for pollution prevention opportunities and reports on the studies to the Waste Abatement Advisory Committee. Already studied by DEHS was the removal of the PCB capacitors from PCB fluorescent light ballasts, in hopes that the remaining ballast could be handled as non-PCB waste. Unfortunately, the study showed that the remaining tar material in the ballast was contaminated to a level that required the entire ballast to be managed as PCB waste. A second ongoing study is the reduction of wastes from histology laboratories. Histology laboratories generate one half (approximately 1,500 gal) of the fuel blend waste solvents generated annually by the University of Minnesota. A combination of product substitution, equipment modification, and distillation recovery is being used to decrease the amount generated.

PURCHASING

A centralized purchasing program for chemicals is strongly recommended by the U.S. EPA and other agencies for control and tracking of chemical emissions, waste minimization, and management of hazardous waste. The University of Minnesota in 1990, however, decentralized the purchasing program to shift the administrative cost burden to individual departments. The new system was not implemented for waste minimization reasons. The purchasing system does shift inventory control to the individual departments and the tracking of chemicals should be easier for the smaller, departmental units. In practice though, the accounting duties were added to departmental staff with full work loads, and a successful tracking program has not been implemented as of April 1994, due to time and budget constraints.

Obtaining data on the amount and type of chemicals purchased by the University will be easier in the future due to new prime vendor contracts being established with four chemical supply companies. The new contracts require vendors to supply the University with an annual report on chemicals purchased, quantities, and University locations receiving those chemicals. Because of the substantial price savings University laboratories will receive under this contract, it is expected that most chemicals, approximately 90% by weight, will be ordered through this system, allowing for better data tracking on purchases.

Other techniques used by the purchasing department (University Stores and Lab Services) at the University of Minnesota were the phasing out of the mercury thermometers in the –20 to 110°C range and selling only non-mercury thermometers (a good economic substitute could not be found for mercury thermometers going up to 300 or 400°C), stocking smaller chemical container sizes, holding waste minimization training sessions in conjunction with DEHS for staff product review committees, publication of product substitutes in purchasing's newsletter, and requiring all vendors who sell gas cylinders, including lecture bottles, through the University prime vendor contract to accept the returned empty, or partially empty, cylinders back from the University.

SEGREGATION AND CHARACTERIZATION

Generators are instructed to segregate all wastestreams and accurately label waste containers as to their exact contents. This allows the University to recycle and redistribute reusable wastes or unused chemicals.

SOURCE REDUCTION

Product substitution (substituting a nonhazardous material for a toxic or hazardous one) is the preferred method of reducing hazardous and LLR wastes. Already mentioned is the use of biodegradable scintillation solutions instead of xylene- or toluene-based solutions. The number of 30-gal, xylene/toluene, scintillation vial drums shipped out for treatment has decreased from over 500 in 1987 to 38 in 1993. Histology laboratories which process human and animal tissue for production of microscope slides have greatly reduced the amount of waste generated by substituting alcohol fixative for formaldehyde fixers (less toxic substitution) and substituting citric acid base solutions for xylene. Detergents, enzymatic cleaners, and "no-chromix" sulfuric acid cleaners have been substituted for potassium (and sodium) dichromate-sulfuric acid cleaning solutions or other chromium containing cleaning solutions in many laboratories and photography darkrooms. Peracetic acid is now used instead of formaldehyde for sterilizing kidney dialysis machines. In 1990, Duplicating Services substituted petroleum solvent-aromatic naphtha solvent-1,1,1 trichloroethane blanket wash for petroleum distillate-perchloroethylene-solvisol bright green gc blanket wash and triethanolamine-aromatic hydrocarbon-mineral spirits-dipropylene glycol monomethyl ether roller and blanket deglazer for methylene chloride deglazer (in both cases, this reduced the toxicity of the materials used by employees and the toxicity of the waste generated). Parts washers in machine shops that are not serviced by recycle vendors are testing citric acid-based and alkaline cleaning solutions as possible substitutes.

Process modification is another method of source reduction used by the University of Minnesota. An example of process modification is the down-sizing of experimental equipment which results in smaller amounts of chemicals used and less waste produced. A classic example is the Kjeldahl apparatus which has been used for 50 years or more to determine the amount of nitrogen in biological samples. Micro- and small-scale apparatus is now in common use, reducing from 10- to 100-fold the amount of corrosive and toxic waste produced. Sources discussing microscale techniques are listed in the Endnotes.

Another example of process modification is the use of instrumentation for chemical analysis, such as gas chromatography, spectroscopy, and nuclear magnetic resonance. Use of automated techniques can substantially reduce the amount of waste generated over traditional wet chemistry techniques. A single lab on the St. Paul campus ran over 16,000 inductively coupled plasma (ICP) emission spectroscopy samples in 1989. Each run provided the sample concentration of 15 to 27 metallic ions. In previous years this lab has run as many as 32,000 ICP samples per year. Traditional wet chemistry analysis would have yielded thousands of liters of hazardous waste, rather than 32 l of corrosive waste generated in 1991. These 32 l are easily neutralized and sewered at the ICP lab (after testing to make sure metal concentrations are within discharge limits). While sophisticated equipment such as the ICP emission spectrometer is fairly common on larger campuses, the cost of such equipment is prohibitive to smaller, out-of-state

University of Minnesota campuses and experiment stations. These smaller campuses and experiment stations must use traditional wet chemistry techniques or somehow arrange to have samples analyzed on one of the larger campuses. With the cost estimates for waste disposal generated by DEHS, some of these smaller campuses may be able to justify the initial capital expense of automated equipment by the savings to the University in decreased waste disposal costs.

Purchase of newer generations of automated equipment can also result in waste reduction over older, less-efficient automated equipment. This has been found to be the case in automated histological tissue processors; the processors purchased 12 years ago produce approximately twice as much waste as processors marketed today.

Modification to processes can also indirectly reduce pollutant emissions. Using the histology labs as an example once again, special exhaust enclosures were designed to reduce energy consumption and employee exposure to chemical vapors. Histology staining dishes require a large amount of bench space, and as a result, employees will remove staining dishes from cramped fume hoods. Down draft tables are typically installed to protect the employees from chemical fumes, but these tables require an air flow of approximately 2000 cubic feet per minute (cfm) for proper operation. The air from the down draft tables cannot be recirculated and must be exhausted directly to the roof vent. Make-up air must be tempered and supplied to the histology labs. The University of Minnesota designed modular exhaust enclosure boxes to protect employees from exposure to the histological chemicals and to reduce air flow (and energy) requirements. These exhaust enclosure boxes reduce air flow requirements by 10 to 40 times over equivalent space used by down draft tables.

CHEMISTRY TEACHING LABORATORIES

An exciting approach to product substitution and source reduction is underway in the Department of Chemistry at the University of Minnesota. Professor Kent Mann, Director of Undergraduate Teaching Laboratories, rewrote the undergraduate teaching laboratory experiments so that they either will not generate any hazardous waste or the experiment will include procedures to deactivate any hazardous residuals. (In Minnesota, this may be done under the RCRA exemption for waste treatment in containers.) For the experiments that he redesigned, the teaching labs produce approximately 6000 l less of heavy metal and oxidizer waste per year.[17] This saves the University approximately $35,000 per year in waste handling, management, and treatment costs. In addition, the labs are a valuable tool for teaching students the chemical principles behind the deactivation procedures and the concept that chemicals must be responsibly managed from "cradle to grave."

Dr. Mann made two interesting comments on the behavior of undergraduate students in chemistry labs: first, students tend to take three to five times the amount of chemicals needed for an experiment if they are allowed to dispense their own solutions from stock bottles and, second, many students are now questioning the morality of performing experiments that generate toxic wastes which may harm the environment, much like students who have argued in the past about the morality of performing experiments on animals. These new teaching lab experiments help to address these two concerns.

Steps taken in the past to reduce the amount of chemicals used in teaching labs have included predispensing stock chemicals or the use of very small diameter nozzles on stock bottles. The small-diameter nozzles dispense very slowly, which discourages students from taking excess chemicals. With the rewritten experiments, students are further penalized for taking excess reagents in that they must deactivate the reagents to render them nonhazardous, lengthening the amount of time they must spend in the laboratory. Having to deactivate the reagents also helps to teach the students that they are responsible for all chemicals in their control.

A resource package for incorporating pollution prevention concepts in higher education curricula was developed under a grant by the Washington State Department of Ecology.[25] The package contains material put together for a 2-day seminar, the purpose of which was to show faculty how to incorporate pollution prevention into curricula for all academic disciplines, not just the hard sciences. The material in the package was developed by the Waste Reduction Institute (WRITAR), Minneapolis, which used University of Minnesota faculty to preview the package and comment on its content.

This resource package, as well as articles and books that may aid research laboratories in the treatment of hazardous waste are listed in the Endnotes.

RECYCLING

Waste chemicals that can be reused are redistributed free of charge to generators throughout the statewide system using the University of Minnesota's hazardous materials transportation vehicle. Almost all of the recycled chemicals are unopened or unused portions of reagent-grade chemicals. The program recycled 6570 kg of chemicals in fiscal year 1990–1991, saving the University an estimated $155,000 in avoided disposal and purchase costs. The redistributed chemicals are transported (using a hazardous materials bill-of-lading) on the University's hazardous waste truck, making the delivery when the truck is sent to pick up hazardous waste from the laboratory. Laboratory glassware and supplies are also redistributed, which saved the University an estimated $14,000 in purchase costs. Parts washer solvents, carburetor cleaner solvents, and lead acid batteries from garages and machine shops are all recycled and serviced through outside vendors. Unused portions of pesticides are redistributed through the statewide agricultural extension service. Fuel-grade solvents are used for energy recovery.

RECLAMATION

Precious metals are collected and reclaimed. Mercury from broken thermometers and other laboratory equipment is collected and sent to a vendor for redistillation. In return, the vendor either gives the University a price break on the purchase of triple-distilled mercury, a direct cash refund, or when mercury prices are low does not charge the University for the mercury. Precious metals used in research are returned to the supplier. One researcher in the Department of Chemistry returns all his osmium and other reactive metals to the supplier at no cost. Silver is recovered from spent photofixer, negatives, and black & white prints.

MANAGEMENT

Generators are trained to inventory supplies and rotate older chemicals out first. They are instructed to order only the quantity of chemicals necessary for an experiment and not to buy large volumes at price discounts, because the cost of disposing of the excess material is greater than any cost savings they may receive. Generators are told to train their employees yearly in waste minimization techniques.

INCENTIVES

To date, the only incentive DEHS can directly offer generators for waste reduction is a letter of commendation to the generator's Department Head, which is copied to the generator, Dean of the College and the Waste Abatement Committee. The letter strongly encourages college and departmental support of the waste minimization activities.

Indirect incentives include nominating individuals and departments for state and federal awards. The Department of Chemistry and the Department of Environmental Health and Safety received the Minnesota Governor's Award for Excellence in Pollution Prevention in 1993 for their efforts in reducing pollution from University laboratories, after being nominated by DEHS.

TRAINING

Generators are trained in the above waste minimization techniques through a number of mechanisms. The primary training method is the *University of Minnesota Hazardous Chemical Waste Management Guidebook, 4th Edition,* which all generators must read before generating hazardous wastes. The Guidebook contains a chapter on waste minimization, the University recycling program, and chemical spill response. The certification clause of the internal manifests, used by the University for shipment of waste chemicals, states that the generator has read and understands the Guidebook and is following its guidelines.

Other training methods include monthly training sessions for laboratory safety officers and managers and the direct training of departments or subdepartmental groups. Monthly, DEHS trains approximately 50 laboratory safety officers in hazardous waste management and waste minimization. Annually, DEHS trains the faculty, staff, graduate students, and employed undergraduate students in the Department of Chemical Engineering and Material Science (over 180 people) in safety, regulatory, and waste minimization concepts. Graduate students and faculty were encouraged to use product substitution, process modification, or use neutralization or deactivation steps in their research projects teaching labs to eliminate hazardous wastes generated by their projects. Water-chilling units are recommended to reduce water consumption in distillation setups and to prevent accidental water damage.

Training for other departments is given upon request and is not regularly scheduled due to lack of personnel and financing. Individual requests for information are usually handled over the phone or by mail. Waste minimization training is also given through the DEHS newsletter which is sent to all generators, and through the purchasing department (University Stores) Open House and Product Show. The Open House is put on by the general supply storehouse and usually attracts several thousand University employees. DEHS has a display booth at all the Open Houses. Attendees are given information on waste minimization, the chemical recycling program, glassware recycling, battery collection, and the *Hazardous Chemical Waste Management Guidebook*.

LOW-LEVEL RADIOACTIVE (LLR) WASTES

LLR waste generation was reduced through a number of management practices. Storing radioactive animal wastes which contain short half-life (<90 days) radionuclides in freezers and holding them until they are no longer regulated as LLR waste has reduced the volume of radioactive animal waste shipped off-site by greater than 95%. Segregation and storage of short half-life solid waste has essentially eliminated the need for off-site shipment of nuclear medical wastes from the University Hospital and Veterinary Hospital. As mentioned earlier, the conversion to biodegradable scintillation solutions and the mechanical crushing of scintillation vials has resulted in a greater than 95% reduction in the volume of radioactive scintillation vial waste shipped to off-site facilities. Prior to recent restrictions on the disposal of LLR waste, the supercompaction of solid radioactive waste drums has resulted in a greater than 50% reduction in the LLR waste buried at NRC-licensed disposal facilities. Four 850-gal storage tanks for holding aqueous radioactive wastes are in use to further decay short half-life radioisotopes, prior to disposal in the sanitary sewer. The four tanks and sewer disposal saves the University $15,000/year in disposal costs and will save the University as much as $105,000/year if the University is cut off from its current disposal site.

FUTURE PLANS

The University of Minnesota has recently received operating permits for and built an $8 million Integrated Waste Management Facility, which opened in February 1995. This facility will allow the University to consolidate and enlarge its chemical recycling program (which is currently housed in four separate buildings) and will allow DEHS to neutralize or deactivate hazardous waste and use recovery techniques such as distillation. Previously, DEHS did not have the facilities or the necessary permits to treat hazardous wastes.

The new facility will also allow the 2900 bottles a month that are received to be washed for glass recycling (previously, they were disposed of as solid waste). Solid waste segregation and recycling will also be fully implemented at the facility. Water chillers will be used on large distillation condensers to conserve water.

Training packages on waste minimization will be developed in the next 2 years.

THE PLANNING, POLICY, AND REGULATORY FRAMEWORK FOR WASTE MINIMIZATION

PLANNING AND POLICY FRAMEWORK

To implement a pollution prevention or waste minimization program, the following steps should be used as a guide:

- Identify wastestreams (audit)
- Obtain administrative/management support — draft policy statements
- Set goals and objectives
- Assign responsibility — staff and fund
- Establish planning and procurement practices
- Prioritize efforts (based on cost, technical problems, and regulatory requirements)
- Train, spread word
- Implement techniques listed above
- Reward employee initiatives
- Assess — set new reduction targets — repeat process

Auditing is helpful for establishing a baseline to which pollution reduction efforts can be compared. A true environmental audit may uncover regulatory deficiencies that must be dealt with. Plan on time and expense being spent to resolve these deficiencies. Common examples are disposal problems and large amounts of chemicals, some potentially shock sensitive, not properly managed.

Administrative support is necessary to let all employees know that pollution prevention is a priority at an organization. A draft policy statement may include language such as that presented in Chapter 6. (The term "waste abatement" refers to nonhazardous, solid waste reduction.)

Financial administrators sometimes balk at putting funds into pollution prevention programs that avoid future costs, because there does not appear to be a reduction of last year's costs. These administrators must be convinced that cost avoidance is a true savings to the organization.

Goals and objectives may be numeric or policy driven. Instead of setting a goal of reducing waste generation by 10%, it is equally valid to set as a goal: the training of all personnel in waste minimization techniques. Because the University of Minnesota has implemented various waste minimization practices over the past decade, it may be difficult to achieve a 10% reduction in waste generation as compared to an organization that has not previously implemented a broad-scale waste minimization program.

Methods to prioritize where resources should be spent, to reduce pollution, are discussed by the EPA Guides,[24] Hawkey,[9] and Karam et al.[11] The University of Minnesota has typically rated wastestreams that represent the greatest volume and are the most costly to treat for disposal as the highest priority for waste minimization resources.

SOME GUIDING PRINCIPLES FOR WASTE MINIMIZATION

REGULATORY OVERSIGHT

Regulatory oversight prior to 1990 consisted of filing generator waste minimization reports every 2 years and certifying in annual reports that the University of Minnesota does have a waste minimization program. This changed in May 1990 with the passage of the Minnesota Toxic Pollution Prevention Act, aimed at facilities that must report under SARA Title III, Section 313, and those regulated as large quantity generators (which covers the Minneapolis and St. Paul campuses). The legislation (1) establishes a state policy of pollution prevention (source reduction and minimizing transfer of pollutants from one environmental medium to another are emphasized); (2) enlarges technical assistance programs; (3) provides matching grants for waste reduction technology studies; (4) requires facilities to develop Toxic Pollution Prevention Plans which establish goals for reducing or eliminating toxic releases; (5) submits annual reports to the state legislature; and (6) has the overall goal of reducing state hazardous waste generation by 40% by the year 2009. The programs are funded by fees assessed on the generators.

On September 16, 1991, Minnesota Governor Arne Carlson signed the Governor's Order on Pollution Prevention in State Government, Executive Order 91-17. The order requires all Minnesota state agencies, including the University of Minnesota, to establish pollution prevention programs that give priority to preventing pollution at the source, include a pollution prevention policy statement, and require annual progress reports to be submitted to the state legislature. This executive order helped to coalesce the University's waste minimization efforts.

COMMENTS

Promoting employee participation is an important goal and priority. The 3M Company established a Pollution Prevention Pays (3P) program over 15 years ago. Russ Susag, 3M's corporate environmental director, strongly advises rewarding employees for contributions to waste minimization. Framed certificates, citing excellence in pollution prevention, presented at annual meetings by corporate vice presidents and a commendation on an employee's work record is 3M's primary way of rewarding an employee for valuable pollution prevention suggestions. Job performance reviews of their division heads are also partially based on the percent hazardous waste reduction for the division. If a division head does not reduce the emissions or waste from his or her unit, the division head's performance raise is affected.

Even if a wastestream cannot be eliminated, try to reduce the toxicity of the material or the amount of energy used by the generation process.

Avoid transferring the hazardous materials from one media to another. For example, solvent loss through evaporation in fume hoods should be avoided. Chemical losses through fume hoods and in sewer systems must be minimized. When product substitution is not feasible, condenser or filter systems should be installed to prevent emissions, particularly on new building projects. The University facility discussed previously has both condenser and carbon adsorption systems installed to reduce both solvent and corrosive fume emissions.

The EPA's Guide to Pollution Prevention: Research and Educational Institutions[24] contains information on other waste minimization techniques, such as the use of waste exchanges, and it lists agencies located throughout the 50 states that can supply waste minimization information and/or grants for waste minimization studies.

It is important for research laboratories, colleges, and universities to document early on how much waste generation has been reduced due to waste minimization programs. The University of Minnesota, unfortunately, has not documented the waste reduction rates of certain wastestreams, simply due to the complexity of polling all the generators throughout the statewide system. For example, DEHS can only estimate the percent reduction due to the use of microtechniques or automation, because no clear baseline data was established for wastestreams before these techniques were put into effect (baseline data being the quantity of waste generated prior to the implementation of a waste reduction technique, the number of labs that are using the technique, and the actual percent reduction these labs are now realizing). While instrumentation decreases the amount of waste produced per analysis, it also encourages researchers to perform more analyses. It is cases like these where the worksheets in the EPA Guides would help to clarify the true extent of waste reduction achieved.

Richard W. MacLean,[15,16] Manager-Corporate Environmental Protection for General Electric, offered some insights into waste minimization that can be applied to research laboratory institutions.

- MacLean recommends that an organization must go beyond a simple cost/benefit analysis (as performed by DEHS) when looking at implementing a waste reduction process. A risk analysis must also be performed to determine failure rates and risk probability for injury and future environmental liability. He argues that waste disposal costs are artificially low because they do not include future liability costs, societal costs, and safety/health costs.
- He states that environmental departments must use the power within an institution to promote change and therefore environmental departments must discover who or what group has this power and projects must be sold to this power. Environmental departments may have traditionally operated outside the mainstream administrative decision process, and therefore may be less skilled at negotiating for limited resources.

- In implementing waste minimization programs, you are not implementing another program, rather you are changing college/university culture.
- Incentives are necessary.
- Disincentives may or may not have the intended effect. For example, charging individual departments or researchers for the total costs of disposing of their waste may result in "midnight dumping."
- Recognize that the "target" for regulations has shifted from environmental staffs to lab researchers and service department managers.
- There is a hierarchy of waste management options. Compromises may have to be made and toxics elimination is not always the best option.
- Issues are complex and policies take time to implement.
- Mr. MacLean references a General Electric publication that includes a method for quantifying long-term liabilities and other less-tangible costs.

As a final note, while difficult to implement at a large educational or research institution, waste minimization is one of the more rewarding aspects of an environmental program. It is a field where one can put into practice the preventive skills that protect health, safety, and the environment. Where often the environmental professional is harangued for being too restrictive and expensive by clients and too lenient and unresponsive by environmental activists, waste minimization (with cost analysis) is increasingly being applauded by all as the environmental goal to achieve.

ENDNOTES

1. Armour, M. A., Browne, L. M., and Weir, G. L., Hazardous Chemicals Information and Disposal Guide, 2nd ed., Department of Chemistry, University of Alberta, Edmonton, Alberta, Canada, T6G 2G2, 1984.
2. Armour, M. A., Browne, L. M., McKenzie, P. A., and Renecker, D. M., Potentially Carcinogenic Chemicals (Including Some Antineoplastic Agents) Information and Disposal Guide, Department of Chemistry, University of Alberta, Edmonton, Alberta, Canada, T6G 2G2, 1986.
3. Backus, B. D., Photodeveloping Wastes, presented at the Seventh Annual College and University Hazardous Waste Conference, Ames, Iowa, July 30 to August 1, 1989.
4. Backus, B. D. and Thompson, F. M., University of Minnesota Waste Minimization Strategies, presented at the Eighth Annual College and University Hazardous Waste Conference, Coeur d'Alene, Idaho, September 30 to October 3, 1990.
5. Bergstrom, W. and Howell, M., Inorganic Chemistry Laboratory Experiments, St. Paul Technical Institute, produced by Independent School District #625 with the Minnesota Technical Assistance Program, University of Minnesota and the Environmental Protection Agency.
6. Bergstrom, W. and Howell, M., Hazardous Waste Reduction for Chemistry Instructional Laboratories, St. Paul Technical Institute, study conducted through the Minnesota Waste Management Board, Minnesota Technical Assistance Program and the Environmental Protection Agency.
7. Brauer, G., *Handbook of Preparative Inorganic Chemistry,* 2nd ed., Academic Press, New York, 1963.

8. Fosnacht, J., Leadership in waste minimization and pollution control, *Am. Lab.,* 1992, 37–40.
9. Hawkey, S. A., An algorithm for setting priorities and selecting target wastes for minimization, *Am. Ind. Hyg. Assoc. J.,* 53, 154–156, 1992.
10. Hered, G. R., *Basic Laboratory Studies in College Chemistry with Semimicro Qualitative Analysis,* 7th ed., D.C. Heath, Lexington, MA, 1984.
11. Karam, J. G., Cin, C. St., and Tilly, J., Economic evaluation of waste minimization options, *Environmental Progress,* 7(3), 192–197, 1988.
12. Kaufman, J. A., Ed., *Waste Disposal in Academic Institutions,* Lewis Publishers, Chelsea, MI, 1990.
13. Lagowski, J. J. and Hampton, M. D., *Introduction and Semimicro Qualitative Analysis,* 7th ed., Prentice Hall, NJ, 1991.
14. Lunn, G. and Sansone, E. B., *Destruction of Hazardous Chemicals in the Laboratory,* John Wiley & Sons, New York, 1990.
15. MacLean, R. W., Economics of Waste Minimization, presented before the Waste Minimization Conference, sponsored by the New Jersey Department of Environmental Protection, Division of Hazardous Waste Management, New Brunswick, NJ, May 10, 1989.
16. MacLean, R. W., Motivating Industry toward Waste Minimization and Clean Technology, presented at the Waste Minimization and Clean Technology: Moving Toward the 21st Century, ISWA and EPA Conference, Geneva, May 30,1989.
17. Mann, K. R., Pollution Prevention in Chemical Laboratory Instruction, Minnesota Office of Waste Management, St. Paul, 1992.
18. Mills, J. L. and Hampton, M. D., *Microscale Laboratory Manual for General Chemistry,* Random House, New York, 1988.
19. National Research Council, *Prudent Practices for the Handling, Storage, Disposal of Hazardous Chemicals in Laboratories,* National Academy of Sciences, Washington, D.C., 1983. [New edition pending.]
20. Neidig, H. A. and Stratton, W. J., Eds., *Modern Experiments for Introductory Chemistry, Second Edition.* Reprinted from *The Journal of Chemical Education.* Easton, 1989.
21. Russo, T., *Microscale for High School General Chemistry,* Kemtec Education Corporation, West Chester, OH, 1986.
22. Szafran, Z., Pike, R. M., and Singh, M. M., *Microscale Inorganic Experiments,* John Wiley & Sons, New York, 1991.
23. Thompson, S., *Chemtrek: Smallscale Experiments for General Chemistry,* Allyn and Bacon, Boston, 1990.
24. U.S. EPA, Guides to Pollution Prevention: Research and Educational Institutions, Risk Reduction Engineering Laboratory and Center For Environmental Research Information, Office of Research and Development, United States Environmental Protection Agency, Cincinnati, OH, EPA/65/7-90/010, June 1990.
25. Washington State Department of Ecology, Incorporating Pollution Prevention Concepts in Higher Education Curricula. A Resource Package, Publication #91-33, Olympia, WA, 1991.

Appendix 1   Chemical and Low Level Radioactive Waste Costs at the University of Minnesota

Chemical Waste Cost Breakdown (kg)

Wastestream	Disposal technique	Quantity generated 1988–1989	Quantity generated 1989–1990	Quantity generated 1990–1991	Total cost 1990–1991	Cost per kilogram 1990–1991
Fuel solvents	Fuel blending	6,060	8,180	2,270	$11,500	$5.07
Bulk solvents	Incineration	30,400	21,100	31,900	$191,000	$6.01
Reactives/mixed hazard classes	Incineration	5,930	5,120	6,880	$104,000	$15.16
Lab packs — treatment	Neutralize/stabilize	17,600	14,800	11,400	$117,000	$10.32
Lab packs — Incineration	Incineration	7,740	12,300	16,100	$252,000	$15.73
Photofixer	Silver reclamation	7,370	15,200	13,500	$59,300	$4.41
PCB material	Incineration/landfill	8,874	11,400	75,900	$255,000	$3.36
Miscellaneous	Best available	1,620	10,700	14,200	$81,600	$5.75
Scintillation vial	Incineration	2,130	479	1,520	$7,150	$4.72
Waste oil	Fuel/recycle	—	—	21,400	$18,000	$0.84
Total		87,800	99,300	173,000	$1,100,000	$5.36

1990–91 Cost Breakdown

	Fuel solvent	Bulk solvent	Reactive/mixture	LP-treatment	LP-incineration	Photofixer	PCB	Misc.	Scintillation Vials	Oil
Total kilograms	2,270	31,900	6,880	11,400	16,100	13,500	75,900	14,200	1,520	21,400
Disposal cost ($)	779	32,700	40,500	42,500	128,000	0	166,000	14,300	0	0
Salary cost ($)	8,790	123,000	40,800	48,600	77,000	48,600	60,700	54,400	5,480	10,500
Supplies cost ($)	186	2,610	13,600	7,260	13,200	1,350	8,730	3,640	697	823
Overhead cost ($)	1,470	20,600	4,440	7,320	10,400	8,680	19,600	9,160	979	6,480
Generator cost ($)	289	12,500	4,800	11,400	24,300	691	332	129	0	195
Total cost ($)	11,500	191,000	104,000	117,000	252,000	59,300	255,000	81,600	7,150	18,000

Low-Level Radioactive Waste Cost Breakdown (kg)

Wastestream	Disposal technique	Quantity generated 1988–1989	Quantity generated 1989–1990	Quantity generated 1990–1991	Total cost 1990–1991	Cost per kilogram 1990–1991
Solid (R)	Ship to U.S. Ecology	18,400	18,800	19,800	$140,000	$7.07
Solid (D)	Decay	504	1,700	2,230	$7,200	$3.23
Aqueous (S)	Sewer	21,200	20,900	22,600	$67,800	$3.00
Aqueous (R)	Ship to U.S. Ecology	4,830	5,140	4,240	$46,800	$11.03
Flammable (R)	Ship to U.S. Ecology	832	472	784	$3,362	$4.29
Vials (D & I)	Decay/incinerate	864	508	579	$15,200	$26.30
Vials (R)	Ship to U.S. Ecology	490	240	68	$3,610	$53.10
Vials (S)	Sewer	4,320	5,610	4,820	$81,300	$16.90
Animal (I)	Incinerate	640	407	272	$751	$2.76
Animal (D)	Decay	4,550	6,190	5,910	$9,900	$1.68
Animal (R)	Ship to U.S. Ecology	290	84	34	$1,510	$44.40
Sealed sources	Long-term storage	0	4,910	0	$0	—
Total		56,800	65,000	61,300	$377,000	$6.15

Appendix 1 Chemical and Low Level Radioactive Waste Costs at the University of Minnesota *(Continued)*

1990-91 Cost Breakdown

	Solid (R)	Solid (D)	Aqueous (S)	Aqueous (R)	Flammable	Vials (D & I)	Vials (R)	Vials (S)
Total kilograms	19,800	2,230	22,600	4,240	784	579	68	4,820
Disposal cost ($)	63,300	0	0	28,900	1,610	8,260	560	0
Salary cost ($)	21,100	1,920	20,500	5,070	461	2,000	922	25,500
Supplies cost ($)	7,900	941	1,000	1,380	250	426	46	1,920
HMS overhead ($)	9,650	881	9,410	2,330	211	916	423	14,300
RPD overhead ($)	12,500	1,140	12,200	3,020	275	1,190	549	13,100
Generator cost ($)	7,310	667	7,120	1,760	160	693	320	7,630
Loan repayment ($)	18,000	1,650	17,600	4,340	395	1,710	790	18,800
Total cost ($)	140,000	7,200	67,800	46,800	3,360	15,200	3,610	81,300

	Animal (I)	Animal (D)	Animal (R)	Sealed Sources
Total kilograms	272	5,910	34	0
Disposal cost ($)	0	0	1,160	0
Salary cost ($)	231	2,610	77	0
Supplies cost ($)	0	1,390	100	0
HMS overhead ($)	106	1,200	35	0
RPD overhead ($)	137	1,560	46	0
Generator cost ($)	80	907	27	0
Loan repayment ($)	197	2,240	66	0
Total cost ($)	751	9,900	1,510	0

Note: The author wishes to thank Gene Christenson and Andrew Phelan for the preparation of these data.

**Part 7:
Conclusions**

CHAPTER **24**

Thoughts on the Future of Pollution Prevention in the Laboratory

Peter C. Ashbrook

CONTENTS

Introduction 425
Why Pollution Prevention — Revisited 426
Importance of Laboratories to Society 427
The Objectives of Pollution Prevention and Waste Minimization 427
General Guidance on Laboratory Pollution Prevention 428
Chemical Redistribution Programs 428
Forces Working against Pollution Prevention 429
The Green Laboratory 429
A Call for Action 430
Conclusion 431
Acknowledgments 431

INTRODUCTION

The intent of this chapter is to give the reader some thoughts about where management of laboratory wastes is heading in the future. As far as regulatory ideas go, I believe that pollution prevention is the end of the line, that is, everyone has come to understand that prevention is easier, cheaper, and more effective than collection and management of wastes. In this respect, hazardous waste management is repeating the approach taken in the control of diseases — it is better to prevent the disease than to treat it after it occurs.

The above statement does not mean that laboratories have seen the end of new regulations. It is possible that tracking of all chemicals, not just waste, may become mandatory. In addition, environmental accounting of chemical usage may become as commonplace as financial accounting. While some tracking and accounting may have value, I question whether most of these requirements would produce much tangible benefit to laboratory workers or to society in general. On the other hand, an increased knowledge of the kinds and amounts of chemicals used — and where they end up after being used — may have intangible benefits in helping to create a better public understanding of the benefits and risks of laboratory work.

Whatever further regulations may arise, pollution prevention should make it much easier to comply with all regulations. Unfortunately, one of the regulatory trends seems to be requirements for more detailed information regarding waste minimization, such as source reduction plans, cost/benefit analyses, waste production indices, and annual reports documenting waste minimization accomplishments. While such documents may make sense for some industrial situations with a small number of wastestreams each with a large volume of waste, they make little sense in the laboratory setting that is likely to have a large number of wastestreams each with a small volume of waste.

Rather than try to predict the future, this chapter will highlight some of the major concepts presented in earlier chapters and identify some trends that may complicate pollution prevention efforts in laboratories. I will also give some thoughts about the concept of what constitutes a "green laboratory."

WHY POLLUTION PREVENTION — REVISITED

The basis of pollution prevention is the efficient use of resources. From this perspective, pollution prevention is not new because our society is constantly looking for ways to do things better, quicker, and more efficiently. Until the Resource Conservation and Recovery Act (RCRA) was passed in 1976, the cost of disposal of chemical waste from laboratories was essentially zero. Wastes either went down the drain or into the trash. No doubt some laboratory wastes still are improperly disposed down the drain or in the trash, but no one really knows how common these practices are because it is impossible to monitor them. Today, many large facilities, such as my institution, experience chemical waste disposal costs that are about the same magnitude as chemical purchase costs. When costs of any item increase by several orders of magnitude, institutions start to notice.

The second reason for interest in pollution prevention is that the concept reflects the maturing of the environmental movement. The initial regulatory approach to any environmental issue has been to collect the pollutant and treat it. It is usually easy and cheap to collect about two thirds of the pollutant. The cost of the next one fourth is more expensive, but usually still manageable. However, the last few percent of the pollutant, which is often still of a large enough magnitude to be a problem, is usually inordinately expensive. It has been only in recent years that control of most environmental problems have reached the stage of dealing with the final few percent of the pollutant. At this point, it becomes obvious that preventing discharges or minimizing discharges would be much cheaper than collecting and treating. This is the stage where both hazardous waste management in general and hazardous waste management in laboratories now find themselves.

A final reason for pollution prevention is the current popularity of "total quality management" or "continual improvement." These concepts are based on the principle that things not only should be done right the first time, they should also be constantly evaluated for possible improvements. This is exactly the concept that laboratories need to embrace to make pollution prevention successful for minimizing wastes.

IMPORTANCE OF LABORATORIES TO SOCIETY

The first thing to remember about pollution prevention in laboratories is that all laboratories have one or more primary missions. The primary missions are not to minimize waste. They are to provide services such as research, teaching, health care screening, environmental monitoring, drug testing, quality control, and the like. If a laboratory is to be successful it must concentrate first and foremost on its primary mission(s). These services are highly valued by society. While I have not seen opinion polls, I would expect that the public would have a very high degree of confidence in and respect for laboratories.

Regardless of whether or not a laboratory is operated for profit, costs are important. Efficient handling of chemicals will have a beneficial impact on the budget. Conversely, if chemical purchases and disposal decisions are made without concern for the environment, health, and/or safety, laboratories can absorb very large, unexpected costs. Environmental regulations are imposing new administrative costs to laboratories. Any procedural modifications that minimize the impact of regulations on laboratories will likely also minimize the administrative costs.

Pollution prevention and waste minimization strategies put a laboratory in a win-win situation. Such activities, when properly evaluated, implemented, and documented, will have a positive budgetary impact and will also facilitate compliance with the environmental regulations demanded by the public.

Therefore, pollution prevention will not be the top priority of a laboratory. However, once the laboratory understands its primary mission(s), pollution prevention would certainly be an integral part in the planning of how the laboratory can efficiently achieve its mission(s).

THE OBJECTIVES OF POLLUTION PREVENTION AND WASTE MINIMIZATION

I suggest that there are three reasons that can justify the implementation of a pollution prevention activity:

- The change will improve safety or benefit the environment
- The change will save money
- The change will save time

If a pollution prevention procedure cannot meet at least one of these three criteria, I question whether it is worth pursuing. While it makes us feel good to be concerned about pollution prevention, let's make sure that any changes we make are really beneficial.

GENERAL GUIDANCE ON LABORATORY POLLUTION PREVENTION

Strategies for laboratory pollution prevention can be summarized as follows:

- Good organization
- Good planning
- Use optimum procedures
- Constant evaluation to optimize procedures

These points are summarized and illustrated in previous chapters in the book. As evidenced by the large number of chapters, there is a wide variety of methods that can be used. Thus, the basic concepts are relatively simple, but the details of implementation provide laboratory workers with many opportunities to use their ingenuity.

It is important to recognize that these strategies should be applied to minimize all laboratory wastes, not merely chemical wastes. Minimization of chemical wastes will likely have the greatest positive economic impact. However, the greatest pollution prevention impact will occur when everyone has a pollution prevention mindset toward all wastes — not just in the laboratory, but at home, in stores, on vacation, wherever.

CHEMICAL REDISTRIBUTION PROGRAMS

One of the cornerstones of new laboratory pollution prevention programs is redistribution of excess chemicals. Such programs are easy to start and require little in the way of investment except for modest amounts of staff time. Redistribution programs have several payoffs because, not only are purchasing and disposal costs minimized, the program also gets laboratory staff thinking about other pollution prevention opportunities.

Theoretically, if an ideal pollution prevention program is in place, the redistribution program should die out. This would occur because laboratories would only be purchasing the chemicals that they need. In reality, I think it is unlikely that redistribution programs will ever die out; however, the apparent success of redistribution programs in monetary terms may decrease over time as laboratories and institutions implement more effective chemical inventory management and pollution prevention techniques.

FORCES WORKING AGAINST POLLUTION PREVENTION

I see several trends that can complicate the success of pollution prevention efforts. First is the increasing concern for doing things safely. I first noticed this trend in hospitals worrying about infection. I think we would all agree that improvements in infection control are good. However, here is the trend I see that is disturbing: the methods used to improve safety involve increased packaging and using single-use items, which means intensive utilization of resources and a large amount of waste generated. I do not have a suggestion about how to improve the situation, but the trend disturbs me.

A second trend, present primarily in research laboratories, is the need to use highly reactive starting materials. As research evolves, scientists begin to look at materials that do not react as easily as those used in the past. To get these less-reactive materials to react, one needs to use more reactive agents. If highly reactive chemicals are used, the users must clearly understand the hazards and should promptly deactivate whatever reactive chemicals are left over.

Lastly is the trend toward convenience items. One sees this at home with meals packaged for individuals so that all they need is to heat them up. Convenience foods require the use of much more packaging than if the same food items are made from scratch. This same concept of convenience has begun to creep into laboratories. For example, you can buy ampules that contain the right amount of hydrochloric acid to create a $1\text{-}M$ solution if one dilutes it to a liter. For that matter, you can buy hydrochloric acid in a wide variety of prepackaged concentrations. These convenience items can increase safety, but they definitely increase waste due to excess packaging.

The purpose of this section is not to condemn packaging; rather it is to point out trends I see as counterproductive to pollution prevention. In some cases, these trends and new products will enhance safety and make laboratories run more efficiently. In other cases, these trends achieve ends that could be achieved in alternate ways with less waste production. When new products or procedures are considered, they should be evaluated, not just for the improvement in the laboratory's services, but also for their implications for waste management.

THE GREEN LABORATORY

The "green" concept has entered our society in a wide variety of areas in recent years to denote products that are supposed to be environmentally friendly. Perhaps the concept could be applied to laboratories. The following are some thoughts about what would be the characteristics of a "green laboratory." Because the concept of the green laboratory is in its infancy, this discussion is meant to introduce ideas rather than to be a definitive statement about what would constitute a green laboratory.

First and foremost would be for the laboratory to develop a written pollution prevention plan. At a minimum, this plan would contain the following elements:

- Top management support for pollution prevention
- Organizational policies on pollution prevention
- Mechanisms(s) for evaluation and continual improvement
- Specific pollution prevention objectives for the coming 6 to 12 months

This plan would be updated periodically — typically on an annual basis. Specific pollution prevention elements could include

- Training new personnel in chemical and environmental safety
- Good housekeeping
- Maintenance of an inventory of all chemicals
- Purchasing chemicals in quantities that can be used within 1 year
- Use redistributed surplus chemicals whenever possible
- Assessment of laboratory air emissions, wastewater discharges, and waste generation to assess how these activities impact the environment
- Written protocols for all procedures, with the inclusion of how wastes are to be handled
- Treat wastes through neutralization or other means as part of the experimental procedure to minimize the hazard of the wastes produced
- Separate wastes into separate categories to the extent practical
- Know the hazards of the chemicals used (e.g., by the use of MSDSs) and search for safer substitutes
- A policy for the prudent use of the sewer for disposal
- Use of benign cleaning agents
- Avoidance of mercury-containing equipment and reagents
- Policies addressing minimization of nonhazardous solid and liquid wastes
- Conservation of energy when feasible
- Work in partnership with the institution's safety professionals

A CALL FOR ACTION

At the risk of seeming too arrogant, I offer the following suggestions for actions that various groups can take.

- Laboratory personnel: Make sure everyone in your laboratory is trained about pollution prevention principles. Define what a green laboratory would be in your case and then develop a written plan to be a green laboratory. One way to coordinate pollution prevention activities is to incorporate them into your chemical hygiene plan.
- Institutions (both management and their environmental health and safety offices: Provide internal resources and incentives for your employees to pursue pollution prevention opportunities. Work with suppliers to facilitate pollution prevention and "green-er" products. Evaluate your own products to make sure that they are as "green" as possible.
- The laboratory community through professional organizations: Voluntarily draft a consensus standard for a green laboratory. This consensus standard would describe appropriate pollution prevention procedures and practices, such as those described above. Provide guidance to implementing an institutional pol-

lution prevention program for laboratories. Draft a model pollution prevention training program. Help create a laboratory pollution prevention clearinghouse for information, developments, and news.
- Regulatory agencies: If additional regulations pertaining to pollution prevention in laboratories is necessary, use the model of the OSHA laboratory standard, which is flexible and performance based. It would make a lot of sense to incorporate pollution prevention into the chemical hygiene plan. Recognize that pollution prevention in laboratories is much different from that in production facilities. Request laboratories to evaluate pollution prevention activities on a periodic basis, but do not follow the SARA model and base laboratory pollution prevention progress on a "production index." Laboratory activities change so frequently and are so varied that the development of a "production index" is meaningless.
- Congress and state legislatures: Recognize that legislation designed for the chemical industry often ends up applying to a wide variety of institutions, including laboratories. Make sure that legislation accounts for these often unintended consequences. Do not politicize pollution prevention by specifying arbitrary reductions in waste generation. Facilitate pollution prevention by establishing incentives and minimizing legislative requirements that result in increased administrative expenses with little or no environmental benefit.
- Everyone: Do not limit your pollution prevention efforts to hazardous wastes. Consider all other wastes, including paper, plastics, glass, and metals. At the same time you are preventing wastes, examine the energy consequences of your activities.

CONCLUSION

It is clear from this book that waste minimization and pollution prevention are continuing processes. Each individual must be responsible for his/her own wastes. Laboratory workers should not take the easy out by arguing that their wastes are too trivial to present a hazard. Any individual can make this argument, but, when large amounts of small quantities of wastes are aggregated, there are hazards. On the other hand, it is important not to overestimate the hazards of laboratory wastes. Although laboratory wastes can present hazards in some cases, these hazards rarely warrant the strict regulation that is applied to industry.

There is no "after" to pollution prevention and waste minimization. These concepts will be with us as long as we live.

ACKNOWLEDGMENTS

As near as I can tell, the first person to articulate the concept of the green laboratory in a public forum was William Raub, the science advisor to the Administrator of the U.S. EPA, at a conference sponsored by the Howard Hughes Medical Center in April 1993. Many, if not most, of the ideas articulated in this chapter came from correspondence and conversations with Peter Reinhardt.

Appendices

APPENDIX A

Friday
May 28, 1993

Part VII

Environmental Protection Agency

Guidance to Hazardous Waste Generators on the Elements of a Waste Minimization Program; Notice

ENVIRONMENTAL PROTECTION AGENCY

[EPA 530-Z-93-007; FRL-4658-5]

Guidance to Hazardous Waste Generators on the Elements of a Waste Minimization Program

AGENCY: Environmental Protection Agency [EPA].

ACTION: Interim final guidance.

SUMMARY: EPA is committed to a national policy for hazardous waste management that places the highest priority on waste minimization. To this end, EPA is today providing interim final guidance to assist hazardous waste generators and owners and operators of hazardous waste treatment, storage, or disposal facilities to comply with the waste minimization certification requirements of sections 3002(b) and 3005(h) of the Resource Conservation and Recovery Act (RCRA), as amended by the Hazardous and Solid Waste Amendments of 1984 (HSWA), 42 U.S.C. 6922(b) and 6925(h).

Section 3002(b) requires generators of hazardous waste to certify on their hazardous waste manifests that they have a waste minimization program in place. Section 3005(h) requires owners and operators of facilities that receive a permit for the treatment, storage, or disposal of hazardous waste on the premises where such waste was generated to make the same certification no less often than annually.

EPA believes waste minimization programs should incorporate, in a way that meets individual organizational needs, the following basic elements common to most good waste minimization programs: (1) Top management support; (2) characterization of waste generation and waste management costs; (3) periodic waste minimization assessments; (4) appropriate cost allocation; (5) encouragement of technology transfer; and (6) program implementation and evaluation. Thus, generators and owners and operators of hazardous waste treatment, storage, and disposal facilities should use these elements to design multimedia pollution prevention programs directed at preventing or reducing wastes, substances, discharges and/or emissions to all environmental media—air, land, surface water and ground water.

EPA is publishing this guidance as an interim final version, and solicits further public comments on it. However, until the guidance is finalized, persons should use it in developing their waste minimization programs in place.

DATES: EPA urges all interested parties to comment on this interim final guidance, in writing, by July 27, 1993.

ADDRESSES: The public must send an original and two copies of their comments to: RCRA Information Center (OS–305), U.S. Environmental Protection Agency, 401 M Street, SW., Washington, DC 20460.

Place the docket number F–93–WMIF–FFFFF on your comments.

Commenters who wish to submit any information they wish to claim as Confidential Business Information must submit an original and two copies, under separate cover, to: Document Control Officer (OS–312), Office of Solid Waste, U.S. Environmental Protection Agency, 401 M Street, SW., Washington, DC 20460.

FOR FURTHER INFORMATION, CONTACT: Becky Cuthbertson, Office of Solid Waste, 703–308–8447, or the RCRA Hotline, toll free at (800) 424–9346. TDD (800) 553–7672.

SUPPLEMENTARY INFORMATION:

Guidance to Hazardous Waste Generators on the Elements of a Waste Minimization Program

I. Purpose

The purpose of today's notice is to provide guidance to hazardous waste generators and owners and operators of hazardous waste treatment, storage, and disposal facilities on what constitutes a waste minimization "program in place," in order to comply with the certification requirements of sections 3002(b) and 3005(h) of the Resource Conservation and Recovery Act (RCRA), as amended by the Hazardous and Solid Waste Amendments of 1984 (HSWA), 42 U.S.C. 6922(b) and 6925(h). Section 3002(b) requires hazardous waste generators who transport their wastes off-site to certify on their hazardous waste manifests that they have programs in place to reduce the volume or quantity and toxicity of hazardous waste generated to the extent economically practicable. Certification of a waste minimization "program in place" is also required as a condition of any permit issued under section 3005(h) for the treatment, storage, or disposal of hazardous waste at facilities that generate and manage hazardous wastes on-site. This guidance fulfills a commitment made by EPA in its 1986 report to Congress [1] entitled The Minimization of Hazardous Waste (EPA/530–SW–86–033, October 1986) to provide additional information to

[1] 51 FR 44683 (December 11, 1986). Notice of Availability of the report to Congress on waste minimization.

generators on the meaning of the certification requirements placed in HSWA.

Additionally, EPA published in the **Federal Register**, on January 26, 1989 (54 FR 3845), a proposed policy statement on source reduction and recycling. This policy commits the Agency to a preventive strategy to reduce or eliminate the generation of environmentally-harmful pollutants which may be released to the air, land, surface water or ground water. We further proposed to incorporate this preventive strategy into EPA's overall mission to protect human health and the environment by making source reduction a priority for every aspect of Agency decision-making and planning, with environmentally-sound recycling as a second and higher priority over treatment and disposal. Today's notice is an important step in implementing this policy with respect to hazardous wastes regulated under RCRA.

EPA has taken the January 26, 1989 proposed pollution prevention policy statement two steps further: By publishing a "Pollution Prevention Strategy" in the February 26, 1991 **Federal Register** (56 FR 7849), and by proposing the creation of a program that would encourage and publicly recognize environmental leadership, and would promote pollution prevention in manufacturing in the January 15, 1993 **Federal Register** (58 FR 4802).

II. Background

A. Statutory Intent and Requirements and Definition of Waste Minimization

In the past, the predominant practice used by manufacturing, commercial and other facilities that generate hazardous waste has been "end of pipe" treatment or land disposal of hazardous and nonhazardous wastes. While this approach has provided substantial progress in improving the quality of the environment, there are limits as to how much environmental improvement can be achieved using methods which manage pollutants after they have been generated.

With the passage of HSWA in 1984, Congress established a significant new policy concerning hazardous waste management. Specifically, Congress declared that the reduction or elimination of hazardous waste generation at the source should take priority over the management of hazardous wastes after they are generated. In particular, section 1003(b), 42 U.S.C. 6902(b), of RCRA the Congress declares it to be the national policy of the United States that, wherever feasible, the generation of hazardous waste is to be reduced or eliminated as expeditiously as possible. Waste that is nevertheless generated should be treated, stored, or disposed of so as to minimize the present and future threat to human health and the environment.

In this declaration, Congress established a clear national priority for eliminating or reducing the generation of hazardous wastes. At the same time, however, the national policy recognized that some wastes will "nevertheless" be generated, and such wastes should be managed in a way that "minimizes" present and future threat to human health and the environment.

In 1990, Congress further clarified the role of pollution prevention in the nation's environmental protection scheme, by passing the Pollution Prevention Act (Pub. L. 101–508, 42 U.S.C. 13101, et seq.). In section 6602(b) of this law, 42 U.S.C. 13101(b), Congress stated that national policy of the United States is that pollution should be prevented or reduced at the source whenever feasible; pollution that cannot be prevented should be recycled in an environmentally safe manner, whenever feasible; pollution that cannot be prevented or recycled should be treated in an environmentally safe manner whenever feasible; and disposal or other release into the environment should be employed only as a last resort and should be conducted in an environmentally safe manner.

Thus, Congress set up a hierarchy of management options in descending order of preference: prevention, environmentally sound recycling, environmentally sound treatment, and environmentally sound disposal.

EPA believes that waste minimization, the term employed by Congress in the RCRA statute, includes (1) source reduction, and (2) environmentally sound recycling. (See later discussion for further clarification of which types of recycling are not waste minimization.)

The first category, source reduction, is defined in section 6603(5)(A) of the Pollution Prevention Act, 42 U.S.C. 13102(5)(a), as any practice which (i) reduces the amount of any hazardous substance, pollutant, or contaminant entering any waste stream or otherwise released prior to recycling (including fugitive emissions) prior to recycling, treatment, or disposal; and

(ii) Reduces the hazards to public health and the environment associated with the release of such substances, pollutants, or contaminants.

The term includes equipment or technology modifications, process or procedure modifications, reformulation or redesign of products, substitution of raw materials, and improvements in housekeeping, maintenance, training, or inventory control.

EPA believes this definition is appropriate for use in identifying opportunities for source reduction under RCRA.

The second category, environmentally sound recycling, is the next preferred alternative for managing those pollutants which cannot be reduced at the source. In the context of hazardous waste management, there are certain practices or activities which the hazardous waste regulations define as "recycling." The definitions for materials that are "recycled" are found in Title 40 of the Code of Federal Regulations, § 261.1(c). A "recycled" material is one which is used, reused, or reclaimed.[2] A material is "used or reused" if it is (i) employed as an ingredient (including use as an intermediate) in an industrial process to make a product (for example, distillation bottoms from one process used as feedstock in another process) * * * or (ii) employed in a particular function or application as an effective substitute for a commercial product.* * *[3]

A material is "reclaimed" if it is "processed to recover a usable product, or if it is regenerated."[4]

On the other hand, the regulations define "treatment" and "disposal" as follows:

Treatment means any method, technique, or process, including neutralization, designed to change the physical, chemical, or biological character or composition of any hazardous waste so as to neutralize such waste, or so as to recover energy or material resources from the waste, or so as to render such waste non-hazardous, or less hazardous; safer to transport, store, or dispose of; or amenable for recovery, amenable for storage, or reduced in volume.[5]

Disposal means the discharge, deposit, injection, dumping, spilling, leaking, or placing of any solid waste or hazardous waste into or on any land or water so that such solid waste or hazardous waste or any constituent thereof may enter the environment or be emitted into the air of discharged into any waters, including ground waters.[6]

Some readers of today's guidance may question whether certain types of recycling are within the concept of waste minimization. EPA believes that recycling activities closely resembling conventional waste management activities do not constitute waste minimization.

Treatment for the purposes of destruction or disposal is not part of waste minimization, but is, rather, an activity that occurs after the opportunities for waste minimization have been pursued.[7] When source reduction and recycling opportunities are exhausted to the extent economically practicable, EPA has set standards for the treatment, storage and disposal of hazardous wastes. Treatment may be either thermal (i.e., incineration), chemical, or biological, especially for organic hazardous wastes. Where destruction methods for treatment are not available or ineffective, immobilization (stabilization) is often effective, especially for inorganic hazardous wastes.

Transfer of hazardous constituents from one environmental medium to another also does not constitute waste minimization. For example, the use of an air stripper to evaporate volatile organic constituents from an aqueous waste only shifts the contaminant from water to air. Furthermore, concentration activities conducted solely for reducing volume does not constitute waste minimization unless, for example, concentration of the waste is an integral setup in the recovery of useful constituents prior to treatment and disposal. Similarly, dilution as a means of toxicity reduction would not be considered waste minimization, unless dilution is a necessary step in a recovery or a recycling operation.

EPA firmly believes that waste minimization will provide additional environmental improvements over "end of pipe" control practices, often with the added benefit of cost savings to generators of hazardous waste and reduced levels of treatment, storage and disposal. Waste minimization has already been shown to result in significant benefits for industry, as evidenced in numerous success stories documented in available literature.

The benefits that accrue to facilities that pursue waste minimization often include:

(1) Minimizing quantities of hazardous waste generated, thereby reducing waste management and compliance costs and improving the protection of human health and the environment;

(2) Reducing or eliminating

---

[2] 40 CFR 261.1(c)(7).
[3] 40 CFR 261.1(c)(5).
[4] 40 CFR 261.1(c)(4).
[5] 40 CFR 260.10. Most types of recycling are in fact classified as treatment (see 48 FR at 14502–14504, April 4, 1983), and some also meet the definition of disposal.
[6] 40 CFR 260.10.

[7] It is, of course, not always easy to distinguish recycling (environmentally sound or otherwise) from conventional treatment. See 56 FR at 7143 (February 21, 1991); 53 FR at 522 (January 8, 1988).

inventories and possible releases of "hazardous chemicals;"
(3) Possible decrease in future Superfund and RCRA liabilities, as well as future toxic tort liabilities;
(4) Improving facility mass/energy efficiency and product yields;
(5) Reducing worker exposure; and
(6) Enhancing organizational reputation and image.

In addition to establishing a national policy to foster waste minimization, HSWA also included several specific requirements that promote implementation of waste minimization at individual facilities. In particular, RCRA section 3002(b) requires generators of hazardous waste who transport wastes off-site to certify on each hazardous waste manifest that they have a program in place to reduce the volume and toxicity of such waste to the degree determined by the generator to be economically practicable. Similarly, certain owners and operators of RCRA permitted treatment, storage and disposal facilities are also required to provide the same certification annually (RCRA Section 3005(h)). These two requirements for certification, taken together, have the effect of insuring that waste minimization programs are put in place for facilities that generate hazardous waste regardless of whether the wastes are managed on-site or off-site. The purpose of today's **Federal Register** notice is to provide guidance to these hazardous waste handlers, who must certify that they have a waste minimization program in place.

Hazardous waste generators and owners/operators of hazardous waste treatment, storage and disposal facilities who manage their own hazardous waste on-site, must also identify in a biennial report to EPA (or the State): (1) The efforts undertaken during the year to reduce the volume and toxicity of waste generated; and (2) the changes in volume and toxicity actually achieved in comparison to previous years.

B. Scope of This Notice

Today's notice provides guidance on the basic elements of a waste minimization "program in place" that, if present, will allow persons to properly certify that they have implemented a program to reduce the volume and toxicity of hazardous waste to the extent "economically practicable." The guidance is directly applicable to generators who generate 1000 or more kilograms per month of hazardous waste ("large quantity" generators) or to owners and operators of hazardous waste treatment, storage, or disposal facilities who manage their own hazardous waste on-site.

Small quantity generators who generate greater than 100 kilograms but less than 1000 kilograms of hazardous waste per month are not subject to the same "program in place" certification requirement as large quantity generators. Instead, they must certify on their hazardous waste manifests that they have "made a good faith effort to minimize" their waste generation. EPA encourages small quantity generators to develop waste minimization programs of their own, to show their good faith efforts.

This notice does not provide guidance on the determination of the phrase "economically practicable". As Congress indicated in its accompanying report to HSWA (S. Rep. No. 98–284, 98th Cong. 1st. Sess., 1983) "economically practicable" is to be defined and determined by the generator. The generator of the hazardous waste, for the purpose of meeting this certification requirement, has the flexibility to determine what is economically practicable for the generator's particular circumstances. Whether this determination is done in a combined fashion for all operations or on a site-specific basis is for the generator to decide.

III. Guidance to Hazardous Waste Generators on the Elements of a Waste Minimization Program, as Required Under RCRA Sections 3002(b) and 3005(h)[8]

Waste minimization programs have been implemented by a wide array of organizations. The elements discussed in this notice reflect the results of EPA interactions with State governments and industry waste minimization program managers. Numerous state governments have already enacted legislation requiring facility specific waste minimization programs (for example, the enactment of the Massachusetts Toxics Use Reduction Act of 1989, Oregon Toxics Use Reduction and Hazardous Waste Reduction Act, and Art. 11.9, Chap. 6.5, Div. 20 of California Health and Safety Code, October 1989.) Other states have legislation pending that may mandate some type of facility specific waste minimization program.

EPA believes that each of the general elements discussed below should be included in a waste minimization program, although the Agency realizes that each element may be implemented in different ways depending on the needs and preferences of individual organizations or facilities. The generator or treatment, storage, or disposal facility should document its program (in writing) so that it is available for interested parties. EPA also believes that the waste minimization program should be signed by that corporate officer who is responsible for ensuring RCRA compliance.

The waste minimization program elements are as follows:

A. *Top management support.* Top management should support an organization-wide effort. There are many ways to accomplish this goal. Some of the methods described below may be suitable for some organizations, while not for others. However, some combination of these techniques or similar ones will demonstrate top management support:

—Make waste minimization a part of the organization policy. Put this policy in writing and distribute it to all departments and individuals. Each individual, regardless of status or rank, should be encouraged to identify opportunities to reduce waste generation. Encourage workers to adopt the policy in day to day operations and encourage new ideas at meetings and other organizational functions. Waste minimization, especially when incorporated into organization policy, should be a process of continuous improvement. Ideally, a waste minimization program should become an integral part of the organization's strategic plan to increase productivity and quality.

—Set explicit goals for reducing the volume and toxicity of waste streams that are achievable within a reasonable time frame. These goals may be quantitative or qualitative. Both can be successful.

—Commit to implementing recommendations identified through assessments, evaluations, waste minimization teams, etc.

—Designate a waste minimization coordinator who is responsible for facilitating effective implementation, monitoring and evaluation of the program. In some cases (particularly in large multi-facility organizations), an organizational waste minimization coordinator may be needed in addition to facility coordinators. In other cases, a single coordinator may have responsibility for more than one facility. In these cases, the coordinator

---

[8] On June 12, 1989, the EPA published a proposed guidance on what constituted a "program in place", and solicited public comments. 33 comments were received in response to the draft guidance; most comments suggested clarifications or expansion of specific points, while some comments disagreed with portions of the proposal. Both the comments and EPA's response to the comments are summarized in the Appendix to this notice.

should be involved or be aware of operations and should be capable of facilitating new ideas at each facility. It is also useful to set up self-managing waste minimization teams chosen from a broad spectrum of operations: engineering, management, research & development, sales & marketing, accounting, purchasing, maintenance and environmental staff personnel. These teams can be used to identify, evaluate and implement waste minimization opportunities.
—Publicize success stories. Set up an environment and select a forum where creative ideas can be heard and tried. These techniques can inspire additional ideas.
—Recognize individual and collective accomplishments. Reward employees that identify cost-effective waste minimization opportunities. These rewards can take the form of collective and/or individual monetary or other incentives for improved productivity/waste minimization.
—Train employees on the waste-generating impacts that result from the way they conduct their work procedures. For example, purchasing and operations departments could develop a plan to purchase raw materials with less toxic impurities or return leftover materials to vendors. This approach can include all departments, such as those in research & development, capital planning, purchasing, production operations, process engineering, sales & marketing and maintenance.

B. *Characterization of waste generation and waste management costs.* Maintain a waste accounting system to track the types and amounts of wastes as well as the types and amounts of the hazardous constituents in wastes, including the rates and dates they are generated. EPA realizes that the precise business framework of each waste generator may be unique. Therefore, each organization must decide the best method to obtain the necessary information to characterize waste generation. Many organizations track their waste production by a variety of means and then normalize the results to account for variations in production rates.

Additionally, a waste generator should determine the true costs associated with waste management and cleanup, including the costs of regulatory oversight compliance, paperwork and reporting requirements, loss of production potential, costs of materials found in the waste stream (perhaps based on the purchase price of those materials), transportation/ treatment/storage/disposal costs, employee exposure and health care, liability insurance, and possible future RCRA or Superfund corrective action costs. Both volume and toxicities of generated hazardous waste should be taken into account. Substantial uncertainty in calculating many of these costs, especially future liability, may exist. Therefore, each organization should find the best method to account for the true costs of waste management and cleanup.

C. *Periodic waste minimization assessments.* Different and equally valid methods exist by which a waste minimization assessment can be performed. Some organizations identify sources of waste by tracking materials that eventually wind up as waste, from point of receipt to the point at which they become a waste. Other organizations perform mass balance calculations to determine input and outputs from processes and/or facilities. Larger organizations may find it useful to establish a team of independent experts outside the organization structure, while some organizations may choose teams comprised of in-house experts.

Most successful waste minimization assessments have common elements that identify sources of waste and calculate the true costs of waste generation and management. Each organization should decide the best method to use in performing a waste minimization assessment that addresses these two general elements:
—Identify opportunities at all points in a process where materials can be prevented from becoming a waste (for example, by using less material, recycling materials in the process, finding substitutes that are less toxic and/or more easily biodegraded, or making equipment/process changes). Individual processes or facilities should be reviewed periodically. In some cases, performing complete facility material balances can be helpful.
—Analyze waste minimization opportunities based on the true costs associated with waste management and cleanup. Analyzing the cost effectiveness of each option is an important factor to consider, especially when the true costs of treatment, storage and disposal are considered.

D. *A cost allocation system.* Where practical and implementable, organizations should appropriately allocate the true costs of waste management to the activities responsible for generating the waste in the first place (e.g., identifying specific operations that generate the waste, rather than charging the waste management costs to "overhead"). Cost allocation can properly highlight the parts of the organization where the greatest opportunities for waste minimization exist; without allocating costs, waste minimization opportunities can be obscured by accounting practices that do not clearly identify the activities generating the hazardous wastes.

E. *Encourage technology transfer.* Many useful and equally valid techniques have been evaluated and documented that are useful in a waste minimization program. It is important to seek or exchange technical information on waste minimization from other parts of the organization/facility, from other companies/facilities, trade associations/affiliates, professional consultants and university or government technical assistance programs. EPA and/or State funded technical assistance programs (e.g., Minnesota Technical Assistance Program—MnTAP, California Waste Minimization Clearinghouse, EPA Pollution Prevention Information Clearinghouse) are becoming increasingly available to assist in finding waste minimization options and technologies.

F. *Program implementation and evaluation.* Implement recommendations identified by the assessment process, evaluations, waste minimization teams, etc. Conduct a periodic review of program effectiveness. Use these reviews to provide feedback and identify potential areas for improvement.

IV. *Additional Resources Available to Generators and Others on Waste Minimization Programs*

EPA and the States have worked cooperatively to put in place a variety of technical information and assistance programs that make information on source reduction and recycling techniques available directly to industry and the public.

EPA has developed information sources that can be used to provide information directly to industry or through State technical assistance programs. EPA maintains a Pollution Prevention Information Clearinghouse (PPIC), which is a reference and referral source for technical, policy, program, legislative and financial information on pollution prevention. PPIC's telephone number is (202) 260–1023; the facsimile number is (202) 260–0178. EPA also publishes a pollution prevention newsletter and produces videos and literature on waste minimization that are available to the public.[9]

Examples of general documents that assist organizations with more detailed guidance on conducting waste minimization assessments and developing pollution prevention programs are the Waste Minimization Opportunity Assessment Manual, EPA 625/7–88/003, July 1988,[10] and the Facility Pollution Prevention Guide, EPA/600/R–92/088.[11] Another general document that introduces the concept of waste minimization is Waste Minimization: Environmental Quality with Economic Benefits, EPA/530–SW–90–044, April 1990.[12] EPA has also developed numerous waste minimization and pollution prevention documents that are tailored to specific manufacturing and other types of processes, and periodically sponsors pollution prevention workshops and conferences.

EPA also promotes technical assistance to industry indirectly by supporting the development of State technical assistance programs. State personnel often have the primary day to day contacts with industry for many RCRA program matters. Examples of State technical assistance programs are; Minnesota Technical Assistance Program—MnTAP and California Waste Minimization Clearinghouse. EPA also provides partial funding for the National Roundtable of State Pollution Prevention Programs, an organization of State technical assistance and regulatory program representatives that meets regularly to discuss technical and programmatic waste minimization issues. The Roundtable uses the PPIC as a central repository for technical exchange and publishes proceedings on state waste minimization activities. EPA's Office of Research and Development also funds several different types of waste minimization research and demonstration projects in a variety of joint ventures with States and industry, and publishes industry-specific pollution prevention guidances.[13]

---

[9] To be added to the newsletter's mailing list, write: Pollution Prevention News, U.S. EPA, PM–222B, 401 M St. SW., Washington, DC 20460.

[10] Available from the National Technical Information Service; telephone (703) 487–4650; the publication number is PB 92–216 985 and the cost is $27.00.

[11] Available by calling the CERI Publications Unit at EPA's Cincinnati, OH office at (513) 569–7562.

[12] Available by calling the RCRA Information Center; telephone (202) 260–9327.

[13] Contact the CERI publications unit at EPA's Cincinnati, OH office, telephone (513) 569–7562, for a list of available pollution prevention publications.

Additionally, at least 29 states reported in their Capacity Assurance Plans (October 1989) that they have in place some type of technical assistance to organizations that seek alternatives to treatment, storage and disposal of waste.

V. Conclusion

EPA is committed to the elimination, reduction, and/or recycling of waste as the first steps in our national waste management strategy. Only through preventing pollution in the first place will our nation be able to ensure both a healthy, vibrant economy that can prevail in a competitive worldwide economy, and a healthy environment that provides us with the resources we need and use in our everyday lives. As a result of the approach Congress has set in both the national policy of RCRA and in the Pollution Prevention Act, generators of waste must shoulder some of the responsibility to implement waste minimization measures, which will assist in prevention of risks to today's and tomorrow's environment. Generators have demonstrated the usefulness and benefits of waste minimization practices. EPA believes that as more organizations implement their waste minimization programs and demonstrate their usefulness and benefits, many other organizations will be encouraged to seek greater opportunities to incorporate waste minimization in their operations. Today's guidance on the elements of effective waste minimization programs may help encourage regulated entities to investigate waste minimization alternatives, implement new programs, or upgrade existing programs. Although the approaches described above are directed toward minimizing hazardous waste, they are also important elements in the design of multi-media source reduction and recycling programs for all forms of pollution.

Dated: May 18, 1993.

Carol M. Browner,
Administrator.

Appendix

Response to Comments on EPA's Draft "Guidance to Hazardous Waste Generators on the Elements of a Waste Minimization Program"

One respondent objected to the nonbinding approach of the guidance, stating that some basic definition of program acceptability should be specifically given. This respondent stated that the approach would encourage only a voluntary effort to implement waste minimization programs. However, most respondents supported the approach and encouraged EPA to retain this approach in the final guidance. These respondents stated that the flexibility inherent in the approach should assist organizations in implementing effective waste minimization programs appropriate to specific circumstances and processes.

While RCRA makes it clear that the waste minimization certification provisions are mandatory and enforceable, the Agency believes that it is the intent of Congress to allow for flexibility in implementing facility specific waste minimization programs. In setting forth the waste minimization approach given in this interim final guidance, EPA believes it has acted in a manner that follows Congressional intent. Because of this, the Agency does not believe it is necessary to describe the approach in the interim final guidance text as "nonbinding" because such a term would be redundant; the guidance is nonbinding by being guidance. However, while the specific elements are guidance, the certification requirements of sections 3002(b) and 3005(h) are mandatory. The nature of the guidance does not reduce in any way these mandatory certification requirements.

Another respondent stated that EPA's definition of waste minimization is too restrictive in allowing only source reduction and recycling activities to define waste minimization. While activities of this nature may be the most desirable, Congress clearly stated the overall goal was to "minimize the present and future threat to human health and the environment." Therefore, better treatment and proper disposal could be considered a part of waste minimization. By not defining treatment and disposal as part of waste minimization, the commenter believed that EPA may be discouraging improvements which could be environmentally beneficial.

The Agency has clearly stated its position that a waste management hierarchy exists where source reduction and environmentally-sound recycling are the primary and secondary priorities of the waste management hierarchy and together define waste minimization. Treatment and disposal are alternatives of last resort to waste minimization, not substitutes for it. EPA disagrees with the respondent's suggestion that defining waste minimization as source reduction and recycling could discourage improvements in treatment and disposal technologies. On the contrary, EPA believes that the main thrust of the RCRA program has been to improve treatment and disposal technology. The Agency believes that the intent of the HSWA National Policy was to move beyond treatment and disposal approaches to prevention approaches. It is on this basis that the Agency concludes that treatment and disposal are not (nor should they be) part of waste minimization.

Guidance Element A: Top Management Support and Facility Coordination:

This element of the proposed guidance stated that top management should ensure that waste minimization is a company-wide effort. Several techniques were proposed that should be used to demonstrate top management support.

Several respondents stated that employee education and feedback as well as management support is important to the success of a waste minimization plan. The Agency agrees that employee education and management support is an important element of any waste minimization program. However, the Agency believes that each organization should decide what the parameters of that support will be, based upon its organizational structure. For example, in some organizations, support may take the form of a directive from top management formally establishing waste minimization teams. In other organizations, support might be in the form of extending the scope of existing quality circles to include waste minimization. What is appropriate for one organization might not be appropriate for others.

Many respondents also recommended that the policy should acknowledge that in some cases individual facility coordinators may be inappropriate, especially for companies with numerous small and/or similar facilities. Respondents suggested that in these cases, a national or regional coordinator may be more appropriate. EPA believes that the key function of a coordinator is to facilitate and maintain plant planning and operations. The most successful programs have an on-site person who deals with day to day tasks necessary to keep the program on track and consistent with organizational goals. Some organizations with multiple facilities also have a coordinator whose function is to facilitate communication and informational flow between facilities and top management and ensure that adequate support is available. Nevertheless, EPA believes each organization should determine how best to fulfill the functions of managing and coordinating waste minimization activities.

Finally, one respondent stated that EPA should recognize that the setting of aggressive goals by upper management to demonstrate commitment may prove counterproductive when these goals are not realized. The Agency believes that the setting of specific, realistic goals is very important to the success of a waste minimization program. However, each organization must determine what these goals are as well as how they are achieved and the timetable for their achievement. These goals can be qualitative and/or quantitative, but can only be successful if management fully supports employee efforts to achieve them. Both types of goals can be successful.

Guidance Element B: Characterizing Waste Generation and Waste Accounting:

This element of the proposed guidance stated that a waste accounting system to track the types, amounts and hazardous constituents of wastes and the dates they are generated should be maintained.

Some respondents recommended that EPA should clarify that waste accounting systems must be unique to each facility and that this uniqueness is a function of the size of the generator as well as waste characteristics and volumes, processes, and other circumstances surrounding waste generation. Therefore, since no two waste accounting systems can be precisely alike, EPA will not mandate any specific type of waste accounting system.

The Agency agrees that each waste accounting system should be facility-specific and should be designed to accommodate each of the parameters mentioned by the respondent. In fact, EPA did not specify particular waste accounting systems in the proposed guidance for precisely those reasons. However, it is important that each facility and/or organization have a system that identifies and characterizes all waste streams and their sources, whatever form the system takes. The Agency believes that there are key parameters that waste accounting systems should address. Among these are identification of all wastes in terms of volume and toxicity as well as sources of all wastes. EPA also believes that it is critical to account for the costs of managing the wastes, including the amounts and costs of raw materials or other by-products found in waste streams and the costs of compliance with the regulations for treatment, storage, and disposal of hazardous wastes.

One respondent indicated that tracking of the rates of waste generation is not mentioned as a program element and that the rates of waste generation are more relevant than the dates of generation as was stated in the draft guidance. The Agency agrees that rates of waste generation are more likely to be relevant than the dates of waste generation when tracking waste generation. However, both are important to providing a clear picture of the sources and quantities of waste. Therefore, the interim final guidance has been changed accordingly.

Guidance Element C: Periodic Waste Minimization Assessments:

This element of the proposed guidance stated that periodic waste minimization assessments should be conducted to identify opportunities for waste minimization and to determine the true costs of waste.

One respondent suggested that the section on periodic waste minimization assessments should contain a flexibility clause stating that there are a number of different ways to accomplish a waste minimization assessment. The respondent stated that some of the methods described in the draft guidance may be suitable for some organizations but not others. In particular, many materials that become wastes do not originate from "loading dock materials" as stated in the draft guidance. Also, some wastes are listed as hazardous because they are residues (by-products) from a specified process or processes and as such would be difficult to track from the "loading dock".

The Agency agrees that there are different ways to complete a waste minimization assessment. In some cases, the actual practice of tracking raw materials through the production process to the point where they become wastes can be exceedingly complex, such as in petrochemical plants where integrally linked processes use multiple raw material inputs. Each organization should determine what level of analysis is necessary to provide adequate information to formulate waste minimization alternatives. The waste minimization team conducting a waste minimization assessment can make this determination.

The interim final guidance has been changed to clarify this point. The interim final guidance stresses that some level of process tracking or materials balance should be used to identify sources and volumes of waste. The interim final guidance stresses that all approaches used should cover five key elements including: waste stream characterization; identification and tracking of wastes; the determination of the true cost of treatment, storage, and disposal; allocation of costs to the activities responsible for waste generation; and identification of opportunities for waste minimization. [Note that information developed in the waste accounting and allocation system is critical to identifying waste minimization opportunities.]

One respondent stated that this section should specifically state that the purchasing of materials and packaging that have been designed to facilitate reuse and recycling should be specified as an identified opportunity for waste minimization.

The Agency agrees that the use of packaging that is designed to facilitate reuse and recycling can be an opportunity in waste minimization. However, numerous suggestions for specific types of waste minimization opportunities were received from respondents. The EPA acknowledges that there are many examples of waste minimization opportunities. However, for the sake of brevity they could not all be included in either the draft guidance or interim final guidance.

Another respondent indicated that EPA should state more forcefully in its interim final guidance that finding substitutes to toxic materials that pose less of a danger to human health and the environment and that are more easily degraded is an important opportunity in waste minimization. The Agency agrees that material substitution is an important aspect of waste minimization, which has been appropriately emphasized in the draft and interim final guidance.

Another respondent suggested that a waste minimization assessment should commence from the "point of receipt" of raw materials rather that "from the loading dock" as written in the draft guidance. The reason for this is that loading docks are used for shipping as well as receiving. The Agency agrees and has changed the language of the interim final guidance accordingly.

Guidance Element D: A Cost Allocation System:

This element of the proposed guidance stated that departments and managers should be charged "fully-loaded" waste management costs for the wastes they generate, factoring in liability, compliance and oversight costs. The guidance encourages organizations to develop and maintain a system for determining and monitoring waste stream characteristics and costs. This information provides a basis for identifying waste minimization opportunities which is discussed further in guidance element F.

Two respondents indicated that the entire Cost Allocation Section should be deleted from the guidance, stating that the guidance is too specific, and that use of the phrase "fully-loaded waste management costs" in the draft guidance implies cost accounting procedures that may not be compatible with existing organizational accounting practices. However, several respondents stated that it was appropriate for EPA to suggest that a waste minimization program include waste management accounting costs, with the understanding that it is inappropriate for EPA to specify the actual methods to be used.

Organizations that have implemented successful waste minimization programs have incorporated cost accounting methods which take into account direct and indirect waste management costs, the costs of lost production, raw materials, treatment, disposal as well as reduced cleanup and liability costs. An understanding of the full costs of waste generation and management is often a critical element for justifying waste minimization decisions.

The Agency does not believe that the cost accounting procedures detailed in the Cost Allocation Section are unduly specific as might have been construed from the phrase "fully-loaded waste management costs". However, this phrase has been deleted from the interim final guidance and the concept has been reworded as "a system to appropriately allocate the true costs of waste management to the activities responsible for generating the waste in the first place" to clarify the Agency's intent. EPA's Waste Minimization Opportunity Assessment Manual (July 1988), and Facility Pollution Prevention Guide (May 1992) provide a sample of a waste accounting system.

Guidance Element E: Encourage Technology Transfer:

This element of the proposed guidance stated that technology transfer on waste minimization should be encouraged from other parts of a company, from other firms, trade associations, State and university technical assistance programs or professional consultants.

Several respondents strongly supported the exchange of waste minimization information among all sources. One respondent stated that variability among facilities requires that judgements on the applicability of technology be made on a facility-specific basis with considerable input from production personnel at the facility. Another respondent indicated that EPA should include specific information on waste minimization resources available to the public from the EPA.

The Agency agrees that the exchange of waste information among all sources is a key factor in the transfer of technology and that production personnel need to play a major role in the application of appropriate technologies. The interim final guidance has additional wording to stress these points. Additionally, a section detailing information on waste minimization programs has been added to the interim final guidance.

Guidance Element F: Program Evaluation:

This element of the proposed guidance stated that a periodic review of program effectiveness should be conducted and that the review be used to provide feedback and identify potential areas for improvement.

In general, the respondents strongly supported periodic program evaluations that can be used to identify areas for improvement and enhance the effectiveness of waste minimization programs.

The Agency continues to support periodic program evaluations as an element in this guidance. To strengthen this section, however, the name has been changed to "Program Implementation and Evaluation" in order to give additional emphasis to implementing as well as evaluating opportunities identified by the assessment process.

[FR Doc. 93–12759 Filed 5–27–93; 8:45 am]
BILLING CODE 6560-50-P

APPENDIX B

National Survey of Laboratory Chemical Disposal and Waste Minimization: Survey Design, Methods, and Analysis

K. Leigh Leonard and Peter A. Reinhardt

CONTENTS

Introduction 443
Survey Objectives 444
Design of the Survey Instrument 444
    Intended Targets and Respondents 445
    Instructions to Respondents 449
Characterization of the Target Populations 451
    Host Institutions of the Howard Hughes Medical Institute 451
    Institutions and Firms with Recipients of ACS's *Network News* 451
    Private Firms that Employ DivCHAS Members 452
Sampling Plan 452
Data Collection Methodology 452
Response Rate 453
Assessing and Controlling Errors 453
    Margin of Error 454
Data Analysis 455
Areas of Future Research 455
Endnotes 456

INTRODUCTION

In 1993 the authors conducted a survey of firms and institutions that generate chemical waste in their laboratories. Chapter 2 of this book discusses some of the results of that survey, particularly those findings that pertain to pollution prevention and waste minimization. Papers that reveal other survey findings are in preparation for publication. This appendix describes how that survey was conducted and the background information behind the findings and conclusions.

SURVEY OBJECTIVES

The survey had four primary objectives:

- Determine what management methods are being used for laboratory chemical waste
- Determine the rationale used by firms and institutions to determine if laboratory waste should be disposed of in the laboratory via the sanitary sewer or by chemical treatment methods
- Determine what methods are in use to minimize laboratory chemical waste
- Examine the costs of waste disposal and savings from waste minimization

DESIGN OF THE SURVEY INSTRUMENT

The survey instrument's design is based on the above objectives and the authors' experience. The University of Wisconsin–Extension Survey Research Laboratory provided consultation on the survey instrument and sampling methodology.[1] The survey was pretested by a small number of selected colleagues who were asked not to respond to the survey, if they received it.

The survey instrument has seven parts (see Figure 1):

1. *Profile* of the respondent, their institution, or firm, including head count, State, and hazardous waste generation rate. Because generation of acute hazardous waste was not specified, the per-month waste generation categories do not necessarily correspond to regulatory categories.
2. *Management* of laboratory chemicals and waste, which queried on-site and off-site use of 16 waste management methods.
3. Disposal of laboratory chemicals to the *sanitary sewer,* including how wastes were selected for the sewer and the factors that inhibit sewer disposal.
4. *Neutralization and chemical treatment* of laboratory waste, including commonly used chemical treatment methods, their legal basis for conducting treatments, the selection of chemicals to be treated, and factors inhibiting neutralization or chemical treatment.
5. *Minimization* of laboratory chemical waste, which requested information on use of 15 waste minimization methods, as well as their most beneficial method, the costs of and savings from waste minimization, and waste minimization plans and goals.
6. Management of *certain waste* laboratory chemicals which asked respondents how they would manage 13 laboratory wastes.
7. *Cost and problems* of laboratory chemical waste which asked for the institution's or firm's annual cost for commercial off-site disposal, and what they consider to be their most difficult laboratory waste problem.

APPENDIX B

National Survey of Laboratory Chemical Disposal and Waste Minimization	
Directions: This survey should be completed by the person who is responsible for, or has the most knowledge of, the laboratory waste disposal practices of your institution or firm. Complete only for laboratories that use chemicals and chemical methods. If you have received more than one survey, *please complete only one survey for each institution or firm.* Base your answers on your institution's or firm's activities in 1992. On average, it takes about 20 minutes to complete the survey. **Deadline: June 4, 1993.** Please return your completed survey by June 4th to save us the cost of mailing reminders. Mail the survey using the enclosed self-addressed stamped envelope to: Safety Department University of Wisconsin-Madison 317 North Randall Avenue Madison, WI 57315-1003	May 10, 1993 Dear Colleague, We need your help with this national survey of chemical disposal and waste minimization practices in laboratories. We are surveying institutions and firms to better understand the variety and benefits of laboratory waste management methods. Survey results will be published in the forthcoming book, *Pollution Prevention and Waste Minimization in Laboratories* (Lewis Publishers). This information will help you and other laboratories select methods to safely and efficiently manage chemical waste. All answers will be kept confidential. The code on the survey tells us who has responded and does not need to be reminded. Please call us if you have questions. We are very interested in any comments you may have; feel free to attach additional sheets. And thank you for participating! Sincerely, K. Leigh Leonard — University of Wisconsin System — 608-263-4419 Peter A. Reinhardt — University of Wisconsin-Madison — 608-262-9735

❑ Check here and return blank survey if you have no chemical labs

❑ Check here if you would like us to send you a summary of the survey results (include your return address)

PROFILE OF YOUR INSTITUTION OR FIRM

1. What *best* describes your responsibilities with respect to laboratories in your institution or firm? (Check one.)
   - ❑ Laboratory staff
   - ❑ Environmental Health and Safety staff
   - ❑ Laboratory Director, Manager or Principle Investigator
   - ❑ Other (specify): _____
   - ❑ Management or Administration

2. What is the *principle* business sector of your institution or firm? (Check one.)
   - ❑ Academic (public or private)
   - ❑ Medical or clinical (public or private)
   - ❑ Government (non-academic)
   - ❑ Other (specify): _____
   - ❑ Other private sector lab not covered by other choices (e.g., R&D, testing, analysis, QA/QC, etc.)

Please answer the remaining questions for *all* laboratories within your institution or firm. You may not have complete knowledge of all your laboratory practices, but please select answers that best represent your knowledge. For institutions or firms with multiple locations, please limit your answers to your general location (e.g., city or state).

3. Approximately how many employees and students (head count) work in *laboratories* at your institution or firm? (Please check your best estimate.)
   - ❑ 1 to 500 employees and students
   - ❑ 1,001 to 5,000 employees and students
   - ❑ 501 to 1,000 employees and students
   - ❑ More than 5,000 employees and students

Figure 1 The survey instrument.

Chapter 2 presents findings from parts 2, 6, and 7.

INTENDED TARGETS AND RESPONDENTS

To ascertain laboratory chemical waste management practices, the survey focused on all the laboratories of an institution or firm within a local proximity. For example, a respondent at a college was asked whether his answers reflected the practices at all the laboratories within the college. This avoids the problem of defining "a laboratory." Some large institutions consider their entire facility to be

4. In what state or province is your institution or firm located? _____

5. What is your institution's or firm's monthly hazardous waste generation rate for *regulated* chemical hazardous waste *from laboratories*. (Please check your best estimate.)
   ☐ Less than 100 kilograms per month   ☐ More than 1,000 kilograms per month
   ☐ 100 to 1,000 kilograms per month

MANAGEMENT OF LABORATORY CHEMICALS AND WASTE

For the purposes of this survey, *laboratory chemicals* mean those fine chemicals, reagents, solutions and gases that are used in laboratory procedures and experiments. It includes both hazardous and nonhazardous substances. Waste from the use of laboratory chemicals include: used organic solvents, aqueous solutions and mineral acids; unwanted stock chemicals, chemically contaminated gloves, pipettes and other disposable labware; and other chemical wastes.

Which of the following waste management methods are used by your institution or firm for laboratory chemicals? Include those methods conducted in the lab or on-site by your institution or firm and those conducted off-site at a nonowned waste facility (e.g., at a commercial facility or by a contractor). (Check all that apply.)	Conducted On-site	Conducted Off-site	Not Used
Landfilling in an EPA or state permitted hazardous waste landfill	☐	☐	☐
Disposal in the normal trash that leads to a sanitary/solid waste landfill	☐	☐	☐
Incineration in an EPA or state permitted hazardous waste incinerator	☐	☐	☐
Fuel blending for energy/heat recovery	☐	☐	☐
Open burning or detonation of explosives	☐	☐	☐
Incineration in a pathological, infectious, medical or animal incinerator	☐	☐	☐
Disposal to the sewer (that leads to a sewage treatment system) Describe system or process: _____	☐	☐	☐
Disposal via another wastewater system Describe system or process: _____	☐	☐	☐
Intentional evaporation for the purpose of disposal	☐	☐	☐
Other intentional releases to the atmosphere for the purpose of disposal	☐	☐	☐
Elementary neutralization (simple acid-base chemistry)	☐	☐	☐
Other chemical treatment methods	☐	☐	☐
Distillation of organic solvents	☐	☐	☐
Exchange or redistribution of unwanted chemicals	☐	☐	☐
Recycling or recovery	☐	☐	☐
Storage of waste for which no treatment or disposal capacity exists	☐	☐	☐

Unless otherwise specified, answer the remaining questions only for laboratory waste management activities and decisionmaking that occur **within the laboratory, at the point of waste generation or on-site of your institution or firm**. For example, in the Neutralization section below, do not consider neutralization that occurs at an off-site commercial hazardous waste facility.

Figure 1 *(Continued.)*

one laboratory, whereas some institutions may consider themselves to have thousands of individual laboratories on their contiguous site.

The survey was sent to individuals who were most likely aware of these practices for their facility: principally environmental health and safety staff or professionals interested in chemical management or safety. The survey instructed the recipient that the "survey should be completed by the person who is responsible for, or has the most knowledge of, the laboratory waste disposal practices of

APPENDIX B

DISPOSAL OF LABORATORY CHEMICALS TO THE SANITARY SEWER

Sanitary Sewer refers to a wastewater disposal system that leads to a sewage treatment works, either privately or publicly owned (POTW), where wastewater is treated by bacterial degradation.

1. Does your institution or firm dispose of *any* laboratory chemicals to the sanitary sewer? (Check one.)
   - ☐ Yes → Continue with question 2.
   - ☐ No → Go to question 4 on this page.

2. What are the three (3) most important factors that your institution or firm considers when selecting laboratory chemicals for disposal to the sanitary sewer? (Check up to **three** factors.)
   - ☐ Chemical is not regulated by environmental laws
   - ☐ Chemical is allowed by sewage treatment works
   - ☐ Chemical is not hazardous or toxic or has low toxicity
   - ☐ Small amount or low concentration
   - ☐ Chemical is degradable by the treatment works
   - ☐ High cost of other disposal options
   - ☐ Other (specify): _____

3. To select laboratory chemicals for disposal to the sanitary sewer, what information sources do you consult (e.g., POTW, environmental regulations, regulatory personnel, treatability tests, references)? Be specific:
   _____
   _____

4. What are the three (3) most important factors that *inhibit* your institution or firm from disposing of laboratory chemicals to the sanitary sewer? (Check up to **three** factors.)
   - ☐ Fear of causing problems at sewage treatment works
   - ☐ Lack of information on waste treatability or degradability
   - ☐ Lack of information on appropriate chemicals for sewer
   - ☐ Poor or uncertain quality of sewage system
   - ☐ Fear of causing public relations problems
   - ☐ Fear of staff mistakes or abuse with sewer disposal
   - ☐ Lack of facilities within your institution or firm for sewer disposal
   - ☐ Sewage treatment works restrictions
   - ☐ Other environmental laws that restrict use
   - ☐ Lack of regulatory guidance on sewer disposal
   - ☐ Occupational risks of sewer disposal
   - ☐ No inhibition: we freely use sewer for disposal
   - ☐ Other (specify): _____

NEUTRALIZATION & CHEMICAL TREATMENT OF LABORATORY WASTE

Chemical Treatment refers to subjecting a waste to a chemical process that reduces the waste's hazard or volume for the purposes of disposal. Treatment products are usually disposed of in the sanitary sewer, normal trash or to a hazardous waste facility. Chemical treatment methods include neutralization, oxidation, precipitation, detoxification, etc.

1. Does your institution or firm dispose of *any* laboratory chemicals via elementary neutralization or chemical treatment in the laboratory or on-site? (Check one.)
   - ☐ Yes → Continue with question 2.
   - ☐ No → Go to question 7 on page 4.

2. Within your institution or firm, *where* does elementary neutralization or chemical treatment occur? (Check all that apply.)
   - ☐ In the laboratory, at the point of waste generation
   - ☐ At one or more unpermitted/unlicensed on-site location(s) that serves several labs
   - ☐ At a RCRA or state permitted on-site facility that serves several labs

Figure 1 *(Continued.)*

[their] institution or firm." In addition, if the respondent did not have complete knowledge of all their laboratory practices, the survey asked to "select answers that best represent [their] knowledge."

Recipients were asked to "complete only one survey for each institution or firm" per location. To capture possible differences arising from state laws and EPA regional policies, firms with locations in several states were instructed to complete a separate survey for each location. These instructions also served to

3. Does your institution or firm chemically treat laboratory waste in ways *other than* elementary neutralization?
   - ☐ Yes → Continue with question 4, next.
   - ☐ No, we only perform elementary neutralizations → Go to question 6 on this page.

4. *Other than* elementary neutralization what are your most commonly used methods to chemically treat laboratory waste? Be specific: _____
_____

5. What is your institution's or firm's *principle* legal basis to chemically treat laboratory waste in ways *other than* elementary neutralization? (Check one.)
   - ☐ Treatment occurs in accumulation container or tank
   - ☐ Laboratory is a totally enclosed treatment facility
   - ☐ RCRA permit or license for chemical treatment
   - ☐ NPDES or state PDES permit for wastewater treatment
   - ☐ Research, development, demonstration or educational permit or exemption
   - ☐ Treatment is part of the waste generation process or experiment
   - ☐ Only unregulated waste is treated
   - ☐ Permit or license by state rule
   - ☐ Other (specify): _____

6. What are the three (3) most important factors that your institution or firm considers when selecting laboratory chemical waste for elementary neutralization or chemical treatment? (Check up to **three** factors.)
   - ☐ Availability of treatment procedures or information
   - ☐ Chemical is not regulated by environmental laws
   - ☐ Off-site disposal is unavailable or difficult
   - ☐ Desire to reduce hazard, volume or weight
   - ☐ Avoidance of handling, transport and off-site risks
   - ☐ Access to permitted wastewater pretreatment system
   - ☐ Treatment products are accepted by the sewage treatment works or solid waste facility
   - ☐ Treatment products can be disposed of at an off-site hazardous waste facility at a lower cost
   - ☐ Too small amount to ship off-site
   - ☐ Ease of treatment procedure
   - ☐ High cost of other disposal options
   - ☐ Ethic of responsibility for waste disposal
   - ☐ Other (specify): _____

7. What are the three (3) most important factors that *inhibit* your institution or firm from disposing of laboratory waste by elementary neutralization or chemical treatment? (Check up to **three** factors.)
   - ☐ Environmental laws that require a treatment permit
   - ☐ Expertise required for chemical treatment
   - ☐ Fear of staff mistakes with chemical treatment
   - ☐ Lack of regulatory guidance on chemical treatment
   - ☐ Inability of sewerage treatment works to accept chemical treatment products
   - ☐ Lack of facilities within your institution or firm for chemical treatment
   - ☐ No inhibition: we freely dispose of laboratory waste by chemical treatment.
   - ☐ Effort and time for chemical treatment
   - ☐ Occupational risks of chemical treatment
   - ☐ Lack of chemical treatment procedures
   - ☐ Other (specify): _____

Figure 1 *(Continued.)*

prevent duplicate responses in the *Network News* data from facility locations that were sent more than one survey.

As a result, the survey findings can be considered to show how institutions and firms manage their laboratory chemical waste. For many of the survey questions, this assumes that all laboratories within a facility location manage their wastes similarly (e.g., according to state law or a facility policy — a reasonable assumption) and that the respondent had sufficient knowledge of waste management practices at all of the institution's or firm's laboratories.

MINIMIZATION OF LABORATORY CHEMICAL WASTE

1. Does your institution or firm make any effort to minimize laboratory chemical waste? (Check one.)
   ☐ Yes → Continue with question 2.   ☐ No → Go to the next section, **Management of Certain Wastes,** page 6.

2. Which of the following waste minimization methods are used at your institution or firm for laboratory chemicals?	Used	Not used	Don't know
Computer simulation or modeling to replace wet lab	☐	☐	☐
Change equipment or procedure to eliminate waste or reduce hazard of waste	☐	☐	☐
Substitute materials with nonhazardous or less hazardous materials	☐	☐	☐
Reduce scale of experiments or procedures (e.g., microscale)	☐	☐	☐
Other process changes to use less materials	☐	☐	☐
Minimize surplus by buying less or in smaller quantities	☐	☐	☐
Minimize surplus by controlling purchases	☐	☐	☐
Minimize surplus by monitoring inventories	☐	☐	☐
Redistribution or exchange of surplus to other laboratories	☐	☐	☐
Distillation of waste solvents for reuse	☐	☐	☐
Other beneficial use of waste (e.g., use of waste base to neutralize acids; use of waste in subsequent experiments)	☐	☐	☐
Other recovery or recycling (e.g., recovery of silver from photo labs, recycling of mercury)	☐	☐	☐
Chemical treatment to reduce hazard or volume	☐	☐	☐
Bulking and commingling similar wastes to reduce volume	☐	☐	☐
Other volume reduction	☐	☐	☐

3. What is your institution's or firm's *single, most beneficial* waste minimization method for laboratory chemical waste? _____

   Why do you consider it to be most beneficial? _____

4. If known, what is the *estimated annual net savings* from your most beneficial waste minimization method? $ _____

5. If known, what are the estimated *annual cost of, and savings from,* laboratory waste minimization within your institution or firm? (Please provide any estimates you have.)

   | Savings $ | − | Cost $ | = | Net Savings or Cost (Specify) $ |

6. Does your institution or firm have a written waste minimization plan or policy that applies to laboratories?
   ☐ Yes → Continue with question 7.   ☐ No → Go to the next section, **Management of Certain Wastes,** page 6.

Figure 1 *(Continued.)*

INSTRUCTIONS TO RESPONDENTS

The survey included the following additional instructions to respondents:

- That the survey responses pertain only to laboratories that use chemical and chemical methods; a check box was provided for those recipients whose facility had no chemical labs
- That the survey answers be based on activities in 1992

7. Does the plan or policy set specific goals for waste minimization?
   ☐ No   ☐ Yes (Please give a sample goal): _____

MANAGEMENT OF CERTAIN WASTE LABORATORY CHEMICALS

Based on the practices of your institution or firm, what would be the *predominant* management method for one kilogram of the following laboratory chemical wastes? (Check only one practice for each waste.)	Disposal to the Normal Trash	Disposal to the Sanitary Sewer	On-site Neutralization or Chemical Treatment	Managed at a Hazardous Waste TSD Facility	Other (specify)	Don't Know
Used glacial acetic acid	☐	☐	☐	☐	☐	☐
Ammonium chloride (surplus stock)	☐	☐	☐	☐	☐	☐
Dilute aqueous solutions of acetone and alcohols used for cleaning glassware	☐	☐	☐	☐	☐	☐
Aqueous solution containing 10% methanol	☐	☐	☐	☐	☐	☐
Calcium carbonate (surplus stock)	☐	☐	☐	☐	☐	☐
Calcium phosphate (surplus stock)	☐	☐	☐	☐	☐	☐
Used chromic-sulfuric acid cleaning solution	☐	☐	☐	☐	☐	☐
Dextrose (surplus stock)	☐	☐	☐	☐	☐	☐
Aqueous solution containing 5% ethidium bromide	☐	☐	☐	☐	☐	☐
Used formalin solution	☐	☐	☐	☐	☐	☐
5 N Hydrochloric acid	☐	☐	☐	☐	☐	☐
Potassium chloride (surplus stock)	☐	☐	☐	☐	☐	☐
Toluene-based liquid scintillation cocktail (<0.05 µCi H-3/g)	☐	☐	☐	☐	☐	☐

COST AND PROBLEM OF LABORATORY CHEMICAL WASTE

1. What is your institution's or firm's approximate annual cost for commercial off-site disposal of chemical waste from laboratories? (A range is acceptable.) $ _____

2. What is your institution's or firm's most difficult laboratory waste problem (i.e., waste type, problem in collection, transport, disposal, cost, compliance, trend, etc.)? _____

   Why is it your biggest problem? _____

Thank You!

Figure 1   *(Continued.)*

The second part of the survey, Management of Laboratory Chemicals and Waste, asks respondents to indicate which methods their institution or firm used to manage laboratory chemical waste, both on-site and off-site. For the remainder of the survey, respondents were asked to limit their answers to those waste management activities that "occur within the laboratory, at the point of waste generation or on-site of your institution or firm."

The survey included definitions of *laboratory chemicals, sanitary sewer* and *chemical treatment* (see Figure 1).

CHARACTERIZATION OF THE TARGET POPULATIONS

The survey's objective was to determine laboratory chemical waste practices nationally. However, to the authors' knowledge, there is no comprehensive list or database of institutions or firms that have chemical laboratories, nor would any combination of lists likely be comprehensive. Access to such lists, if deemed appropriate, would be difficult and expensive, and beyond the authors' modest resources. As a result, we chose to survey three target populations that are subsets of original target (all institutions and firms who generate laboratory chemical waste):[2]

- Host institutions of the Howard Hughes Medical Institute (HHMI)
- Institutions and firms that generate laboratory chemical waste that employ subscribers of the American Chemical Society's *Network News* newsletter (ACS *NN*)
- Private firms that generate laboratory chemical waste that employ members of the American Chemical Society's Division of Chemical Heath and Safety (DivCHAS)

Characteristics of these target populations are described below.

HOST INSTITUTIONS OF THE HOWARD HUGHES MEDICAL INSTITUTE

In 1993, the Howard Hughes Medical Institute (HHMI) funded medical research at 53 large research institutions in the U.S. which are designated as Host Institutions. To conduct such research, chemicals are used in laboratories, which results in the generation of laboratory chemical waste. For nearly all of these institutions, environmental, health, and safety staff have been given the responsibility to manage or monitor laboratory chemical waste. On April 5, 1993, the HHMI Office of Laboratory Safety mailed the National Survey of Laboratory Chemical Disposal and Waste Minimization survey form and a cover letter to the environmental health and safety directors of each Host Institution.

INSTITUTIONS AND FIRMS WITH RECIPIENTS OF ACS'S *NETWORK NEWS*

The American Chemical Society is the nation's largest professional society, with a large proportion of members who work with chemicals in laboratories. Its office of Government Relations and Science Policy publishes *Network News,* a free newsletter that generally reports on chemical, environmental, safety, and compliance issues facing educational institutions, laboratories, and other settings where chemical professionals work. The National Survey of Laboratory Chemical Disposal and Waste Minimization survey form was included in the Spring 1993 issue, which was sent to all 327 subscribers.

PRIVATE FIRMS THAT EMPLOY DIVCHAS MEMBERS

The Division of Chemical Heath and Safety (DivCHAS) is made up of members of the American Chemical Society who pay nominal additional dues to receive a newsletter and participate in the Division's health and safety programming. With permission of the Division, the authors identified 352 members listed in the 1992 Division directory who appeared to be employed by for-profit private sector firms (i.e., organizations that were not educational institutions, hospitals, or government facilities). From that list, the authors randomly selected a sample of 201 individuals, who were mailed the National Survey of Laboratory Chemical Disposal and Waste Minimization survey form on May 10, 1993.

SAMPLING PLAN

In their Regulatory Impact Assessment for their laboratory standard, the U.S. Occupational Safety and Health Administration (OSHA) classified laboratories into three populations: academic, clinical, and industrial.[3] Similarly, our survey categorized laboratories into four populations:

1. Academic (public or private)
2. Medical or clinical
3. Government (nonacademic)
4. Other private sector labs not covered by other choices

HHMI and *Network News* populations have a large percentage of academic institutions (almost exclusively postsecondary schools). Unlike the other categories, however, private sector laboratories are more likely to be concerned about profits and, perhaps, the financial aspects of waste management and minimization. Determining the costs and benefits of waste management for private laboratory firms (in contrast to academic, medical, and government laboratories) was a major objective of sampling the DivCHAS population.

For HHMI and *Network News* populations, the sample consisted of the entire target population. Because of limited resources, only 201 of 352 private members of DivCHAS were sampled.

DATA COLLECTION METHODOLOGY

Data collection methodology varied by target population. As explained below, survey research is most accurate when high response rates are achieved. Response rates are highest when mailings are personalized and sent by first class mail. Each survey form was coded on the lower right corner for its target population.

For HHMI, data collection consisted of a first mailing to all Host Institutions, which included a survey form and a personalized cover letter from the HHMI

APPENDIX B	453

Office of Laboratory Safety. This was sent first class and included a first class self-addressed stamped envelope. Nonrespondents were sent a second mailing that was identical except that the personalized cover letter was from the authors.

For *Network News,* data collection consisted of a survey form being mailed to all subscribers as part of the Spring 1993 newsletter. Note that the first page of the survey included a short, generic cover note, explaining the purpose and benefits of the survey (see Figure 1). A reminder to return the survey was printed in the subsequent newsletter.

Data collection for DivCHAS consisted of up to four mailings. So that subsequent mailings would go to nonrespondents, each survey was coded to identify the recipient, which was in addition to a "CHAS" coding. First, the randomly selected sample was sent, by first class mail, a survey form and a first class self-addressed stamped envelope. A handwritten picture postcard was sent as a reminder to nonrespondents. The third mailing was identical to the first, except that a personalized cover letter was included, which encouraged their participation in the study. Nonrespondents were then sent a second handwritten picture postcard.

RESPONSE RATE

The response rate was calculated to be:

$$\frac{\text{Number of completed surveys returned}}{\text{Total surveys sent } - \text{ undeliverables}} = \text{Response rate}$$

Undeliverables include those surveys returned, but marked "no laboratories," "addressee no longer here," "no such address," and "moved." Total surveys sent included surveys returned blank and refusals to participate (some recipients had a company policy to not participate in surveys).

Response rates for each target population are given in Table 1.

ASSESSING AND CONTROLLING ERRORS

There are four primary types of errors in survey research:

1. *Coverage error.* This type of error is minimized when everyone in the desired population has a nonzero chance of being selected in the sample. For the HHMI and ACS NN populations, the entire population was sampled. Although much care was taken to avoid this error with DivCHAS, it is possible that all members employed in private sector labs were not identified by the authors and thereby did not have a chance of being selected for the sample.
2. *Sample error.* This depends upon the size of the sample. In general, larger samples minimizes the margin of error. Sample sizes are given in Table 1.

Table 1 Survey Samples and Response Rates

	Target population		
	HHMI	DivCHAS	ACS NN
Total population	53	352	327 [a]
Population surveyed	53 (all)	175 [b] (randomly selected)	327 (all)
No. surveys returned	46	74	60
Response rate (%)	87	42	18
Margin of error (%)	±15	±12	±12

[a] Because some institutions and firms receive multiple *Network News* subscriptions, the total ACS *NN* population includes some duplicates. Respondents were asked to return only one survey per institution or firm location, and it appears that those instructions were followed and the 60 returned ACS *NN* surveys include no duplicates. As a result, the ACS *NN* response rate may be higher, with a commensurate decrease in the population's margin of error.

[b] 201 DivCHAS randomly selected members were initially surveyed, prior to subtracting one duplicate, and those members who had moved, etc. (undeliverables).

3. *Nonresponse error.* This error is the result of differences in the characteristics of respondents and nonrespondents. For example, people who where concerned about the legality of their waste disposal practices may have been less likely to respond than those who were not. Some bias may have resulted because the target populations tended to include individuals who had an interest in safety or were safety professionals. Nonresponse error is minimized when the response rate is high. In addition to follow-up reminders, we tried to minimize nonresponse bias and improve our response rate by explaining, in the survey's cover note, the benefit of this information to all laboratory waste managers, and by offering to return a summary of our survey results to those respondents who checked the box.

4. *Measurement error.* This is a measure of the effectiveness of the survey instrument and the accuracy of the respondents. Response bias can result when respondents misunderstand, answer a different question, lie, or do not have a complete knowledge of their organization's laboratory hazardous waste management. We hoped to minimize these errors by careful survey design and by pretesting the survey. The survey asked of waste management practices in 1992. Because a year is a relatively long period, and because some surveys were not returned until late 1993, some respondents may have forgotten the details of their laboratory waste management during 1992 (i.e., recall bias). On the other hand, the respondent's memory should be augmented by information documented in policies, procedures, and reports, and because an organization's waste management practices are routine and change slowly.

MARGIN OF ERROR

Based on sample sizes and response rates, the survey's margin of error is ±15% for HHMI and ±12% for DivCHAS and ACS *NN* at a confidence level of 95%, as calculated by the Wisconsin Survey Research Laboratory. However, because high response rates minimizes nonresponse error, the HHMI results should be considered a fairly accurate reflection of the HHMI population. Conversely, the significantly lower response rate for ACS *NN* raises the possibility of nonresponse error.

DATA ANALYSIS

Answers from the completed surveys were keyed into a computer and the data were analyzed using Paradox 3.5 database software. Analysis was done by target population.

Percentages given are the percent of institutions and firms in each category. In calculating percentages, blanks and "Don't Know" answers were treated as a separate category and were not subtracted from the total.

AREAS OF FUTURE RESEARCH

In contrast to this formal survey, Yale University, EM Science, Inc., and New York University have circulated questionnaires to colleges and universities to gather anecdotal information on waste minimization in laboratories. Although these projects can be a very useful way to exchange information, they cannot provide an accurate overview of a population's behavior. A formal, rigorous survey such as ours, scientifically conducted and analyzed, can be the basis for policymaking, regulatory decisions, and an organization's strategic planning.

We believe that laboratories (and the research, teaching, and medical and environmental testing that they do) are vital to our society. Chemical waste management appears to be a significant burden and cost to laboratories. The results of this survey provides some insight into these problems and their solutions. As a result, several areas seem ripe for future research:

- Ideally, future surveys of laboratories would cover a broader population of institutions and firms, and thereby more closely represent a sample of all U.S. laboratories. Such a survey would be very costly, however.
- Although some analysis of subgroups are described in Chapter 2, a larger sample and lower error rates would permit a more thorough analysis, including examinations of divisions of subgroups (such as private sector small quantity generators).
- A survey to identify the types and quantities of laboratory wastes would be interesting, as well as information on specific laboratory activities that generate waste. This would shed light on why certain waste management and minimization practices were in use.
- Future surveys should evaluate the effectiveness of EPA's Waste Minimization Guidelines as they apply to laboratories.
- A future survey might be useful to review other pollution prevention initiatives, such as reduction of sewer effluents and fume hood emissions.

ENDNOTES

1. The authors wish to extend their deepest appreciation to Linda J. Penaloza, Associate Director, and Jane Campbell, Senior Research Specialist, of the University of Wisconsin–Extension, Continuing Education Extension, Wisconsin Survey Research Laboratory for their consultation, advice and assistance in survey design, data analysis, statistical interpretation of findings, and presentation of results.
2. The authors wish to thank W. Emmett Barkley and Cheryl A. Warfield of the Howard Hughes Medical Institute Office of Laboratory Safety for their support and assistance in conducting the survey. We thank the American Chemical Society Division of Chemical Health and Safety for permission to use their member directory. We very much appreciate the support of David Schleicher of the ACS Office of Government Relations and Science Policy for including the survey in their newsletter.
3. U.S. Occupational Safety and Health Administration, Occupational Exposures to Hazardous Chemicals in Laboratories; Final Rule, *Fed. Reg.,* 55 (21), 3306, 1990.

APPENDIX C

POLLUTION PREVENTION DIRECTORY

Publication Date: September 1994
Publisher: U.S. Environmental Protection Agency, Pollution Prevention Division
Publication Number: EPA 742-B-94-005

This comprehensive (103-page) reference book can be obtained free of charge from EPA's Pollution Prevention Division. It describes in detail pollution prevention programs and grants administered by EPA, other federal programs, and states. In addition, it lists clearinghouses, databases, periodicals, directories, hotlines, centers, and associations that focus on pollution prevention activities. The *Directory* can be obtained by contacting:

Pollution Prevention Information Clearinghouse
Environmental Protection Agency
U.S. EPA (7409)
401 M Street SW (3404)
Washington, D.C. 20406
Tel: (202) 260-1023
Fax: (202) 260-0178

THE POLLUTION PREVENTION YELLOW PAGES

Publication Date: September 1994
Publisher: National Pollution Prevention Roundtable

This reference was compiled from the results of a comprehensive survey (conducted by the Roundtable) of state and local government pollution prevention programs. The 346-page reference is tightly organized by state. Within each state, entries are alphabetized by organization name. Entries are also indexed by contact names and areas of expertise. Detailed information is included, such as program type, its goals, and its budget. The Yellow Pages cost $25 at the time this book was published. It may be obtained by contacting

National P2 Roundtable
218 D Street S.E.
Washington, D.C. 20003
Tel: (202) 543-P2P2
Fax: (202) 543-3844

Glossary

Glossary

beneficial use: when applied to recycling of waste, the recovery of heat, energy, or materials of use to society.

bulk: (1) as a verb, to combine smaller amounts together or to commingle; (2) as an adjective, it implies that the quantity of a material is relatively large.

commingle: to combine wastes in small containers into larger containers.

de minimis: a trivial amount; an amount so small that it need not be considered (from the Latin *de minimis non curat lex* — the law does not deal with trifles).

disposal: when applied to chemical waste, a method for getting rid of the waste such that the material cannot easily be used for any purpose.

distillation: the process of purifying one or more compounds by first heating a mixture until some of the material vaporizes, then condensing out the vapors.

effluent: a gaseous or liquid discharge, generally regulated under the Clean Water Act.

emissions: when applied to chemical waste, releases (usually gaseous or liquid) to the environment, generally regulated under the Clean Air Act.

energy recovery: the incineration or combustion of materials in such a manner that beneficial energy, usually in the form of heat or electricity, is recovered or utilized.

environmentally sound recycling: a method of recycling that greatly minimizes adverse effects on the environment (note that this does not include processes that divert pollution to another media).

environmental receptors: animals and/or plants that may be impacted by pollution.

generator: (1) from a regulatory point of view, an industry or other establishment that produces waste in the course of conducting business; (2) more generally as applied to waste, any person whose process or activity produces waste.

hazardous material: (1) broadly, a material that presents a chemical, radioactive, or biological hazard; (2) when applied to chemicals only, a material that exhibits hazardous characteristics (the material may or may not be a waste).

hazardous waste: (1) from a regulatory point of view, hazardous waste is a solid waste that is ignitable, corrosive, reactive, or meets one or more criteria pertaining to toxicity; (2) any waste that presents danger to environmental receptors.

heat recovery: a process by which the heat from an incineration or combustion process is recovered for a beneficial purpose.

incineration: an engineered process in which waste is combusted at high temperatures and under such conditions that it is thermally degraded.

in-process recycling: an engineered method for recycling wastes to an earlier stage of the process that generated them, where the wastes can be used beneficially.

laboratory: (1) OSHA definition, a laboratory is an operation in which (a) multiple chemicals are routinely used, (b) multiple processes are routinely used, (c) a single person can manipulate both the chemicals and the processes, and (d) where personnel protective equipment and engineering controls common to laboratories are found; (2) generally, an area in which teaching, research, and/or analytical work is conducted by individuals using small amounts of chemicals, typically 4 kg or less, at a time.

laboratory use: the use of small amounts of chemicals, typically 4 kg or less, at a time by individuals for teaching, research, and/or analytical purposes.

manifest: from a regulatory point of view, a descriptive legal document required to accompany the shipment of hazardous wastes to an off-site treatment, storage, or disposal facility.

media: air, water, or land — the portion of the environment that pollution can impact.

nonpoint source: as applied to pollutants, a source of emission that is diffuse and widely dispersed in area and, hence, is typically difficult to control.

point source: as applied to pollutants, a source of emission that is well defined and discrete, such as a pipe or stack and, hence, is typically relatively easy to control.

pollutant: a material or effect that has a detrimental impact on health and/or the environment.

pollution control: a method or methods for preventing or limiting the releases of pollutants to the environment.

pollution prevention: (1) broadly, activities that keep pollutants from being created in any media; (2) according to EPA, source reduction and environmentally sound recycling.

pollution prevention audit: a method for identifying and evaluating pollution prevention activities.

production index: measurement used as a baseline to indicate the amount of pollution created relative to production.

RCRA: the Resource Conservation and Recovery Act, the legislation passed by Congress in 1976 which authorizes the EPA to regulate hazardous wastes.

recycling: a method by which a beneficial use or product is obtained from a waste material.

redistribution: a process by which a usable material, such as a surplus chemical, that identified as a waste by one user is given to another user to be used beneficially.

reuse: a process by which a waste material that has been disposed as a waste is given to another user to be used on an "as-is" basis for a beneficial purpose.

solid waste: (1) from a regulatory point of view, any solid, semisolid, gaseous, sludge, or liquid waste (the entire universe of all waste — hazardous waste and other specialized categories are subsets of solid waste); (2) more generally, a waste that is not a gas or a liquid; (3) sometimes used to denote those solid wastes that are not hazardous waste.

source reduction: a method of reducing the amount of waste generated at the point that waste is produced; this may be accomplished by methods such as process modification, process improvement, material substitution, and product modification.

source separation: a method by which waste components are separated into more homogenous subgroups at the point where the waste is generated (this often facilitates waste minimization and more efficient waste management).

speculative accumulation: the accumulation of a certain waste in the hope that enough can be aggregated in a timely manner so that markets for the waste will develop or that the recycled products will rise in value.

storage: what is done with wastes between the time the waste is produced until the time it is disposed (waste that is temporarily accumulated at the generation point or being transported is generally not considered to be in storage).

substitution: as applied to pollution prevention, the process of replacing a starting material with a different starting material for the purpose of producing less hazardous or a smaller amount of waste.

surplus chemicals: in a laboratory, unused chemicals in the laboratory's possession that are no longer needed, but which may be suitable for use by others.

totally enclosed treatment facility: from a regulatory point of view, "a facility for the treatment of hazardous waste which is directly connected to an industrial production process and which is constructed and operated in a manner which prevents the release of any hazardous waste or any constituent thereof into the environment during treatment."

toxic release inventory (TRI), Form R: a form required by some states pertaining to chemical emissions that laboratories.

treatment: (1) according to RCRA: "any method, technique, or process, including neutralization, designed to change the physical, chemical, or biological character or composition of any hazardous waste so as to neutralize such waste, or so as to recover energy or material resources from the waste, or so as to render such waste non-hazardous, or less hazardous; safer to transport, store, or dispose of; or amenable for storage, or reduced in volume." (2) a method applied to a waste to alter the material or to recover useful components of the waste.

TSD: a treatment, storage, or disposal facility as defined and regulated under RCRA.

volume reduction: a method by which the volume of waste is decreased (e.g., precipitation of metals from aqueous solutions).

waste: (1) a product of an activity that has no value, and generally has liabilities associated with it; (2) from a regulatory point of view, a waste is any discarded material that is abandoned, recycled, or considered to be inherently waste-like.

waste audit: a process for assessing the kinds and amounts of wastes produced, the sources of these wastes, and the methods used to manage these wastes.

waste generation: a process that creates waste.

waste management: all of the activities applied to a waste from the time the waste is produced to the time it has been disposed or recycled; such activities include generation, handling, accumulation, storage, transport, treatment, and disposal.

waste minimization: (1) according to the U.S. EPA, any method that reduces the volume or toxicity of waste requiring disposal; (2) more generally, any method that reduces the amount of waste.

waste minimization audit: a process for identifying and evaluating both existing and potential waste minimization activities.

waste minimization certification: required by the U.S. EPA on all manifests whereby the generator certifies that they have a program to minimize hazardous wastes to the extent practical; note that small quantity generators need merely to state that they have made a good-faith effort to minimize waste generation and select the best waste management method that is available and that they can afford.

waste minimization hierarchy: the ordering of various waste minimization methods by order of desirability on the basis of environmental impact or closeness to generation and waste management pathway.

waste reduction: a process that reduces that amount of waste produced.

waste segregation: the separation, or effort to maintain separation, of various wastes or waste components into more homogenous subsets to facilitate subsequent management.

Index

A

Acetone, 36
  waste recovery analysis, 369
Acetonitrile, waste recovery analysis, 352, 355
Acids
  chromosulfuric, 160
  neutralization, 105, 155, 183
  spill treatment, 288–289
  waste reduction, 47
Acts, legislative, See Laws, environmental
Administrations, See also Educational institutions; executive management
  waste minimization programs, 71
    commitment establishment, 74–75, 91–94, 408, 415
    cost savings, 190–191
    infrastructure, 137
Agracetus program, waste minimization audit, 382
  benefits, 387–388
  chemical usage, 380–381
  company background, 380
  goals, 378
  management controls, 386–387
  program implementation, 379–380
  recycling, 385–386
  source reduction methods, 382–388
  wastestreams, 381–382
Air emissions
  characterization, 125
  control, 125, 129–131
  estimations, purpose, 127–128
  identification methods, 128
  management program, 135–138
  measurements, 128–129, 138
  policies, 136–138
  regulations
    environmental health, 132
    occupational health, 131–132
  research needs, 138
  sources, 126-127
  types, 126
Air pollutants, Clean Air Act provisions, 5
Alcohol mixtures, solvent recovery, 347, 375
Alkenes, hydroperoxide destruction, 278–279
Ambient air
  defined, 133–134
  hypothetical state proposals, 135
American Chemical Society (ACS)
  Chemical Health and Safety Division, waste management study, 17
    annual costs, 24
    cost savings, 28
    management practices not used, 22
    management practices used, 20–21
    minimization practices, 25–28
    respondent profile, 19
    sewer and normal trash use, 23–24
  Laboratory Equity and Waste Minimization Act
    description, 116
    provisions, 117–118
Amines, aromatic, disposal reactions, 283–284
Animal carcasses, radioisotope presence, waste management, 198
Ash, incineration, radioactive monitoring, 199, 201
Audits, waste minimization, 63–64, 415
Autoclaves
  infectious waste sterilization, 179
  limitations, 223
Automated walkway systems
  examples, 174
  laboratory personnel safety, 174
Azeotrope
  defined, 248
  minimum boiling, 248
Azeotropic behavior, defined, 248
Azides
  inorganic, 285–286
  organic, 282-283

465

B

Bar codes, chemical inventory management, 161, 302
"Baseline" conundrum, waste minimization reforms, 60
Bases
  elementary neutralization, 105, 277
  spill treatment, 288–289
Batteries, disposal, 400
Biosafety, disposable item reuse and, 172–173
Board of Regents, involvement in waste minimization programs, 91
Bromine, disposal reaction, 284
Bulking, defined, 108

C

California Hazardous Waste Source Reduction Act, effect on waste generators, 58
CDC, See Centers for Disease Control
Centers for Disease Control (CDC), infectious waste categories, 167
CERCLA, See Comprehensive Environmental Response, Compensation, and Liability Act
CFCs, See Chlorofluorocarbons
Champions
  places to find, 90–91
  role in waste minimization programs, 71
CHAS, See Chemical Health and Safety Division, ACS
Chemical brokers
  ideal criteria, 300
  problems associated with, 299–301
Chemical Health and Safety Division, ACS (CHAS), waste management study, 17
  annual costs, 24
  cost savings, 28
  management practices not used, 22
  management practices used, 20–21
  minimization practices, 25–28
  respondent profile, 19
  sewer and normal trash use, 23–24
Chemical inventory management
  bar codes, 161, 302
  procurement process, 302–304
  stockroom, 394
  waste management practice, 30, 38, 48-49
Chemical Manufacturers Association (CMA)
  pollution prevention programs, 9
  Responsible Care program, 306
Chemicals
  commonly used types in laboratories, 48
  computerized databases, 115, 161, 302
  inorganic, See Inorganic chemicals
  organic, See Organic chemicals
  purchasing programs, 393–394, 409
  redistribution programs, See University of Arizona
  sanitary sewer drainage, 142
  spills
    air contamination, 127
    protection methods, 151
  stock, 106
  storage practices, 136
  surplus amounts
    management controls, 386–387
    redistribution, 27, 38, 105–106, 114, 299–301, 428
    take-backs, 301
  waste minimization, 48
Chemical treatment
  elementary neutralization, 19, 22, 39, 59
    acids, 47, 105, 155, 183
    bases, 105
  fuel blending, 19, 22, 100, 155–156
    description, 155
    off-site types, 155
  precipitation and evaporation, 155
  waste minimization, 29, 108–109
Chemical waste
  management practices, on-site and off-site, 19–21, 183
  regulatory differences, 116
  source separation, 169–170
Chemistry
  automatic analyzers, 175
  waste minimization advancements, 175–176
Chemistry industry, microscale chemistry applications, 217–219
Chlorofluorocarbons (CFCs), legislative control, 7
Chromatography
  gas
    liquid purifications, 214
    as process modification, 410
    solvent recovery analysis, 331–332
  high-performance liquid
    solvent recovery analysis, 331–332, 352–366
  solvent use, 240
Chromic acid, waste reduction, 47
Chromosulfuric acid, reduction methods, 160
Clean Air Act
  air pollutant classifications, 132–133
  principles, 6
  provisions, 133

INDEX 467

state requirements, 134
suggestions to improve, 64
Cleaning supplies
  alternative types, 397
  chromic acid, 47
  as source of waste, 42, 410
Clean Water Act, See also Federal Water Pollution Control Act
  goals, 142–143
  principles, 5–6
  suggestions to improve, 64
CMA, See Chemical Manufacturers Association
Combined Wastewater Formula, 147
Commingling
  advantages, 27
  aqueous wastes, 324
  at central facility, 323–325
  cost savings, 320–321
  defined, 320
  in laboratories, 322–323
  organic solvents, 108, 237
  safety hazards, 321–322
  waste management, 30, 233
  for waste minimization, 321
Compaction treatment
  infectious waste, 183
  at UIUC, 200–201
Companies, See specific companies
Composition–boiling point curves, vapor–liquid
  defined, 243
  example, 244
Comprehensive Environmental Response, Compensation, and Liability Act (CERCLA)
  effect on wastestream liability, 61
  principles, 7
Congress
  Pollution Prevention Act passage, 8
  waste minimization legislation, 117
Containers, waste
  assessment, 312
  for commingling, 323
  criteria, 234–235
  identification, 235
  recycling, 395–397
  types, 227–228
Costs
  disposal, 225
  source separation
    assessment and analysis, 226–227
    collection and containment, 227
    laboratory personnel and waste handler training, 228

overview, 231
waste assessment and analysis, 226–227
waste management practices, 24, 415
liability, 84
Cost savings
  commingling, 320–321
  fuel blending, 155
  microscale chemistry, 216
  solvent recovery, 374–375
  from waste management practices, 30, 92, 190–191
Cross-media pollution, considerations, 64–65
Crystallizations, in microscale chemistry laboratories, 213–214
Cyanides, disposal reaction, 284

D

Databases, computerized
  chemical supply control, 115, 161, 302
  redistribution promotion, 81
  wastestream assessment process, 80–81
DCM, See Dichlormethane
Decay, radioactive
  half-life, 412
  storage methods, 156–157
Department of Environmental Health and Safety (DEHS), waste minimization programs, 407–408
Department of Risk Management & Safety (RM&S), UA chemical redistribution program, 312
Department of Transportation (DOT), container disposal regulations, 396–397
Dichlormethane (DCM), waste recovery analysis, 366–367
Discharge permits, types, 145
Disposable items
  examples, 185
  product substitution, 172–173
  reuse, 185–186
Disposal, waste
  to atmosphere, 401
  costs, 225
Disposal reactions
  inorganic chemicals
    bromine, 284
    cyanides, 284–285
    inorganic azides, 285–286
    iodine, 284
    oxidizing agents, 284
  organic chemicals
    aromatic amines, 283-284
    diethyl sulfate, 280

dimethyl sulfate, 280
N-Bromosuccinimide, 279
N-Chlorosuccinimide, 279
nitriles, 282
N,N-Dimethylacetamide, 280
N,N-Dimethylformamide, 280
organic azides, 282–283
oxalic acid, 279
oxalyl chloride, 279
picric acid, 281
sodium oxalate, 279
Disposal sites, hazardous waste cleanup, liability, 61
Distillation, 22, See also Redistillation
 cuts, 259
 fractionating columns
  characteristics, 255–259
  types, 248
 solvents, 38, 47
  commercial units, 264
  fractionated, 243, 246
  on-site, 21
  prevalence, 20
  purpose, 242
  recovery analysis methods, 331–333
  simple, 242
  theoretical background, 240–248
 use as waste minimization practice, 26
DOT, See Department of Transportation

E

Economic evaluation, waste minimization program implementation in laboratories, 84–85
Economic isolation, barrier of, defined, 113
Education in waste minimization practices for laboratory personnel, 86
Educational institutions, See also specific universities
 waste minimization programs
  administrative support, 91–94
  common variables, 104
  feasibility determinations, 83
  finding "champions," 90–91
  implementation, 74–75, 93
  maintaining progress in, 94–95
  reasons to start, 89–90
  sociopolitical climate importance, 93
EE&PC, See Environmental Engineering & Pollution Control
Efficiency of fractionating columns, 256, 259
Elementary neutralization, 19, 22, 39, 59
 acids, 47, 105, 155, 183

 bases, 105
 on-site management, 21
Emissions
 air
  characterization, 125
  control, 125, 129–131
  estimations, 127–128
  identification methods, 128
  management program, 135–138
  measurements, 128–129, 138
  policies, 136–138
  regulations, 131–132
  research needs, 138
  sources, 126–127
  types, 126
 fugitive
  from chemical storage container leaks, 126
  defined, 126
  environmental contamination, 127
  measurements, 129
Emissions targets, defined, 126
Energy savings, conversion to microscale chemistry, 218
Engineering controls
 chemical storage cabinets, 130–131
 general ventilation system, 130
 local ventilation devices, 130
Environmental disasters, examples, 4–5
Environmental Engineering & Pollution Control (EE&PC), 3M pollution programs, compliance monitoring, 391
Environmental laws
 Clean Air Act
  air pollutant classifications, 132–133
  principles, 6
  provisions, 133
  state requirements, 134
  suggestions to improve, 64
  waste minimization, 58–59
 Clean Water Act
  goals, 142–143
  principles, 5–6
  suggestions to improve, 64
 Comprehensive Environmental Response, Compensation, and Liability Act, principles, 7
 Federal Water Pollution Control Act, provisions, 142–143
 Laboratory Equity and Waste Minimization Act
  description, 116
  provisions, 117–118

INDEX 469

Minnesota Toxic Pollution Prevention Act, principles, 416
Resource Conservation and Recovery Act, 6–7
  generator certifications amendment, 56–58, 406
  Hazardous and Solid Waste Amendments, 6–7
  laboratory waste exemptions from, 59
  principles, 6–7
  sanitary sewer restrictions, 142
  suggestions to improve, 64
  treatment facility restrictions, 64
Toxic Substances Control Act, principles, 7
Environmentally sound recycling, defined, 10
Environmental Protection Agency (EPA)
  audits, environmental compliance, 63
  Boiler-Industrial-Furnace rules, 155
  container disposal regulations, 396–397
  hazardous waste definitions, 158, 224
  hierarchy, 10–11
  infectious waste categories, 167
  pollution prevention
    benefits, 8–9
    defined, 9–10
  Total Cost Assessment method, elements, 83–84
  Toxic Characteristic Leachate Procedure, 227
  waste minimization
    defined, 10
    generator certifications, 56
    hierarchy, 70, 100
    priority approaches, 38–40
    recommendations, 35
Environmental regulations, See also Environmental laws
  terminology
    ambient air standards, 133
    emission-based standards, 133
    hazardous air pollutants, 132
    source, 132
Environmental wastes, recovery analysis
  dichlormethane, 366–367
  freon, 367
EPA, See Environmental Protection Agency
Ethanol, solvent recovery analysis, 347
Ethers, hydroperoxide destruction, 278–279
Evaporation, 155, 278
Executive management, See also Administrations
  waste minimization, 73–74, 86
Exhausted emissions, defined, 126

F

Feasibility analysis for waste minimization program implementation, 37, 82–85
Federal Water Pollution Control Act (FWPCA), 5, See also Clean Water Act
  provisions, 142–143
Fees, disposal, for waste minimization promotion, 112
Fenske equation, theoretical plate calculations, 262
Flooding, defined, 259
Floor drains, chemical spill protection, 151
Formalin, solvent recovery, 347, 351, 375
Fractional distillation, defined, 243, 246
Fractionating columns
  characteristics
    efficiency, 256
    equilibration time, 259
    holdup, 256
    pressure drop, 259
    reflux ratio, 258–259
    throughput, 257–258
  defined, 246
  types, 248, 254–255
Freon, waste recovery analysis, 367
Fuel blending, 19–20, 22, 100
  cost savings, 155
  description, 155
  off-site types, 155
Fugitive emissions
  from chemical storage container leaks, 126
  defined, 126
  environmental contamination, 127
  measurements, 129
Fume hoods
  air contamination contributor, 126, 130, 136, 307
  chemical spill protection, 151
FWPCA, See Federal Water Pollution Control Act

G

Gas chromatography (GC), liquid purifications, 214
Generators, hazardous waste
  defined, 34
  divisions in waste management research studies, 19
  fee and tax imposition, 62
  hazardous waste, liability establishment
    disposal sites, 61
    incidents handling, 62

large quantity, 34, 379
legislative mandates against, 56–58
small quantity, 34, 379
waste minimization programs
  EPA certification mandates, 109, 158
  responsibilities, 53
Government Relations and Science Policy department (GRASP), waste minimization legislation, 116–117
Greenhouses
  hazardous waste production, 381
  product substitution opportunities, 383–384
"Green" laboratory
  defined, 429
  pollution prevention, 430
"Green Lights Program," 397
Group permit, proposed legislation, 117–118

H

Hazardous and Solid Waste Amendments (HSWA), to Resource Conservation and Recovery Act, principles, 6–7
Hazardous Materials Transportation Act (HMTA), principles, 7
Hazardous Waste Minimization and Source Reduction Task Force at Mayo laboratories, 159–161
Hazardous wastes
  defined, 34
  legislative protection against, 6
  liability establishment
    disposal site cleanups, 61
    incident response, 62
  minimization, See Waste minimization
Health and safety teams, waste minimization programs, 95
Health regulations, air emissions
  environmental, 132
  occupational, 131–132
Hematology, waste minimization advancements, 175–176
Hexafluoroisopropanol (HFIP), 240
  waste recovery analysis, 362, 366
HHMI, See Howard Hughes Medical Institute
High-performance liquid chromatography (HPLC)
  solvent recovery analysis, 331–332
    acetonitrile wastes, 352, 355
    hexafluoroisopropanol, 362, 366
    isopropanol, 356
    methanol, 356
    nonpolar solvents, 361
    tetrahydrofuran, 362
  solvent use, 240

HMTA, See Hazardous Materials Transportation Act
Holdup
  defined, 242, 257
  dynamic, 262–263
Hospital wastes, solvent recovery analysis, 340–341
Howard Hughes Medical Institute (HHMI), National Survey participation, 17, 451
  annual costs, 24
  cost savings, 28
  management practices not used, 22
  management practices used, 20–21
  minimization practices, 25–28
  respondent profile, 19
  sewer and normal trash use, 23–24
HPLC, See High-performance liquid chromatography
HSWA, See Hazardous and Solid Waste Amendments
Hydrocarbons, in recovered freon, 367, 369
Hydroperoxides destruction, 278–279

I

Illinois Department of Nuclear Safety (IDNS), radioactive waste classifications, 197
Immunology, waste minimization advancements, 175
Incineration, 100
  infectious waste treatment, 182, 208
  limitations, 223–224
  liquid scintillation cocktails, 198, 200
  mercury, 223
  prevalence, 22
  solvents, 186
Infectious waste
  categories, 167
  on-site treatment
    in facility, 179, 182–183
    in laboratory, 179
  reduction programs, strategies, 188–189
  shredding treatment, 182–183
  source separation, 169
  testing methods, 174–175
Information dissemination, role in waste minimization programs, 71–72
Inorganic chemicals, disposal reactions
  bromine, 284
  cyanides, 284–285
  inorganic azides, 285–286
  iodine, 284
  oxidizing agents, 284
In-process recycling, defined, 10

INDEX

Instructional laboratory, See Teaching laboratory
Inventory management of chemicals
  bar codes, 161, 302
  procurement process
    as information source, 303
    as internal control point, 303–304
    limitations, 304–306
  stockroom, 394
  waste management practice, 30, 38, 48–49
Iodine, disposal reaction, 284
Isopropanol, waste recovery analysis, 356
Isotopes, radioactive
  half-lives, 184
  waste management, 199

L

Laboratories, research, See also specific laboratories
  air emissions
    control, 125, 129–131
    environmental impact, 124–125, 127
    estimations, 127–129
    identification methods, 128
    management program, 135–138
    measurements, 128–129, 138
    policies, 136–138
    regulations, 131–132
    research needs, 138
    sources, 126–127
    types, 125–126
  instructional, See Teaching laboratory
  legislative effects, See also Environmental laws
    Clean Air Act, 6
    Clean Water Act, 5–6
    Hazardous Materials Transportation Act, 7
    Toxic Substances Control Act, 7
  medical, See Medical laboratory
  personnel, See Laboratory personnel
  purpose, 427
  solvents, 36
    acetone, 369
    toluene, 369, 371
  source separation
    error monitoring, 236
    sanitary sewer and normal trash use, 234
    training, 233–234
    waste collection containers, 234–235
  teaching, See Teaching laboratory
  waste management studies
    management practices, 19–21
    sampling approach, 17
    survey instrument, 16

  waste minimization, See Waste minimization
  wastestream assessment process, 78–82
Laboratory Equity and Waste Minimization Act
  description, 116
  provisions, 117–118
Laboratory personnel, 71
  air emission risks, 127–128
  biosafety concerns, 172–173, 208
  infective disease exposure, 166–168
  safety issues, automated walkaway systems, 174
  source separation training, 233–234
  waste minimization programs, 109
    awareness programs, 189–190
    communication methods, 111
    economic isolation barrier, 113
    education, 86, 130, 136–137
    participation, 32, 72–73
    training, 190, 228, 413–414, 430
Landfills, surplus chemicals disposal, 106, 208
Laws, environmental
  Clean Air Act, 6
    air pollutant classifications, 132–133
    principles, 6
    provisions, 133
    state requirements, 134
    suggestions to improve, 64
    waste minimization, 58–59
  Clean Water Act
    goals, 142–143
    principles, 5–6
    suggestions to improve, 64
  Comprehensive Environmental Response, Compensation, and Liability Act
    effect on wastestream liability, 61
    principles, 7
  Federal Water Pollution Control Act, provisions, 142–143
  Laboratory Equity and Waste Minimization Act
    description, 116
    provisions, 117–118
  Minnesota Toxic Pollution Prevention Act, principles, 416
  Resource Conservation and Recovery Act, 6–7
    generator certifications amendment, 56–58, 406
    Hazardous and Solid Waste amendments, 6–7
    laboratory waste exemptions from, 59
    principles, 6–7
    sanitary sewer restrictions, 142

suggestions to improve, 64
    treatment facility restrictions, 64
  Toxic Substances Control Act, principles, 7
Lead, recycling practices, 105
Legislation, environmental, See Laws, environmental
Liquid chromatography system, waste minimization benefits, 174
Liquid scintillation cocktail
  biodegradable, 407
  incineration, 198, 200
  vial disposal, 200, 414
  waste management, 197–198
  waste reduction, substitution method, 100, 414
LLR, See Radioactive waste, low-level

M

MACT, See Maximum achievable control technology
Management
  air emissions, 135–138
  executive, See Executive management
  waste minimization programs, 166–173
Manhole, control for wastewater sampling, 149
Manifest, defined, 34
Market, effect on chemical procurement, 304–305
Mass balance, 36, 49, 189
Material safety data sheets (MSDS)
  at Agracetus, 385
  OSHA requirements, 300
Materials management and waste management, relationship, 161
Maximum achievable control technology (MACT)
  EPA classifications, 133
  hypothetical scenarios using, 134–135
Mayo laboratories
  Hazardous Waste Minimization and Source Reduction Task Force, 159–161
  radionuclide storing feasibility study, 156
  waste minimization program, 154
Mechanical treatment, infectious waste, 182
Medical laboratory
  defined, 164
  disposable product use, 172
  product substitution, 176
  technological advances in testing methodology
    chemistry, 175–176
    hematology, 175–176
    immunology, 175
    microbiology, 174–175
    nonradioactive methods, 176–177
    paperless information recording, 177
    product substitutions, 176
    serology, 175
  waste minimization, 165
    on-site treatment, 177–185
    program implementation, 188–191
    recycling, 187
    reusability, 185–187
Medical waste
  regulated, 167
  types, 165
Mercury
  product substitution, 176, 383
  reclaiming, 395
  recycling practices, 105, 412–413
  reduction methods, 47–48, 160
  solutions, disposal reactions, 287–288
  as waste source, 42, 225
Metal carbonyls, disposal reactions, 288
Methanol, waste recovery analysis, 356
Microbiology, waste minimization advancements, 174–175
Microscale chemistry, See also Process modification
  advantages, 27, 215–216
  conversion costs, 218–219
  disadvantages, 215
  equipment
    glassware assembly, 211–212
    heating, 212
    reaction vessels, 211
    stirring, 212
  implementation impediments, 209
  organic, 210
  in teaching laboratory, See Teaching laboratory
  techniques
    changed reaction conditions, 215
    chemical measurements, 212–213
    crystallization, 213–214
    liquid transfer and filtering, 213
    separation, 214
  for waste minimization, 102, 104
    concept introduction, 209–211
    defined, 208
    prevalence, 210–211
Microspheres, radioactive, management, 157–158
Minnesota, hazardous waste definitions, 158
Minnesota Toxic Pollution Prevention Act, principles, 416

INDEX

Mixing of wastes, 222–223, 237
MSDS, See Material safety data sheets
Multihazardous waste
  defined, 184
  disposal concerns, 184, 224

N

National Laboratory Waste Survey, source separation errors, 229
National Pollutant Discharge Elimination System (NPDES), Clean Water Act enforcement, 143
National Survey of Laboratory Chemical Disposal and Waste Minimization
  conclusions, 28–30
  data analysis, 455
  data collection methodology, 452–453
  DivCHAS members, 17, 452
  error assessment and control, 453–454
  future research areas, 455
  Howard Hughes Medical Institute participation, 17, 451
  instructions to respondents, 449–450
  intended targets and respondents, 445–448
  margin of error, 454–455
  Network News, ACS, 17, 451
  objectives, 444
  response rate calculations, 453
  sampling plan, 452
  survey groups profile, 18–19, 451
  survey instrument, 16–18, 444
  waste management practices
    costs, 24
    off-site, 20
    on-site, 20–21
    sewer and normal trash use, 23–24
    type not used, 21–22
  waste minimization practices, 25–28
Naturally occurring radioactive materials (NORM), waste disposal, 202
Network News (NET), waste management study, 17
Neutralization, elementary, 19, 22, 39, 59
  acids, 47, 105, 155, 183, 277
  bases, 105, 277
Neutralization tanks
  construction of, 150
  wastewater regulation, 150, 170
"No Chromix," 382, 408
Nonpolar solvents, waste recovery analysis, 361
NORM, See Naturally occurring radioactive materials

Normal trash
  chemical waste disposal, nonhazardous, 23–24
  defined, 223
NPDES, See National Pollutant Discharge Elimination System
Nuclear Regulatory Commission (NRC)
  radioactive waste regulations, 197, 224
  waste-disposal regulations, 184

O

Occupational safety, wastewater disposal, 185, 230–231
Occupational Safety and Health Administration (OSHA)
  bloodborne pathogens standards, 224, 226–227
  infectious waste categories, 167
  laboratory standards, 128
  material safety data sheets, 300
  occupational regulations, 131
  regulated waste definition, 168
  worker safety regulations, 116
Odor threshold, chemical determinations, 129
Off-site treatment
  on-site and, comparison, 178
  types, 19–20
  vendor services, efficiency improvement, 298–299
Oil, engine, waste segregation, 385
On-site treatment
  off-site and, comparison, 178
  types, 19, 21
Organic chemicals, disposal reactions
  diethyl sulfate, 280
  dimethyl sulfate, 280
  N-Bromosuccinimide, 279
  N-Chlorosuccinimide, 279
  nitriles, 282
  N,N-Dimethylacetamide, 280
  N,N-Dimethylformamide, 280
  oxalic acid, 279
  oxalyl chloride, 279
  picric acid, 281
  sodium oxalate, 279
OSHA, See Occupational Safety and Health Administration

P

Packaging materials
  recycling, 187
  reusability, 187
  strategies to reduce, 189

PCIWO, See Pima County Industrial
    Wastewater Ordinance
PEL, See Permissible exposure limits
Permissible exposure limits (PEL), defined, 131
Personnel, laboratory, See Laboratory
    personnel
Picric acid, disposal reactions, 281
Pima County Industrial Wastewater Ordinance
    (PCIWO)
  chemical discharge regulations, 142
    permit types, 145
  POTW protection, 147–148
  regulation of University of Arizona,
    143–147
Plates, theoretical
  calculations, 262
  defined, 246
Pollution, cross-media, 64–65
Pollution prevention, See also Waste
    minimization
  champions
    places to find, 90–91
    role in waste minimization programs, 71
  corporate participation, 9
  environmental pressures, 426–427
  EPA's definition, 9–10
  in laboratories, 428
  mandates, 58–60
  noncategorical activities, 10
  objectives, 427–428
  obstacles to, 429
  preventive measures
    feasibility evaluation, 37
    informational and compliance programs, 34
    planning and organization, 35
    waste minimization assessment, 34–35
    waste minimization implementation, 37
    wastestream assessment, 35–37
  programs, See 3M R&D laboratories
  reasons to start, 89–90, 426–427
Pollution Prevention Act (PPA)
  laboratory exclusions from, 58
  principles, 8
  source reduction mandates, 57
Pollution Prevention Directory, 457
Pollution Prevention Yellow Pages, 457–458
Potentially responsible parties (PRP)
  defined, 7
  liability, 61, 388
POTW, See Publicly owned treatment works
PPA, See Pollution Prevention Act
Precipitation, 155
  of metals, 286
Process modification, See also Microscale
    chemistry
  defined, 410
  examples, 410–411
Procurement, chemical
  as information source, 303
  as internal control point, 303–304
  limitations
    market factors, 304–305
    market inertia, 305–306
    purchasing leverage, 305
Product stewardship, purpose, 306
Product substitution
  disposable items, 172
  infectious wastes, 172
  waste minimization technique, 100, 108,
    382–384, 410–411
PRP, See Potentially responsible parties
Public image, institution, waste management
    errors and, relationship, 229–230
Publicly owned treatment works (POTW)
  chemical discharge, 143
  hydraulic loading, 146
  prohibited wastes, 145–146
Purchasing control, chemicals, waste
    minimization strategies, 29–30, 189,
    393–394, 409
Purchasing leverage, defined, 305
Pureform 2000 Recovery System, 272, 337

Q

Quality assurance and quality control (QA/QC),
    benefits of disposable items, 172

R

Radioactive decay, storage methods, 156–158
Radioactive waste
  low-level (LLR)
    classifications, 197–198
    recycling, 202–203
    waste minimization, 195–196, 406, 414
  on-site treatment, 183–184
  programs, See University of Illinois at
    Urbana-Champaign, radiation safety
    program
  source separation, 170
  waste minimization, nonradioactive assays,
    176–177
Radioimmunoassays (RIA)
  nonradioactive use, 176
  radiolabeled microsphere management, 157
  testing elimination, 156–157
  waste from decay process, 156
Radioisotopes
  half-lives, 184

INDEX 475

waste management, 199
Radionuclides, 156
RCRA, See Resource Conservation and Recovery Act
Reactions, disposal
  inorganic chemicals
    bromine, 284
    cyanides, 284–285
    inorganic azides, 285–286
    iodine, 284
    oxidizing agents, 284
  organic chemicals
    aromatic amines, 283–284
    diethyl sulfate, 280
    dimethyl sulfate, 280
    N-Bromosuccinimide, 279
    N-Chlorosuccinimide, 279
    nitriles, 282
    N,N-Dimethylacetamide, 280
    N,N-Dimethylformamide, 280
    organic azides, 282–283
    oxalic acid, 279
    oxalyl chloride, 279
    picric acid, 281
    sodium oxalate, 279
Reagents, low-toxicity, waste minimization benefits, 173
Recovery as on-site management practice, 20–21
Recycling
  defined, 38
  environmentally sound, defined, 10
  in-process, defined, 10
  on-site hazardous, 301–302
  waste minimization, 226, 394
Recycling methods
  bottles, 91, 170
  cans, 91, 170
  glass, 395
  packaging materials, 187
  paper, 91, 170, 187
  plastic, 395
Redistillation, 39, 183, See also Distillation
  solvents, 186–187
Redistribution
  role of computerized databases in, 81
  surplus chemicals, 38, 105–106, 428
    barriers to, 114
    off-site, 299–301
    program, See University of Arizona, chemical redistribution program
  waste minimization practice, 27, 29–30
Reflux ratio
  defined, 258–259
  total, 261–262

Research institutions, See Laboratories, research
Research personnel, See Laboratory personnel
Resource Conservation and Recovery Act (RCRA)
  amendments
    generator certifications, 56–58, 406
    Hazardous and Solid Waste, 6–7
  laboratory waste exemptions from, 59
  principles, 6–7
  purpose, 371, 374
  sanitary sewer restrictions, 142
  suggestions to improve, 64
  treatment facility restrictions, 64
Reusability
  disposable items, 172, 185–186
  effluent recirculation, 186
  packaging materials, 187
  solvents, 186
  xylene, 104, 240
RIA, See Radioimmunoassays
Rivers and Harbors Act (1899), 5
RM&S, See Department of Risk Management & Safety

S

Safety departments, role in waste minimization practices, 104–111
Sanitary sewer system
  chemical waste disposal, 23, 142, 208
  effluent recirculation, 186
  3M guidelines, 400–401
  wastewater, 170, 185
SARA, See Superfund Amendments and Reauthorization Act
S.C. Johnson, pollution prevention programs, 9
Scale reduction, See Microscale chemistry
Scientists, See Laboratory personnel
Scintillation cocktails, liquid
  biodegradable, 407
  incineration, 198, 200
  product substitution, 100, 383, 414
  vial disposal, 200, 414
  waste management, 197–198
Serology, waste minimization advancements, 175
Sewer, sanitary
  chemical waste disposal, 23, 142, 400–401
  effluent recirculation, 186
  3M guidelines, 400–401
  wastewater, 170
Short-term exposure limit, OSHA guidelines, 131

Shredding treatment, infectious waste, 182–183
Significant Industrial User (SIU), defined, 145
Simple distillation
  defined, 242
  vapor-liquid composition curve, 243–244
Site visits, wastestream assessment step, 80
SIU, See Significant Industrial User
Solid-phase extraction, waste reduction, 44, 47
Solutions, disposal reactions
  chromium ions, 287–288
  heavy metal ion, 286–287
  mercury ions, 287–288
  metal carbonyls, 288
Solvents
  commingling, 108
  distillation, 38, 47
    commercial units, 264
    fractionated, 243, 246
    on-site, 21
    prevalence, 20
    purpose, 242
    simple, 242
    spinning band, See Spinning band stills
    theoretical background, 240–248
  ethanol, 347
  flammable, 223–224
  halogenated, 322
  liquid scintillation cocktail, 197–198
  nonpolar, waste recovery analysis, 361
  pollution prevention, 44, 47
  recovery analysis
    fluorescence, 332
    gas chromatography, 331–332
    high-performance liquid chromatography, 332, 352-366
    practical aspects, 371, 374–375
    ultraviolet, 332
    water, 331
  recycling, 187
    safety considerations, 263–264
  reusability, 186
  toluene, 100, 197, 369, 410
  xylene
    commingling, 324
    distillation and reuse practices, 104, 240
    fuel blending, 155
    in liquid scintillation cocktail, 197, 410
    solvent recovery, 341, 344, 346, 375
    substitutes, solvent recovery, 346–347, 375
SOP, See Standard operating procedures
Source category, defined, 132
Source reduction, 70, 154
  defined, 10

  in Pollution Prevention Act, 57
  waste minimization, 38, 158, 208, 410–411
Source separation, See also Waste separation
  benefits, 223, 231
  costs, waste
    assessment and analysis, 226–227
    collection and containment, 227
    overview, 231
    training of laboratory personnel and waste handlers, 228
  decision balances, 230–233
  defined, 110, 222
  errors
    monitoring, 236
    public relations and, relationship, 229–230
  implementation in laboratories
    error monitoring, 236
    laboratory personnel training, 233–234
    sanitary sewer and normal trash use, 234
    waste collection containers, 234–235
  radioactive waste, 170
  radioisotopes, 201
  safety concerns, 224
  waste management options, 226
  wastestreams, 110
    assessment and analysis, 226–227
    benefits, 171–172
    collection and containment, 227
    commingling, 233
    criteria, 231, 233
    implementation, 168–171, 190
Spills, chemical
  air contamination, 127
  protection methods, 151
Spinning band stills
  characteristics, 260
  description, 260–261
  error types, 260
  holdup determinations, 262–263
  operating feature, 261
  plate use, 261–262
  pressure drop determinations, 263
  safety considerations, 263–264
  solvent recovery tests
    alcohol mixtures, 347
    equipment, 333–337
    ethanol, 347
    formalin, 347, 351
    hospital wastes, 340–341
    xylene, 341, 344, 346
    xylene substitutes, 346–347
  theoretical plate use, 261–262

Standard operating procedures for waste
    minimization programs, 86
Steam sterilizers for infectious waste, 179, 182
Stills, See also Distillation
  spinning band
    characteristics, 260
    description, 260–261
    error types, 260
    holdup determinations, 262–263
    pressure drop determinations, 263
    solvent recovery tests, 333–351
    theoretical plate use, 261–262
  waste minimization, 104, 115–116, 214
Stock chemicals, defined, 106
Sulfuric dichromate, 160
Superfund Amendments and Reauthorization
    Act (SARA), waste minimization
    prevention, 57

T

Take-backs, product, excess chemicals, 301
Task force, waste minimization
  at Mayo laboratories, 159–161
  program plan design and implementation,
    76–77
  responsibilities, 75–76
TCA, See Total Cost Assessment
TCLP, See Toxic Characteristic Leachate
    Procedure
Teaching laboratory
  microscale chemistry, 102–104
    chemical use reduction, 215
    conversion costs, 218–219
    cost savings, 216–217
    fire and explosion risk reduction, 215
    safety, 215
  waste minimization practices
    at University of Minnesota, 411–412
Tetrahydrofuran (THF)
  recycling, 264
  waste recovery analysis, 362
Theoretical plates
  calculations, 262
  defined, 246
THF, See Tetrahydrofuran
3M
  Pollution Prevention Pays program, 9, 390,
    416
  R&D laboratories, pollution prevention
    programs
    compliance monitoring, 391
    equipment cleaning, 397
    hazard reviews, 398–399

  hierarchy and cardinal rules, 391
  waste disposal, 398–401
  waste minimization, 392–398
  waste reduction strategies, 393
  wastestream profile, 392
  waste minimization statements, 217–218
Throughput, defined, 257–258
Toluene, 100
  in liquid scintillation cocktail, 197, 410
  waste recovery analysis, 369
Total Cost Assessment (TCA), waste
    minimization economic analysis
    method, 83–84
Total quality management
  microscale chemistry, 208
  principles, 427
Toxic Characteristic Leachate Procedure
    (TCLP), principles, 227
Toxic substances, legislative control, 7
Toxic Substances Control Act (TSCA)
  chemical disposal regulations, 300–400
  principles, 7
TQM, See Total quality management
Training, laboratory personnel
  source separation, 228, 233–234
  waste minimization, 190, 228, 413–414, 430
Trash, normal
  chemical waste disposal, nonhazardous,
    23–24
  defined, 223
Treatment
  chemical, 154–155
    elementary neutralization, 19, 22, 39, 47,
      59, 105, 155, 183
    fuel blending, 19, 22, 100, 155–156
    precipitation and evaporation, 155
    waste minimization, 29, 108–109
  RCRA restrictions, 59, 64
  waste minimization technique, 39
1,1,2,-Trichlorotrifluoroethane, See Freon
TSCA, See Toxic Substances Control Act

U

UA, See University of Arizona
UIUC, See University of Illinois at
    Urbana-Champaign
Ultraviolet analysis, solvent recovery, 332
University of Arizona (UA)
  chemical redistribution program
    criteria for success, 314–315
    impediments, 315–317
    overview, 312–313
    quality control, 316

publicly owned treatment works use, 146
wastewater management program
  chemical regulations, 143–147
  compliance training, 148–149
  industrial source identification process, 148
  problems associated with, 150–151
  sampling and analysis, 149–150
  spill protection, 151
University of Illinois at Urbana-Champaign (UIUC)
  radiation safety program
    conclusions and recommendations, 203–204
    incineration, 199
    low-level radioactive waste classifications, 197–198
    overview, 196–197
    Radiation Safety Office, 196–197
    statistics, 199–203
  waste minimization studies
    follow-up, 49–52
    laboratory selection and focus, 48–49
    results and conclusions, 40–49
    structure, 40
University of Minnesota
  "Pollution Prevention and Waste Abatement" policy, 93–94
  waste minimization program
    assessment and methods, 407–408
    chemistry teaching laboratories, 411–412
    economic assessment, 408
    future plans, 414–415
    incentives, 413
    management, 413
    methods, 406–407
    planning, 408–409
    purchasing, 409
    of radioactive wastes, 414
    reclamation, 412–413
    recycling, 412
    segregation and characterization, 409–410
    source reduction, 410–411
    training, 413–414
University of Wisconsin–Madison, waste minimization practices
  common methods, 102–103
  departmental communication, 111
  disposal fee proposals, 112–113
  materials substitution, 100
  options not chosen, 112–114
  proposed initiatives, 114–116
  safety department involvement, 104–111
US EPA, See Environmental Protection Agency

V

Vapor-liquid composition curves
  defined, 243
  example, 244
Vendors, waste minimization, 296–298
Ventilation devices
  fume hoods
    air contamination contributor, 126, 130, 136
    chemical spill protection, 151
  local
    air contamination contributor, 126
    proper design, 127

W

Waste containers
  assessment, 312
  for commingling, 323
  criteria, 234–235
  identification, 235
  recycling, 395–397
  types, 227–228
Waste handlers, separation policies, 236–237
Waste management, See also Waste minimization
  regulatory agency differences, 116
  research studies
    average annual cost, 24
    generator categories, 19
    management practices used and not used, 20–22
    results, 18–28
    sampling approach, 17
    survey instrument, 16
    waste minimization practices, 25–28
Waste manager, role in waste minimization programs, 32
Waste minimization, See also Pollution prevention; waste management
  administrative support, 71
  commitment establishment, 74–75, 91–94, 408, 415
  cost savings, 190–191
  infrastructure, 137
  audits, 63–64, 415
  benefits, 306–308, 387–388, 406
  checklist to reduce, 49–52
  commingling, 321
  common methods, 101
  cost savings, 30, 92
  economist's view, 113–114
  EPA's definition, 10
  hierarchy, 98–100

INDEX

impediments, 98
informational and compliance programs, 34
laboratory personnel involvement, 31–32, 416
  awareness programs, 189–190
  training, 190, 413–414, 430
legal incentives, 61
legal requirements, mandates, 56–58
management practices
  product substitution, 172–173
  source separation, 168–172
  waste identification, 166–168
microscale chemistry
  concept introduction, 209–211
  defined, 208
  equipment, 211–212
  prevalence, 210–211
  techniques, 213–215
noncategorical activities, 10
objectives, 427–428
product substitution
  disposable items, 172
  infectious wastes, 172
  waste minimization technique, 100, 108, 382–384, 410–411
program implementation, See also specific university programs
  benefit emphasis, 109–110
  essential elements, 71–72
  feasibility evaluations, 37, 82–85
  goal setting, 76
  legal considerations, 63
  managerial support, 73–75
  planning and organization, 35, 72–78, 188, 415, 417–418, 428, 430–431
  policy statement, 75, 415–416
  prioritization process, 38–40, 81–82
  program evaluation, 87–88
  program initiation, 37, 85–87
  reasons to start, 89–90, 101–102, 377–378
  wastestream assessment process, 78–82
radioactive decay, storage methods, 156–157
reduction methods
  chemical treatment, 154–155
  documentation of, 417
  management practices, 166–173
  recycling, 187
  reuse, 185–187
  source, 157–158
regulatory reforms, 415–416, See also Environmental laws
  "actual" vs. "allowable" reductions, 60
  "baseline" conundrum, 60
research studies
  beneficial techniques, 27–28
  cost savings, 27–28
  frequency, 25–26
  popularity, 30–49
  structure, 40
social costs, 113
solvent recovery, distillation, 240–248
source separation, See Source separation
task forces
  at Mayo laboratories, 159–161
  program plan design and implementation, 76–77
  responsibilities, 75–76
technological advances in laboratory testing
  immunology, 175
  microbiological, 174–175
  serology, 175
treatment methods
  infectious wastes, 177–183
  multihazardous wastes, 184
  off-site, 19–20
  on-site, 19–20, 177–179
  radioactive wastes, 183–184
vendors, 296–298
Wastes
environmental, recovery analysis, 366–367
hazardous
  defined, 34
  legislative protection against, 6
  liability establishment, 61–62
infectious
  categories, 167
  on-site treatment, 179, 182–183
  reduction programs, 188–189
  shredding treatment, 182–183
  source separation, 169
  testing methods, 174–175
metal-containing, disposal reactions, 286–289
radioactive
  low-level (LLR), 195–198, 202–203, 406, 414
  on-site treatment, 183–184
  programs, See University of Illinois at Urbana-Champaign, radiation safety program
  source separation, 170
  waste minimization, nonradioactive assays, 176–177
Waste segregation
  defined, 222
  disposal method selection, 223–224
  separation policies, 236–237
  for waste minimization, 384–385
Waste separation, See also Source separation
  assessment and analysis, 226–227

definitions of, 222–223
reasons, 224
Wastestreams, See also Wastewater
   assessment process, 35–37, 78–82
   common problems, 25
   defined, 222
   discharge regulations, 59
   management program
      compliance training, 148–149
      implementation and operation, 148–151
      industrial source identification, 148
      problems associated with, 150–151
      sampling and analysis, 149–150
      spill protection, 151
   minimization strategies, 44, 47–48
   source separation, 110
      assessment and analysis, 226–227
      benefits, 171–172
      collection and containment, 227
      commingling, 233
      criteria, 231, 233
      implementation, 168–171
Waste volume reduction, See Waste minimization
Wastewater, See also Wastestreams
   occupational safety concerns, 185
   on-site treatment, 185
   sanitary sewer system, 170
   source separation, 170
WW/PP, See Waste minimization, program implementation

X

Xylene
   commingling, 324
   distillation and reuse practices, 104, 240
   fuel blending, 155
   in liquid scintillation cocktail, 197, 410
   solvent recovery, 341, 344, 346, 375
   substitutes, solvent recovery, 346–347, 375